Guide to Convolutional Neural Networks

Hamed Habibi Aghdam
Elnaz Jahani Heravi

Guide to Convolutional Neural Networks

A Practical Application to Traffic-Sign Detection and Classification

 Springer

Hamed Habibi Aghdam
University Rovira i Virgili
Tarragona
Spain

Elnaz Jahani Heravi
University Rovira i Virgili
Tarragona
Spain

ISBN 978-3-319-86190-6   ISBN 978-3-319-57550-6   (eBook)
DOI 10.1007/978-3-319-57550-6

Printed on acid-free paper

This Springer imprint is published by Springer Nature
The registered company is Springer International Publishing AG
The registered company address is: Gewerbestrasse 11, 6330 Cham, Switzerland

To my wife, Elnaz, who possess the most accurate and reliable optimization method and guides me toward global optima of life.

Hamed Habibi Aghdam

Preface

General paradigm in solving a computer vision problem is to represent a raw image using a more informative vector called feature vector and train a classifier on top of feature vectors collected from training set. From classification perspective, there are several off-the-shelf methods such as gradient boosting, random forest and support vector machines that are able to accurately model nonlinear decision boundaries. Hence, solving a computer vision problem mainly depends on the feature extraction algorithm.

Feature extraction methods such as scale invariant feature transform, histogram of oriented gradients, bank of Gabor filters, local binary pattern, bag of features and Fisher vectors are some of the methods that performed well compared with their predecessors. These methods mainly create the feature vector in several steps. For example, scale invariant feature transform and histogram of oriented gradients first compute gradient of the image. Then, they pool gradient magnitudes over different regions and concatenate them in order to create the final feature vector. Similarly, bag of feature and Fisher vectors start with extracting a feature vector such as histogram of oriented gradient on regions around bunch salient points on image. Then, these features are pooled again in order to create higher level feature vectors.

Despite the great efforts in computer vision community, the above hand-engineered features were not able to properly model large classes of natural objects. Advent of convolutional neural networks, large datasets and parallel computing hardware changed the course of computer vision. Instead of designing feature vectors by hand, convolutional neural networks learn a composite feature transformation function that makes classes of objects linearly separable in the feature space.

Recently, convolutional neural networks have surpassed human in different tasks such as classification of natural objects and classification of traffic signs. After their great success, convolutional neural networks have become the first choice for learning features from training data.

One of the fields that have been greatly influenced by convolutional neural networks is automotive industry. Tasks such as pedestrian detection, car detection, traffic sign recognition, traffic light recognition and road scene understanding are rarely done using hand-crafted features anymore.

Designing, implementing and evaluating are crucial steps in developing a successful computer vision-based method. In order to design a neural network, one must have the basic knowledge about the underlying process of neural network and training algorithms. Implementing a neural network requires a deep knowledge about libraries that can be used for this purpose. Moreover, neural network must be evaluated quantitatively and qualitatively before using them in practical applications.

Instead of going into details of mathematical concepts, this book tries to adequately explain fundamentals of neural network and show how to implement and assess them in practice. Specifically, Chap. 2 covers basic concepts related to classification and it derives the idea of feature learning using neural network starting from linear classifiers. Then, Chap. 3 shows how to derive convolutional neural networks from fully connected neural networks. It also reviews classical network architectures and mentions different techniques for evaluating neural networks.

Next, Chap. 4 thoroughly talks about a practical library for implementing convolutional neural networks. It also explains how to use Python interface of this library in order to create and evaluate neural networks. The next two chapters explain practical examples about detection and classification of traffic signs using convolutional neural networks. Finally, the last chapter introduces a few techniques for visualizing neural networks using Python interface.

Graduate/undergraduate students as well as machine vision practitioners can use the book to gain a hand-on knowledge in the field of convolutional neural networks. Exercises have been designed such that they will help readers to acquire deeper knowledge in the field. Last but not least, Python scripts have been provided so reader will be able to reproduce the results and practice the topics of this book easily.

Books Website

Most of codes explained in this book are available in https://github.com/pcnn/. The codes are written in Python 2.7 and they require *numpy* and *matplotlib* libraries. You can download and try the codes on your own.

Tarragona, Spain   Hamed Habibi Aghdam

Contents

1	**Traffic Sign Detection and Recognition**	1
	1.1 Introduction	1
	1.2 Challenges	2
	1.3 Previous Work	5
	1.3.1 Template Matching	5
	1.3.2 Hand-Crafted Features	5
	1.3.3 Feature Learning	7
	1.3.4 ConvNets	10
	1.4 Summary	12
	References	12
2	**Pattern Classification**	15
	2.1 Formulation	16
	2.1.1 K-Nearest Neighbor	17
	2.2 Linear Classifier	20
	2.2.1 Training a Linear Classifier	22
	2.2.2 Hinge Loss	30
	2.2.3 Logistic Regression	34
	2.2.4 Comparing Loss Function	37
	2.3 Multiclass Classification	41
	2.3.1 One Versus One	41
	2.3.2 One Versus Rest	44
	2.3.3 Multiclass Hinge Loss	46
	2.3.4 Multinomial Logistic Function	48
	2.4 Feature Extraction	51
	2.5 Learning $\Phi(\mathbf{x})$	58
	2.6 Artificial Neural Networks	61
	2.6.1 Backpropagation	65
	2.6.2 Activation Functions	71
	2.6.3 Role of Bias	78
	2.6.4 Initialization	79
	2.6.5 How to Apply on Images	79

2.7   Summary   81
2.8   Exercises   82
References   83

3   Convolutional Neural Networks   85
3.1   Deriving Convolution from a Fully Connected Layer   85
        3.1.1   Role of Convolution   90
        3.1.2   Backpropagation of Convolution Layers   92
        3.1.3   Stride in Convolution.   94
3.2   Pooling.   95
        3.2.1   Backpropagation in Pooling Layer   97
3.3   LeNet.   98
3.4   AlexNet   100
3.5   Designing a ConvNet.   101
        3.5.1   ConvNet Architecture.   102
        3.5.2   Software Libraries   103
        3.5.3   Evaluating a ConvNet   105
3.6   Training a ConvNet   111
        3.6.1   Loss Function   112
        3.6.2   Initialization   113
        3.6.3   Regularization.   115
        3.6.4   Learning Rate Annealing   121
3.7   Analyzing Quantitative Results   124
3.8   Other Types of Layers   126
        3.8.1   Local Response Normalization   126
        3.8.2   Spatial Pyramid Pooling.   127
        3.8.3   Mixed Pooling   127
        3.8.4   Batch Normalization   127
3.9   Summary   128
3.10   Exercises   128
References   129

4   Caffe Library   131
4.1   Introduction   131
4.2   Installing Caffe   132
4.3   Designing Using Text Files.   132
        4.3.1   Providing Data   137
        4.3.2   Convolution Layers   139
        4.3.3   Initializing Parameters   141
        4.3.4   Activation Layer   142
        4.3.5   Pooling Layer.   144
        4.3.6   Fully Connected Layer.   145
        4.3.7   Dropout Layer.   146
        4.3.8   Classification and Loss Layers   146

4.4		Training a Network	152
4.5		Designing in Python	154
4.6		Drawing Architecture of Network	157
4.7		Training Using Python	157
4.8		Evaluating Using Python	158
4.9		Save and Restore Networks	161
4.10		Python Layer in Caffe	162
4.11		Summary	164
4.12		Exercises	164
		Reference	166

5   Classification of Traffic Signs 167

5.1		Introduction	167
5.2		Related Work	169
	5.2.1	Template Matching	170
	5.2.2	Hand-Crafted Features	170
	5.2.3	Sparse Coding	171
	5.2.4	Discussion	171
	5.2.5	ConvNets	172
5.3		Preparing Dataset	173
	5.3.1	Splitting Data	174
	5.3.2	Augmenting Dataset	177
	5.3.3	Static Versus One-the-Fly Augmenting	185
	5.3.4	Imbalanced Dataset	185
	5.3.5	Preparing the GTSRB Dataset	187
5.4		Analyzing Training/Validation Curves	188
5.5		ConvNets for Classification of Traffic Signs	189
5.6		Ensemble of ConvNets	199
	5.6.1	Combining Models	200
	5.6.2	Training Different Models	201
	5.6.3	Creating Ensemble	202
5.7		Evaluating Networks	203
	5.7.1	Misclassified Images	208
	5.7.2	Cross-Dataset Analysis and Transfer Learning	209
	5.7.3	Stability of ConvNet	214
	5.7.4	Analyzing by Visualization	217
5.8		Analyzing by Visualizing	217
	5.8.1	Visualizing Sensitivity	218
	5.8.2	Visualizing the Minimum Perception	219
	5.8.3	Visualizing Activations	220
5.9		More Accurate ConvNet	222
	5.9.1	Evaluation	224
	5.9.2	Stability Against Noise	226
	5.9.3	Visualization	229

5.10   Summary   230
5.11   Exercises   231
References   232

6   Detecting Traffic Signs   235
6.1   Introduction   235
6.2   ConvNet for Detecting Traffic Signs   236
6.3   Implementing Sliding Window Within the ConvNet   239
6.4   Evaluation   243
6.5   Summary   246
6.6   Exercises   246
References   246

7   Visualizing Neural Networks   247
7.1   Introduction   247
7.2   Data-Oriented Techniques   248
7.2.1   Tracking Activation   248
7.2.2   Covering Mask   248
7.2.3   Embedding   249
7.3   Gradient-Based Techniques   249
7.3.1   Activation Maximization   250
7.3.2   Activation Saliency   253
7.4   Inverting Representation   254
7.5   Summary   257
7.6   Exercises   257
References   258

Appendix A: Gradient Descend   259

Glossary   275

Index   279

Acronyms

Adagrad	Adaptive gradient
ADAS	Advanced driver assistant system
ANN	Artificial neural network
BTSC	Belgium traffic sign classification
CNN	Convolutional neural network
ConvNet	Convolutional neural network
CPU	Central processing unit
DAG	Directed acyclic graph
ELU	Exponential linear unit
FN	False-negative
FNN	Feedforward neural network
FP	False-positive
GD	Gradient descend
GPU	Graphic processing unit
GTSRB	German traffic sign recognition benchmark
HOG	Histogram of oriented gradients
HSI	Hue-saturation-intensity
HSV	Hue-saturation-value
KNN	K-nearest neighbor
Leaky ReLU	Leaky rectified linear unit
LRN	Local response normalization
OVO	One versus one
OVR	One versus rest
PCA	Principal component analysis
PPM	Portable pixel map
PReLU	Parameterized rectified linear unit
ReLU	Rectified linear unit
RMSProp	Root mean square propagation
RNN	Recurrent neural network
RReLU	Randomized rectified linear unit
SGD	Stochastic gradient descend
SNR	Signal-to-noise ratio
SPP	Spatial pyramid pooling

TN	True-negative
TP	True-positive
t-SNE	t-distributed stochastic neighbor embedding
TTC	Time to completion

List of Figures

Fig. 1.1   Common pipeline for recognizing traffic signs   2
Fig. 1.2   Some of the challenges in classification of traffic signs.
           The signs have been collected in Germany and Belgium   4
Fig. 1.3   Fine differences between two traffic signs.   4
Fig. 1.4   Traditional approach for classification of objects   6
Fig. 1.5   Dictionary learnt by Aghdam et al. (2015) from 43 classes of
           traffic signs   10
Fig. 2.1   A dataset of two-dimensional vectors representing two
           classes of objects   17
Fig. 2.2   K-nearest neighbor looks for the K closets points
           in the training set to the query point   18
Fig. 2.3   K-nearest neighbor applied on every point on the plane for
           different values of K   19
Fig. 2.4   Geometry of linear models   21
Fig. 2.5   The intuition behind squared loss function is to minimized
           the squared difference between the actual response and
           predicted value. *Left* and *right* plots show two lines with
           different w_1 and b. The *line* in the *right* plot is *fitted* better
           than the line in the left plot since its prediction error is lower
           in total.   23
Fig. 2.6   Status of the gradient descend in four different iterations. The
           parameter vector **w** changes greatly in the first iterations.
           However, as it gets closer to the minimum of the squared loss
           function, it changes slightly   26
Fig. 2.7   Geometrical intuition behind least square loss function is to
           minimize the sum of unnormalized distances between the
           training samples x_i and their corresponding hypothetical
           line   27
Fig. 2.8   Square loss function may fit inaccurately on training data if
           there are noisy samples in the dataset   27
Fig. 2.9   The sign function can be accurately approximated
           using $\tanh(kx)$ when $k \gg 1$ 29
Fig. 2.10 The sign loss function is able to deal with noisy datasets and
           separated clusters problem mentioned previously.   30

Fig. 2.11   Derivative of tanh(kx) function saturates as $|x|$ increases.
            Also, the ratio of saturation growth rapidly when $k > 1$ 30
Fig. 2.12   Hinge loss increases the margin of samples while it is
            trying to reduce the classification error. Refer to text for
            more details   31
Fig. 2.13   Training a linear classifier using the hinge loss function on
            two different datasets   33
Fig. 2.14   Plot of the sigmoid function (*left*) and logarithm of the
            sigmoid function (*right*). The domain of the sigmoid function
            is real numbers and its range is $[0, 1]$ 35
Fig. 2.15   Logistic regression is able to deal with separated
            clusters   37
Fig. 2.16   Tanh squared loss and zero-one loss functions are not
            convex. In contrast, the squared loss, the hinge loss, and its
            variant and the logistic loss functions are convex   38
Fig. 2.17   Logistic regression tries to reduce the logistic loss even after
            finding a hyperplane which discriminates the classes
            perfectly   39
Fig. 2.18   Using the hinge loss function, the magnitude of **w** changes
            until all the samples are classified correctly and they
            do not fall into the critical region   40
Fig. 2.19   When classes are not linearly separable, $\|\mathbf{w}\|$ may have an
            upper bound in logistic loss function   41
Fig. 2.20   A samples dataset including four different classes.
            Each class is shown using a unique color and shape   42
Fig. 2.21   Training six classifiers on the four class classification
            problem. One versus one technique considers all unordered
            pairs of classes in the dataset and fits a separate binary
            classifier on each pair. A input **x** is classified by computing
            the majority of votes produced by each of binary classifiers.
            The *bottom plot* shows the class
            of every point on the plane into one of four classes   43
Fig. 2.22   One versus rest approach creates a binary dataset by
            changing the label of the class-of-interest to 1 and the label
            of the other classes to -1. Creating binary datasets is
            repeated for all classes. Then, a binary classifier is trained on
            each of these datasets. An unseen sample is classified based
            on the classification score of the binary classifiers   45
Fig. 2.23   A two-dimensional space divided into four regions using
            four linear models fitted using the multiclass hinge loss
            function. The plot on the *right* shows the linear models
            (lines in two-dimensional case) in the space   48
Fig. 2.24   Computational graph of the softmax loss on one sample   51

Fig. 2.25   The two-dimensional space divided into four regions using
             four linear models fitted using the softmax loss function. The
             plot on the *right* shows the linear models (lines in
             two-dimensional case) in the space   52
Fig. 2.26   A linear classifier is not able to accurately discriminate
             the samples in a nonlinear dataset   52
Fig. 2.27   Transforming samples from the original space (*left*) into
             another space (*right*) by applying on each sample. The
             bottom colormaps show how the original space is
             transformed using this function   54
Fig. 2.28   Samples become linearly separable in the new space. As the
             result, a linear classifier is able to accurately discriminate
             these samples. If we transform the linear model from the new
             space into the original space, the linear decision
             boundary become a nonlinear boundary   55
Fig. 2.29   43 classes of traffic in obtained from the GTSRB dataset
             (Stallkamp et al. 2012)   55
Fig. 2.30   Weights of a linear model trained directly on raw pixel
             intensities can be visualized by reshaping the vectors so they
             have the same shape as the input image. Then, each channel
             of the reshaped matrix can be shown using a colormap   56
Fig. 2.31   Computational graph for (2.78). Gradient of each node
             with respect to its parent is shown on the edges   60
Fig. 2.32   By minimizing (2.78) the model learns to jointly transform
             and classify the vectors. The *first row* shows the distribution
             of the training samples in the two-dimensional space. The
             second and *third rows* show the status of the model in three
             different iterations starting from the left plots   61
Fig. 2.33   Simplified diagram of a biological neuron   62
Fig. 2.34   Diagram of an artificial neuron   63
Fig. 2.35   A feedforward neural network can be seen as a directed
             acyclic graph where the inputs are passed through different
             layer until it reaches to the end   63
Fig. 2.36   Computational graph corresponding to a feedforward
             network for classification of three classes. The network
             accepts two-dimensional inputs and it has two hidden layers.
             The hidden layers consist of four and three neurons,
             respectively. Each neuron has two inputs including the
             weights and inputs from previous layer. The derivative of
             each node with respect to each input is shown on thee
             edges   65

Fig. 2.37   Forward mode differentiation starts from the end node
            to the starting node. At each node, it sums the output edges
            of the node where the value of each edge is computed by
            multiplying the edge with the derivative of the child node.
            Each *rectangle* with different color and line style shows
            which part of the partial derivative is computed until that
            point   67
Fig. 2.38   A sample computational graph with a loss function. To cut
            the clutter, activations functions have been fused
            with the soma function of the neuron. Also, the derivatives
            on edges are illustrated using small letters. For example,
            g denotes $\frac{\delta \mathbf{H}_0^2}{\delta \mathbf{H}_1^1}$ 69
Fig. 2.39   Sigmoid activation function and its derivative.   72
Fig. 2.40   Tangent hyperbolic activation function and its derivative   73
Fig. 2.41   The softsign activation function and its derivative   74
Fig. 2.42   The rectified linear unit activation function and its
            derivative.   74
Fig. 2.43   The leaky rectified linear unit activation function and its
            derivative.   75
Fig. 2.44   The softplus activation function and its derivative   76
Fig. 2.45   The exponential linear unit activation function and its
            derivative.   77
Fig. 2.46   The softplus activation function and its derivative   78
Fig. 2.47   The weights affect the magnitude of the function for a fixed
            value of bias and $\mathbf{x}$ (*left*). The bias term shifts the function
            to left or right for a fixed value of $\mathbf{w}$ and $\mathbf{x}$ (*right*)   78
Fig. 2.48   A deeper network requires less neurons to approximate
            a function   81
Fig. 3.1   Every neuron in a fully connected layers is connected
            to every pixel in a grayscale image   86
Fig. 3.2   We can hypothetically arrange the neurons in blocks. Here,
            the neurons in the hidden layer have been arranged into
            50 blocks of size 12×12   87
Fig. 3.3   Neurons in each block can be connected locally to the input
            image. In this figure, each neuron is connected to a 5×5
            region in the image   87
Fig. 3.4   Neurons in one block can share the same set of weights
            leading to reduction in the number of parameters   88
Fig. 3.5   The above convolution layer is composed of 49 filters of size
            $5\times$. The output of the layer is obtained by convolving
            each filter on the image   89
Fig. 3.6   Normally, convolution filters in a ConvNet are
            three-dimensional array where the first two dimensions are
            arbitrary numbers and the third dimension is always equal to
            the number out channels in the previous layer   90

Fig. 3.7   From ConvNet point of view, an RGB image is a
three-channel input. The image is taken from www.
flickr.com.   91

Fig. 3.8   Two layers from middle of a neural network indicating the
one-dimensional convolution. The weight $\mathbf{W}^2$ is shared
among the neurons of $\mathbf{H}^2$. Also, δ_i shows the gradient of loss
functions with respect to H_i^2   92

Fig. 3.9   A pooling layer reduces the dimensionality of each feature
map separately   96

Fig. 3.10   A one-dimensional max-pooling layer where the neurons
in $\mathbf{H}^2$ compute the maximum of their inputs.   97

Fig. 3.11   Representing LeNet-5 using a DAG   98

Fig. 3.12   Representing AlexNet using a DAG   100

Fig. 3.13   Designing a ConvNet is an iterative process. Finding a good
architecture may require several iterations
of design–implement–evaluate   101

Fig. 3.14   A dataset is usually partitioned into three different parts
namely *training set*, *development set* and *test set*.   105

Fig. 3.15   For a binary classification problem, confusion matrix
is a 2×2 matrix.   108

Fig. 3.16   Confusion matrix in multiclass classification problems   109

Fig. 3.17   A linear model is highly biased toward data meaning that it is
not able to model nonlinearities in the data   115

Fig. 3.18   A nonlinear model is less biased but it may model any small
nonlinearity in data   116

Fig. 3.19   A nonlinear model may still overfit on a training set
with many samples   116

Fig. 3.20   A neural network with greater weights is capable of
modeling sudden changes in the output. The right decision
boundary is obtained by multiplying the third layer of the
neural network in left with 10.   118

Fig. 3.21   If dropout is activated on a layer, each neuron in the layer
will be attached to a blocker. The blocker blocks information
flow in the forward pass as well as the backward pass
(i.e., backpropagation) with probability p   120

Fig. 3.22   If the learning rate is kept fixed it may jump over local
minimum (*left*). But, annealing the learning rate helps the
optimization algorithm to converge to a local minimum   122

Fig. 3.23   Exponential learning rate annealing   122

Fig. 3.24   Inverse learning rate annealing   123

Fig. 3.25   Step learning rate annealing   124

Fig. 4.1   The Caffe library uses different third-party libraries and it
provides interfaces for C++, Python, and MATLAB
programming languages   132

Fig. 4.2   The NetParameter is indirectly connected to many other
            messages in the Caffe library   135
Fig. 4.3   A computational graph (neural network) with three layers   136
Fig. 4.4   Architecture of the network designed by the protobuf text.
            Dark rectangles show nodes. *Octagon* illustrates the name
            of the top element. The number of *outgoing arrows* in a node
            is equal to the length of top array of the node. Similarly, the
            number of *incoming arrows* to a node shows the length of
            bottom array of the node. The *ellipses* show the tops that
            are not connected to another node.   141
Fig. 4.5   Diagram of the network after adding a ReLU activation.   144
Fig. 4.6   Architecture of network after adding a pooling layer   145
Fig. 4.7   Architecture of network after adding a pooling layer   146
Fig. 4.8   Diagram of network after adding two fully connected layers
            and two dropout layers   147
Fig. 4.9   Final architecture of the network. The architecture is similar
            to the architecture of LeNet-5 in nature. The differences are
            in activations functions, dropout layer, and connection in
            middle layers   150
Fig. 5.1   Some of the challenges in classification of traffic signs.
            The signs have been collected in Germany and Belgium   168
Fig. 5.2   Sample images from the GTSRB dataset   173
Fig. 5.3   The image in the *middle* is the flipped version of the image in
            the *left*. The image in the *right* another sample from dataset.
            Euclidean distance from the *left* image to the *middle* image
            is equal to 25,012.461 and the Euclidean distance from the
            left image to the *right* image is equal to 27,639.447   177
Fig. 5.4   The original image in *top* is modified using Gaussian filtering
            (*first row*), motion blur (*second* and *third rows*), median
            filtering (*fourth row*), and sharpening (*fifth row*) with
            different values of parameters   181
Fig. 5.5   Augmenting the sample in Fig. 5.4 using random cropping
            (*first row*), hue scaling (*second row*), value scaling
            (*third row*), Gaussian noise (*fourth row*), Gaussian noise
            shared between channels (*fifth row*), and dropout (*sixth row*)
            methods with different configuration of parameters   184
Fig. 5.6   Accuracy of model on training and validation set tells us
            whether or not a model is acceptable or it suffers from high
            bias or high variance   188
Fig. 5.7   A ConvNet consists of two *convolution-hyperbolic
            activation-pooling* blocks without fully connected layers.
            Ignoring the activation layers, this network is composed of
            five layers   190

Fig. 5.8   Training, validation curve of the network illustrated
in Fig. 7 193

Fig. 5.9   Architecture of the network that won the GTSRB
competition (Ciresan et al. 2012a). 194

Fig. 5.10   Training/validation curve of the network illustrated in
Fig. 5.9 195

Fig. 5.11   Architecture of network in Aghdam et al. (2016a) along with
visualization of the first fully connected layer as well as the
last two pooling layers using the t-SNE method. *Light blue,
green, yellow* and *dark blue* shapes indicate convolution,
activation, pooling, and fully connected layers, respectively.
In addition, each *purple* shape shows a linear transformation
function. Each class is shown with a unique color in the
scatter plots 196

Fig. 5.12   Training/validation curve on the network illustrated in
Fig. 5.11 198

Fig. 5.13   Compact version of the network illustrated in Fig. 5.11 after
dropping the first fully connected layer and the subsequent
Leaky ReLU layer. 199

Fig. 5.14   Incorrectly classified images. The *blue* and *red* numbers
below each image show the actual and predicted class labels,
respectively. The traffic sign corresponding to each class
label is illustrated in Table 5.5 209

Fig. 5.15   Sample images from the BTSC dataset 210

Fig. 5.16   Incorrectly classified images from the BTSC dataset.
The *blue* and *red* numbers below each image show the actual
and predicted class labels, respectively. The traffic sign
corresponding to each class label is illustrated in Table 5.5 . . . 211

Fig. 5.17   The result of fine-tuning the ConvNet on the BTSC dataset
that is trained using GTSRB dataset. Horizontal axis shows
the layer n at which the network starts the weight adjustment.
In other words, weights of the layers before the layer n are
fixed (frozen). The weights of layer n and all layers after
layer n are adjusted on the BTSC dataset. We repeated the
fine-tuning procedure 4 times for each $n \in \{1, \ldots, 5\}$,
separately. *Red circles* show the accuracy of each trial and
blue squares illustrate the mean accuracy. The t-SNE
visualizations of the best network for $n = 3, 4, 5$ are also
illustrated. The t-SNE visualization is computed on the
$LReLU_4$ layer 213

Fig. 5.18   Minimum additive noise which causes the traffic sign to be
misclassified by the minimum different compared with the
highest score 216

Fig. 5.19   Plot of the SNRs of the noisy images found by optimizing
(5.7). The mean SNR and its variance are illustrated 216

Fig. 5.20   Visualization of the transformation and the first convolution
            layers   217
Fig. 5.21   Classification score of traffic signs averaged over
            20 instances per each traffic sign. The *warmer color* indicates
            a higher score and the colder color shows a lower score.
            The corresponding window of element (m, n) in the score
            matrix is shown for one instance. It should be noted that the
            (m, n) is the *top-left* corner of the window not its center and
            the size of the window is 20% of the image size in all the
            results   218
Fig. 5.22   Classification score of traffic signs averaged over
            20 instances per each traffic sign. The warmer color indicates
            a higher score. The corresponding window of element (m, n)
            in the score matrix is shown for one instance. It should be
            noted that the (m, n) is the *top-left* corner of the window not
            its center and the size of the window is 40% of the image size
            in all the results   219
Fig. 5.23   Classification score of traffic signs averaged over
            20 instances per each traffic sign. The warmer color indicates
            a higher score. The corresponding window of element (m, n)
            in the score matrix is shown for one instance. It should be
            noted that the (m, n) is the *top-left* corner of the window not
            its center and the size of the window is 40% of the image size
            in all the results   220
Fig. 5.24   Receptive field of some neurons in the last pooling layer   221
Fig. 5.25   Average image computed over each of 250 channels using
            the 100 images with highest value in position $(0, 0)$
            of the last pooling layer. The corresponding receptive field of
            this position is shown using a cyan rectangle   221
Fig. 5.26   The modified ConvNet architecture compare
            with Fig. 5.11   223
Fig. 5.27   Relation between the batch size and time-to-completion
            of the ConvNet   224
Fig. 5.28   Misclassified traffic sings. The *blue* and the *red* number
            indicate the actual and predicted class labels, respectively   226
Fig. 5.29   Lipschitz constant (*top*) and the correlation between
            $d(x, x + \mathcal{N}(0, \sigma))$ and $d(\mathscr{C}_{fc_2}(x), \mathscr{C}_{fc_2}(x + \mathcal{N}(0, \sigma)))$
            (*bottom*) computed on 100 samples from every category in
            the GTSRB dataset. The *red circles* are the noisy instances
            that are incorrectly classified. The size of each circle is
            associated with the values of σ in the Gaussian noise   228
Fig. 5.30   Visualizing the *relu4* (*left*) and the *pooling3* (*right*) layers in
            the classification ConvNet using the t-SNE method. Each
            class is shown using a different color.   229

Fig. 5.31 Histogram of leaking parameters 230
Fig. 6.1 The detection module must be applied on a high-resolution
image 236
Fig. 6.2 The ConvNet for detecting traffic signs. The *blue,green,* and
yellow color indicate a convolution, LReLU and pooling
layer, respectively. $C(c, n, k)$ denotes n convolution kernel of
size $k \times k \times c$ and $P(k, s)$ denotes a max-pooling layer
with pooling size $k \times k$ and stride s. Finally, the number in
the LReLU units indicate the leak coefficient of the activation
function 237
Fig. 6.3 Applying the trained ConvNet for hard-negative mining. 238
Fig. 6.4 Implementing the sliding window detector within the
ConvNet 240
Fig. 6.5 Architecture of the sliding window ConvNet 241
Fig. 6.6 Detection score computed by applying the fully
convolutional sliding network to 5 scales of the
high-resolution image 241
Fig. 6.7 Time to completion of the sliding ConvNet for different
strides. *Left* time to completion per resolution and *Right*
cumulative time to completion 242
Fig. 6.8 Distribution of traffic signs in different scales computed
using the training data 243
Fig. 6.9 *Top* precision-recall curve of the detection ConvNet along
with models obtained by HOG and LBP features. *Bottom*
Numerical values (%) of precision and recall for the
detection ConvNet. 244
Fig. 6.10 Output of the detection ConvNet before and after
post-processing the bounding boxes. A *darker* bounding box
indicate that it is detected in a lower scale image 245
Fig. 7.1 Visualizing classes of traffic signs by maximizing the
classification score on each class. The *top-left* image
corresponds to class 0. The class labels increase from *left* to
right and *top* to *bottom* 252
Fig. 7.2 Visualizing class saliency using a random sample from each
class. The order of images is similar Fig. 7.1 253
Fig. 7.3 Visualizing expected class saliency using 100 samples from
each class. The order of images is similar to Fig. 7.1. 254
Fig. 7.4 Reconstructing a traffic sign using representation of different
layers 257

Traffic Sign Detection and Recognition

1.1 Introduction

Assume you are driving at speed of 90 km/h in a one-way road and you are about to join a new road. Even though there was a "*danger: two way road*" sign in the junction, you have not seen the sign and you keep driving in opposite lane of the new road. This is a hazardous situation which may end up with a fatal accident because the driver assumes he or she is still driving in a two-way road. This was only a simple example in which failing to detect traffic sign may cause irreversible consequences. This danger gets even more serious with inexperienced drivers and senior drivers, specially, in unfamiliar roads.

According to National Safety Council, medically consulted motor-vehicle injuries for the first 6 months of 2015 were estimated to be about 2,254,000.[1] Also, World Health Organization reported that[2] there have been about 1,250,000 fatalities in 2015 due to car accidents. Moreover, another study shows that human error accounts solely for 57% of all accidents and it is a contributing factor in over 90% of accidents. The above example is one of the scenarios which may occur because of failing to identify traffic signs.

Furthermore, self-driving cars are going be commonly used in near future. They must also conform with the road rules in order not to endanger other users of road, Likewise, smart-cars try to assist human drivers and make driving more safe and comfortable. Advanced Driver Assistant System (ADAS) is a crucial component on these cars. One of the main tasks of this module is to recognize traffic signs. This helps a human driver to be aware of all traffic signs and have a more safe driving experience.

---

[1] www.nsc.org/NewsDocuments/2015/6-month-fatality-increase.pdf.
[2] www.who.int/violence_injury_prevention/road_safety_status/2015/GSRRS2015_data/en/.

© Springer International Publishing AG 2017
H. Habibi Aghdam and E. Jahani Heravi, *Guide to Convolutional Neural Networks*, DOI 10.1007/978-3-319-57550-6_1

1.2 Challenges

A Traffic signs recognition module is composed of two main steps including *detection* and *classification*. This is shown in Fig. 1.1. The detection stage scans image of scene in a multi-scale fashion and looks for location of traffic signs on the image. In this stage, the system usually does not distinguish one traffic sign from another. Instead, it decides whether or not a region includes a traffic sign regardless of its type. The output of detection stage is a set of regions in the image containing traffic signs. As it is shown in the figure, detection module might make mistakes and generate a few *false-positive* traffic signs. In other words, there could be a few regions in the output of detection module without any traffic sign. These outputs have been marked using a red (dashed) rectangle in the figure.

Next, classification module analyzes each region separately and determines type of each traffic sign. For example, there is one "no turning to left" sign, one "roundabout" sign, and one "give way" sign in the figure. There are also three "pedestrian crossing"

Fig. 1.1 Common pipeline
for recognizing traffic signs

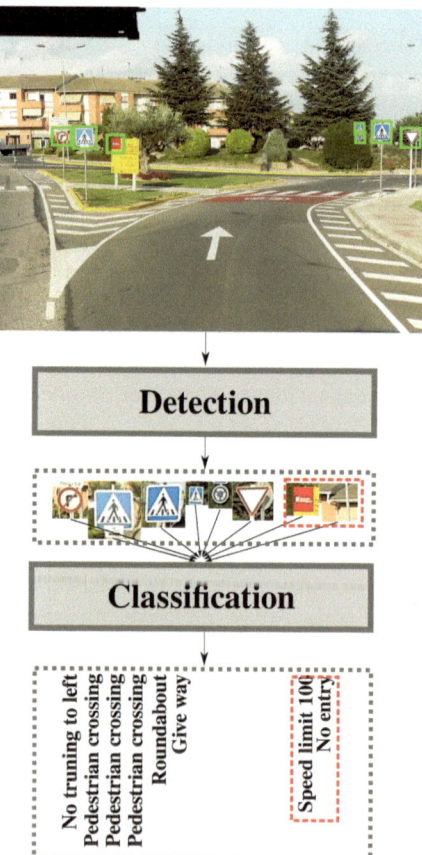

signs. Moreover, even though there is no traffic sign inside the false-positive regions, the classification module labels them into one of traffic sign classes. In this example, the false-positive regions have been classified as "speed limit 100" and "no entry" signs.

Dealing with false-positive regions generated by detection module is one of major challenges in developing a practical traffic sign recognition system. For instance, a self-driving car may suddenly brake in the above hypothetical example because it has detected a no entry sign. Consequently, one of the practical challenges in developing a detection module is to have zero false-positive region. Also, it has to detect all traffic signs in the image. Technically, its true-positive rate must be 100%. Satisfying these two criteria is not trivial in practical applications.

There are two major goals in designing traffic signs. First, they must be easily distinguishable from rest of objects in scene and, second, their meaning must be easily perceivable and independent from spoken language. To this end, traffic signs are designed with a simple geometrical shape such as triangle, circle, rectangle, or polygon. To be easily detectable from rest of objects, traffic signs are painted using basic colors such as red, blue, yellow, black, and white. Finally, the meaning of traffic signs is mainly carried out by pictographs in center of traffic signs. It should be noted that some signs heavily depend on text-based information. However, we can still think of the texts in traffic signs as pictographs.

Although classification of traffic signs is an easy task for a human, there are some challenges in developing an algorithm for this purpose. Some of these challenges are illustrated in Fig. 1.2. First, image of traffic signs might be captured from different perspectives. This may nonlinearly deform the shape of traffic signs.

Second, weather condition can dramatically affect the appearance of traffic signs. An example is illustrated in Fig. 1.2 where the "no stopping" sign is covered by snow. Third, traffic signs are being impaired during time and some artifacts may appear on signs which might have a negative impact on their classification. Fourth, traffic signs might be partially occluded by another signs or objects. Fifth, the pictograph area might be manipulated by human which in some cases might change the shape of the pictograph. Another important challenge is illumination variation caused by weather condition or daylight changes. The last and more important issue shown in this figure is pictograph differences of the same traffic sign from one country to another. More specifically, we observe that the "danger: bicycle crossing" sign posses important differences between two countries.

Referring to the Vienna Convention on Road Traffic Signs, we can find roughly 230 pictorial traffic signs. Here, text-based signs and variations on pictorial signs are counted. For example, the speed limit sign can have 24 variations including 12 variation for indicating speed limits and 12 variations for end of speed limit. Likewise, traffic signs such as recommended speed, minimum speed, minimum distance with front car, etc. may have several variations. Hence, traffic sign recognition is a large multi-class classification problem. This makes the problem even more challenging.

Note that some of the signs such as "crossroad" and "side road" signs differ only by very fine details. This is shown in Fig. 1.3 where both signs differ only in small

Fig. 1.2 Some of the challenges in classification of traffic signs. The signs have been collected in Germany and Belgium

Fig. 1.3 Fine differences between two traffic signs

part of pictograph. Looking at Fig. 1.1, we see signs which are only 30 m away from the camera occupy very small region in the image. Sometimes, these regions can be as small as 20×20 pixels. For this reason, identifying fine details become very difficult on these signs.

In sum, traffic sign classification is a specific case of object classification where the objects are more rigid and two dimensional. Also, their discriminating parts are well defined. However, there are many challenges in developing a practical system for detection and classification of traffic signs.

1.3   Previous Work

1.3.1   Template Matching

Arguably, the most trivial way for recognizing objects is *template matching*. In this method, a set of templates is stored on system. Given a new image, the template is matched with every location on the image and a score is computed for each location. The score might be computed using *cross correlation, sum of squared differences, normalized cross correlation*, or *normalized sum of squared differences*. Piccioli et al. (1996) stored a set of traffic signs as the templates. Then, the above approach was used in order to classify the input image. Note that template-matching approach can be used for both detection and classification.

In practice, there are many problems with this approach. First, it is sensitive to perspective, illumination and deformation. Second, it is not able to deal with low quality signs. Third, it might need a large dataset of templates to cover various kinds of samples for each traffic sign. For this reason, selecting appropriate templates is a tedious task.

On the one hand, template matching compares raw pixel intensities between the template and the source image. On the other hand, pixel intensities greatly depend on perspective, illumination, and deformation. As the result, a slight change in illumination may affect the matching score, significantly. To tackle with this problem, we usually apply some algorithms on the image in order to extract more useful information from it. In other words, in the case of grayscale images, a *feature extraction* algorithm accepts a $W \times H$ image and transforms the $W \times H$ dimensional vector into a D-dimensional vector in which the D-dimensional vector carries more useful information about the image and it is more tolerant to perspective changes, illumination, and deformation. Based on this idea, Gao et al. (2006) extracted shape features from both template and source image and matched these feature vectors instead of raw pixel intensity values. In this work, matching features were done using the Euclidean distance function. This is equivalent to the sum of square differences function. The main problem with this matching function was that every feature was equally important. To cope with this problem, Ruta et al. (2010) learned a similarity measure for matching the query sign with templates.

1.3.2   Hand-Crafted Features

The template matching procedure can be decomposed into two steps. In the first step, a template and an image patch are represented using more informative vectors called feature vectors. In the second step, feature vectors are compared in order to find class of the image patch. This approach is illustrated in Fig. 1.4. Traditionally, the second step is done using techniques of machine learning. We will explain the basics of this step in Sect. 2.1. However, roughly speaking, extracting a feature vector from an image can be done using *hand-crafted* or *automatic* methods.

Fig. 1.4 Traditional approach for classification of objects

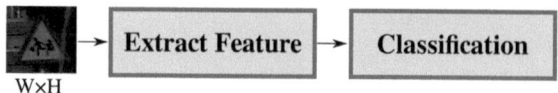

Hand-crafted methods are commonly designed by a human expert. They may apply series of transformations and computations in order to build a feature vector. For example, Paclík et al. (2000) generated a binary image depending on color of traffic sign. Then, moment invariant features were extracted from the binary image to form the feature vector. This method could be very sensitive to noise since a clean image and its degraded version may have two different binary images. Consequently, the moments of the binary images might vary significantly. Maldonado-Bascon et al. (2007) transformed the image into the HSI color space and calculated histogram of Hue and Saturation components. Although this feature vector can distinguish general category of traffic signs (for example, mandatory vs. danger signs), they might act poorly on modeling traffic signs of the same category. This is due to the fact that traffic signs of the same category have the same color and shape. For instance, all danger signs are triangle with a red margin. Therefore, the only difference would be the pictograph of signs. Since all pictographs are black, they will fall into the same bin on this histogram. As the result, theoretically, this bin will be the main source of information for classifying signs of same category.

In another method, Maldonado Bascón et al. (2010) classified traffic signs using only the pictograph of each sign. To this end, they first segment the pictograph from the image of traffic sign. Although the region of pictograph is binary, accurate segmentation of a pictograph is not a trivial task since automatic thresholding methods such as Otsu might fail taking into account the illumination variation and unexpected noise in real-world applications. For this reason, Maldonado Bascón et al. (2010) trained SVM where the input is a 31 × 31 block of pixels in a grayscale version of pictograph. In a more complicated approach, Baró et al. (2009) proposed an Error Correcting Output Code framework for classification of 31 traffic signs and compared their method with various approaches.

Zaklouta et al. (2011), Zaklouta and Stanciulescu (2012), and Zaklouta and Stanciulescu (2014) utilized more sophisticated feature extraction algorithm called Histogram of Oriented Gradient (HOG). Broadly speaking, the first step in extracting HOG feature is to compute the gradients of the image in x and y directions. Then, the image is divided into non-overlapping regions called *cells*. A histogram is computed for each cell. Bins of the histogram show the orientation of the gradient vector. Value of each bin is computed by accumulating the gradient magnitudes of the pixels in each cell. Next, blocks are formed using neighbor cells. Blocks may have overlap with each other. Histogram of a block is obtained by concatenating histograms of the cells within the block. Finally, histogram of each block is normalized and final feature vector is obtained by concatenating the histogram of all blocks.

This method is formulated using size of each cell, size of each block, number of bins in histograms of cell, and type of normalization. These parameters are called *hyperparameters*. Depending on the value of these parameters we can obtain different feature vectors with different lengths on the same image. HOG is known to be a powerful hand-crafted feature extraction algorithm. However, objects might not be linearly separable in the feature space. For this reason, Zaklouta and Stanciulescu (2014) trained a Random Forest and a SVM for classifying traffic sings using HOG features. Likewise, Greenhalgh and Mirmehdi (2012), Moiseev et al. (2013), Mathias et al. (2013), Huang et al. (2013), and Sun et al. (2014) extracted the HOG features. The difference between these works mainly lies on their classification model (e.g., SVM, Cascade SVM, Extreme Learning Machine, Nearest Neighbor, and LDA). However, in contrast to the other works, Huang et al. (2013) used a two-level classification model. In the first level, the image is classified into one of super-classes. Each super-class contains several traffic signs with similar shape/color. Then, the perspective of the input image is adjusted based on its super-class and another classification model is applied on the adjusted image. The main problem of this method is sensitivity of the final classification to the adjustment procedure.

Mathias et al. (2013) proposed a more complicated procedure for extracting features. Specifically, the first extracted HOG features with several configurations of hyperparameters. In addition, they extracted more feature vectors using different methods. Finally, they concatenated all these vectors and built the final feature vector. Notwithstanding, there are a few problems with this method. Their feature vector is a 9000-dimensional vector constructed by applying five different methods. This high-dimensional vector is later projected to a lower dimensional space using a transformation matrix.

1.3.3 Feature Learning

A hand-crafted feature extraction method is developed by an expert and it applies series of transformations and computations in order to extract the final vector. The choice of these steps completely depends on the expert. One problem with the hand-crafted features is their limited representation power. This causes that some classes of objects overlap with other classes which adversely affect the classification performance. Two common approaches for partially alleviating this problem are to develop a new feature extraction algorithm and to combine various methods. The problems with these approaches are that devising a new hand-crafted feature extraction method is not trivial and combining different methods might not separate the overlapping classes.

The basic idea behind feature learning is *to learn features from data*. To be more specific, given a dataset of traffic signs, we want to learn a mapping $\mathcal{M} : \mathbb{R}^d \to \mathbb{R}^n$ which accepts $d = W \times H$-dimensional vectors and returns an n-dimensional vector. Here, the input is a flattened image that is obtained by putting the rows of image next to each other and creating a one-dimensional array. The mapping $\mathcal{M}$ can be any arbitrary function that is linear or nonlinear. In the simplest scenario, $\mathcal{M}$ can be a linear function such as

$$\mathcal{M}(x) = W^+(x^T - \bar{x}), \qquad (1.1)$$

where $W \in \mathbb{R}^{d \times n}$ is a weight matrix, $x \in \mathbb{R}^d$ is the flattened image, and $\bar{x} \in \mathbb{R}^d$ is the flattened mean image. Moreover, $W^+ = (W^T W)^{-1} W^T$ denotes the MoorePenrose pseudoinverse of W. Given the matrix W we can map every image into a n-dimensional space using this linear transformation. Now, the question is how to find the values of W?

In order to obtain W, we must devise an objective and try to get as close as possible to the objective by changing the values of W. For example, assume our objective is to project x into a five-dimensional space where the projection is done arbitrarily. It is clear that any $W \in \mathbb{R}^{3 \times d}$ will serve our purpose. Denoting $\mathcal{M}(x)$ with z, our aim might be to project data on a $n \leq d$ space while maximizing the variance of z. The W that is found using this objective function is called *principal component analysis*. Bishop (2006) has explained that to find W that maximizes this objective function, we must compute the covariance matrix of data and find eigenvectors and eigenvalues of the covariance matrix. Then, the eigenvectors are sorted according to their eigenvalues in descending order and the first n eigenvectors are picked to form W.

Now, given any $W \times H$ image, we plug it in (1.1) to compute z. Then, the n-dimensional vector z is used as the feature vector. This method is previously used by Sirovich and Kirby (1987) for modeling human faces. Fleyeh and Davami (2011) also projected the image into the principal component space and found class of the image by computing the Euclidean distance of the projected image with the images in the database.

If we multiply both sides with W and rearrange (1.1) we will obtain

$$x^T = Wz + \bar{x}^T. \qquad (1.2)$$

Assume that $\bar{x}^T = \mathbf{0}$. Technically, we say our data is *zero-centered*. According to this equation, we can *reconstruct* x using W and its mapping z. Each column in W is a d-dimensional vector which can be seen as a template learnt from data. With this intuition, the first row in W shows set of values of first pixel in our dictionary of templates. Likewise, n^{th} row in W is set of values of nth pixel in the templates. Consequently, the vector z shows how to linearly combine these templates in order to reconstruct the original image. As the value of n increases, the reconstruction error decreases.

The value of W depends directly on the data that we have used during the training stage. In other words, using the training data, a system learns to extract features. However, we do not take into account the class of objects in finding W. In general, methods that do not consider the class of object are called *unsupervised* methods.

One limitation of principal component analysis is that $n \leq d$. Also, z is a real vector which is likely to be non-sparse. We can simplify (1.2) by omitting the second term.

Now, our objective is to find W and z by minimizing the constrained reconstruction error:

$$E = \sum_{i=1}^{N} \|x_i^T - W z_i\|_2^2 \quad s.t. \quad \|z\|_1 < \mu, \qquad (1.3)$$

where μ is a user-defined value and N is the number of training images. W and z_i also have the same meaning as we mentioned above. The $\mathscr{L}_1$ constrained in the above equation forces z_i to be *sparse*. A vector is called sparse when most of its elements are zero. Minimizing the above objective function requires an alternative optimization of W and z_i. This method is called *sparse coding*. Interested readers can find more details in Mairal et al. (2014). It should be noted that there are other formulations for objective function and the constraint.

There are two advantages with the sparse coding approach compared with principal component analysis. First, the number of columns in W (i.e., n) is not restricted to be smaller than d. Second, z_i is a sparse vector. Sparse coding has been also used to encode images of traffic signs.

Hsu and Huang (2001) coded each traffic sign using the *Matching Pursuit* algorithm. During testing, the input image is projected to different sets of filter bases to find the best match. Lu et al. (2012) proposed a graph embedding approach for classifying traffic signs. They preserved the sparse representation in the original space using $L_{1,2}$ norm. Liu et al. (2014) constructed the dictionary by applying k-means clustering on the training data. Then, each data is coded using a novel coding input similar to Local Linear Coding approach (Wang et al. 2010). Moreover, Aghdam et al. (2015) proposed a method based on visual attributes and Bayesian network. In this method, each traffic sign is described in terms of visual attributes. In order to detect visual attributes, the input image is divided into several regions and each region is coded using *elastic net* sparse coding method. Finally, attributes are detected using a random forest classifier. The detected attributes are further refined using a Bayesian network. Figure 1.5 illustrates a dictionary learnt by Aghdam et al. (2015) from 43 classes of traffic signs.

There are other unsupervised feature learning techniques. Among them, *autoencoders*, *deep belief networks*, and *independent component analysis* have been extensively studied and used in the computer vision community. One of the major disadvantages of unsupervised feature learning methods is that they do not consider the class of objects during the learning process. More accurate results have been obtained using supervised feature learning methods. As we will discuss in Chap. 3, convolutional neural networks (ConvNet) have shown a great success in classification and detection of objects.

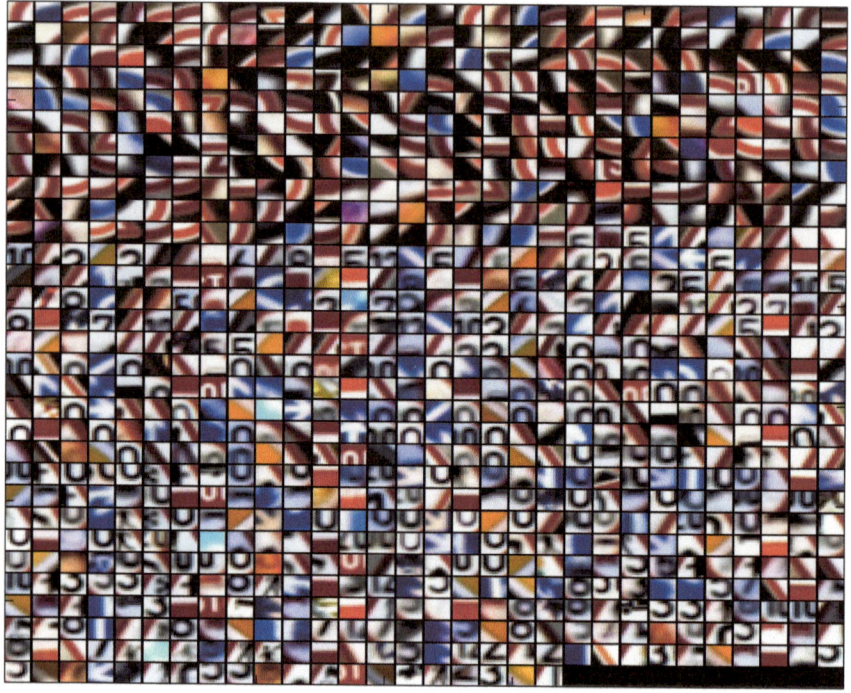

Fig. 1.5 Dictionary learnt by Aghdam et al. (2015) from 43 classes of traffic signs

1.3.4 ConvNets

[3]ConvNets were first utilized by Sermanet and Lecun (2011) and Ciresan et al. (2012) in the field of traffic sign classification during the German Traffic Sign Recognition Benchmark (GTSRB) competition where the ensemble of ConvNets designed by Ciresan et al. (2012) surpassed human performance and won the competition by correctly classifying 99.46% of test images. Moreover, the ConvNet of Sermanet and Lecun (2011) ended up in the second place with a considerable difference compared with the third place which was awarded for a method based on the traditional classification approach. The classification accuracies of the runner up and the third place were 98.97 and 97.88%, respectively.

Ciresan et al. (2012) constructs an ensemble of 25 ConvNets each consists of 1,543,443 parameters. Sermanet and Lecun (2011) creates a single network defined by 1,437,791 parameters. Furthermore, while the winner ConvNet uses the *hyperbolic* activation function, the runner-up ConvNet utilizes the *rectified sigmoid* as the activation function. It is a common practice in ConvNets to make a prediction by calculating the average score of slightly transformed versions of the query image.

---

[3]We shall explain all technical details of this section in the rest of this book.

However, it is not clearly mentioned in Sermanet and Lecun (2011) that how do they make a prediction. In particular, it is not clear that the runner-up ConvNet classifies solely the input image or it classifies different versions of the input and fuses the scores to obtain the final result.

Regardless, both methods suffer from the high number of arithmetic operations. To be more specific, they use highly computational activation functions. To alleviate these problems, Jin et al. (2014) proposed a new architecture including 1,162,284 parameters and utilizing the *rectified linear unit* (ReLU) activations (Krizhevsky et al. 2012). In addition, there is a Local Response Normalization layer after each activation layer. They built an ensemble of 20 ConvNets and classified 99.65% of test images correctly. Although the number of parameters is reduced using this architecture compared with the two networks, the ensemble is constructed using 20 ConvNets which is not still computationally efficient in real-world applications. It is worth mentioning that a ReLU layer and a Local Response Normalization layer together needs approximately the same number of arithmetic operations as a single hyperbolic layer. As the result, the run-time efficiency of the network proposed in Jin et al. (2014) might be close to Ciresan et al. (2012).

Recently, Zeng et al. (2015) trained a ConvNet to extract features of the image and replaced the classification layer of their ConvNet with an Extreme Learning Machine (ELM) and achieved 99.40% accuracy on the GTSRB dataset. There are two issues with their approach. First, the output of last convolution layer is a 200-dimensional vector which is connected to 12,000 neurons in the ELM layer. This layer is solely defined by $200 \times 12,000 + 12,000 \times 43 = 2,916,000$ parameters which makes it impractical. Besides, it is not clear why their ConvNet reduces the dimension of the feature vector from $250 \times 16 = 4000$ in Layer 7 to 200 in Layer 8 and then map their lower dimensional vector to 12,000 dimensions in the ELM layer (Zeng et al. 2015, Table 1). One reason might be to cope with calculation of the matrix inverse during training of the ELM layer. Finally, since the input connections of the ELM layer are determined randomly, it is probable that their ConvNet does not generalize well on other datasets.

The common point about all the above ConvNets is that they are only suitable for the *classification* module and they cannot be directly used in the task of *detection*. This is due to the fact that applying these ConvNets on high-resolution images is not computationally feasible. On the other hand, accuracy of the *classification* module also depends on the *detection* module. In other words, any false-positive results produced by the detection module will be entered into the classification module and it will be classified as one of traffic signs. Ideally, the false-positive rate of the detection module must be zero and its true-positive rate must be 1. Achieving this goal usually requires more complex image representation and classification models. However, as the complexity of these models increases, the detection module needs more time to complete its task.

The ConvNets proposed for traffic sign classification can be explained from three perspectives including *scalability*, *stability*, and *run-time*. From generalization point of view, none of the four ConvNets have assessed the performance on other datasets. It is crucial to study how the networks perform when the signs slightly change from one country to another. More importantly, the transferring power of the network must be estimated by fine-tuning the same architecture on a new dataset with various numbers of classes. By this way, we are able to estimate the *scalability* of the networks. From *stability* perspective, it is crucial to find out how tolerant is the network against noise and occlusion. This might be done through generating a few noisy images and fetch them to the network. However, this approach does not find the minimum noisy image which is misclassified by the network. Finally, the *run-time efficiency* of the ConvNet must be examined. This is due to the fact that the ConvNet has to consume as few CPU cycles as possible to let other functions of ADAS perform in real time.

1.4 Summary

In this chapter, we formulated the problem of traffic sign recognition in two stages namely detection and classification. The detection stage is responsible for locating regions of image containing traffic signs and the classification stage is responsible for finding class of traffic signs. Related work in the field of traffic sign detection and classification is also reviewed. We mentioned several methods based on hand-crafted features and then introduced the idea behind feature learning. Then, we explained some of the works based on convolutional neural networks.

References

Aghdam HH, Heravi EJ, Puig D (2015) A unified framework for coarse-to-fine recognition of traffic signs using Bayesian network and visual attributes. In: Proceedings of the 10th international conference on computer vision theory and applications, pp 87–96. doi:10.5220/0005303500870096

Baró X, Escalera S, Vitrià J, Pujol O, Radeva P (2009) Traffic sign recognition using evolutionary adaboost detection and forest-ECOC classification. IEEE Trans Intell Transp Syst 10(1):113–126. doi:10.1109/TITS.2008.2011702

Bishop CM (2006) Pattern recognition and machine learning. Information science and statistics. Springer, New York

Ciresan D, Meier U, Schmidhuber J (2012) Multi-column deep neural networks for image classification. In: 2012 IEEE conference on computer vision and pattern recognition. IEEE, pp 3642–3649. doi:10.1109/CVPR.2012.6248110, arXiv:1202.2745v1

Fleyeh H, Davami E (2011) Eigen-based traffic sign recognition. IET Intell Transp Syst 5(3):190. doi:10.1049/iet-its.2010.0159

Gao XW, Podladchikova L, Shaposhnikov D, Hong K, Shevtsova N (2006) Recognition of traffic signs based on their colour and shape features extracted using human vision models. J Visual Commun Image Represent 17(4):675–685. doi:10.1016/j.jvcir.2005.10.003

Greenhalgh J, Mirmehdi M (2012) Real-Time Detection and Recognition of Road Traffic Signs. Ieee Transactions on Intelligent Transportation Systems 13(4):1498–1506. doi:10.1109/tits.2012.2208909

Hsu SH, Huang CL (2001) Road sign detection and recognition using matching pursuit method. Image Vis Comput 19(3):119–129. doi:10.1016/S0262-8856(00)00050-0

Huang GB, Mao KZ, Siew CK, Huang DS (2013) A hierarchical method for traffic sign classification with support vector machines. In: The 2013 international joint conference on neural networks (IJCNN). IEEE, pp 1–6. doi:10.1109/IJCNN.2013.6706803

Jin J, Fu K, Zhang C (2014) Traffic sign recognition with hinge loss trained convolutional neural networks. IEEE Trans Intell Transp Syst 15(5):1991–2000. doi:10.1109/TITS.2014.2308281

Krizhevsky A, Sutskever I, Hinton G (2012) Imagenet classification with deep convolutional neural networks. In: Advances in neural information processing systems. Curran Associates, Inc., pp 1097–1105

Liu H, Liu Y, Sun F (2014) Traffic sign recognition using group sparse coding. Inf Sci 266:75–89. doi:10.1016/j.ins.2014.01.010

Lu K, Ding Z, Ge S (2012) Sparse-representation-based graph embedding for traffic sign recognition. IEEE Trans Intell Transp Syst 13(4):1515–1524. doi:10.1109/TITS.2012.2220965

Mairal J, Bach F, Ponce J (2014) Sparse modeling for image and vision processing. Found Trends Comput Graph Vis 8(2–3):85–283. doi:10.1561/0600000058

Maldonado Bascón S, Acevedo Rodríguez J, Lafuente Arroyo S, Fernndez Caballero A, López-Ferreras F (2010) An optimization on pictogram identification for the road-sign recognition task using SVMs. Comput Vis Image Underst 114(3):373–383. doi:10.1016/j.cviu.2009.12.002

Maldonado-Bascon S, Lafuente-Arroyo S, Gil-Jimenez P, Gomez-Moreno H, Lopez-Ferreras F (2007) Road-sign detection and recognition based on support vector machines. IEEE Trans Intell Transp Syst 8(2):264–278. doi:10.1109/TITS.2007.895311

Mathias M, Timofte R, Benenson R, Van Gool L (2013) Traffic sign recognition - How far are we from the solution? In: Proceedings of the international joint conference on neural networks. doi:10.1109/IJCNN.2013.6707049

Moiseev B, Konev A, Chigorin A, Konushin A (2013) Evaluation of traffic sign recognition methods trained on synthetically generated data. In: 15th international conference on advanced concepts for intelligent vision systems (ACIVS). Springer, Poznań, pp 576–583

Paclík P, Novovičová J, Pudil P, Somol P (2000) Road sign classification using Laplace kernel classifier. Pattern Recognit Lett 21(13–14):1165–1173. doi:10.1016/S0167-8655(00)00078-7

Piccioli G, De Micheli E, Parodi P, Campani M (1996) Robust method for road sign detection and recognition. Image Vis Comput 14(3):209–223. doi:10.1016/0262-8856(95)01057-2

Ruta A, Li Y, Liu X (2010) Robust class similarity measure for traffic sign recognition. IEEE Trans Intell Transp Syst 11(4):846–855. doi:10.1109/TITS.2010.2051427

Sermanet P, Lecun Y (2011) Traffic sign recognition with multi-scale convolutional networks. In: Proceedings of the international joint conference on neural networks, pp 2809–2813. doi:10.1109/IJCNN.2011.6033589

Sirovich L, Kirby M (1987) Low-dimensional procedure for the characterization of human faces. J Opt Soc Am A 4(3):519–524. doi:10.1364/JOSAA.4.000519, http://josaa.osa.org/abstract.cfm?URI=josaa-4-3-519

Sun ZL, Wang H, Lau WS, Seet G, Wang D (2014) Application of BW-ELM model on traffic sign recognition. Neurocomputing 128:153–159. doi:10.1016/j.neucom.2012.11.057

Wang J, Yang J, Yu K, Lv F, Huang T, Gong Y (2010) Locality-constrained linear coding for image classification. In: 2010 IEEE computer society conference on computer vision and pattern recognition. IEEE, pp 3360–3367. doi:10.1109/CVPR.2010.5540018

Zaklouta F, Stanciulescu B (2012) Real-time traffic-sign recognition using tree classifiers. IEEE Trans Intell Transp Syst 13(4):1507–1514. doi:10.1109/TITS.2012.2225618

Zaklouta F, Stanciulescu B (2014) Real-time traffic sign recognition in three stages. Robot Auton Syst 62(1):16–24. doi:10.1016/j.robot.2012.07.019

Zaklouta F, Stanciulescu B, Hamdoun O (2011) Traffic sign classification using K-d trees and random forests. In: Proceedings of the international joint conference on neural networks, pp 2151–2155. doi:10.1109/IJCNN.2011.6033494

Zeng Y, Xu X, Fang Y, Zhao K (2015) Traffic sign recognition using deep convolutional networks and extreme learning machine. In: Intelligence science and big data engineering. Image and video data engineering (IScIDE). Springer, pp 272–280

Pattern Classification

Machine learning problems can be broadly classified into *supervised learning*, *unsupervised learning* and *reinforcement learning*. In supervised learning, we have set of *feature vectors* and their corresponding *target* values. The aim of supervised learning is to learn a model to accurately predict targets given unseen feature vectors. In other words, the computer must learn a mapping from feature vectors to target values. The feature vectors might be called independent variable and the target values might be called dependent variable. Learning is done using and objective function which directly depends on target values. For example, classification of traffic signs is a supervised learning problem.

In unsupervised setting, we only have a set of feature vectors without any target value. The main goal of unsupervised learning is to learn structure of data. Here, because target values do not exist, there is not a specific way to evaluate learnt models. For instance, assume we have a dataset with 10,000 records in which each data is a vector consists of [driver's age, driver's gender, driver's education level, driving experience, type of car, model of car, car manufacturer, GPS point of accident, temperature, humidity, weather condition, daylight, time, day of week, type of road]. The goal might be to divide this dataset into 20 categories. Then, we can analyze categories to see how many records fall into each category and what is common among these records. Using this information, we might be able to say in which conditions car accidents happen more frequently. As we can see in this example, there is not a clear way to tell how well the records are categorized.

Reinforcement learning usually happens in dynamic environments where series of actions lead the system into a point of getting a reward or punishment. For example, consider a system that is learning to drive a car. The system starts to driver and several seconds later it hits an obstacle. Series of actions has caused the system to hit the obstacle. Notwithstanding, there is no information to tell us how good was the action which the systems performed at a specific time. Instead, the system is

© Springer International Publishing AG 2017
H. Habibi Aghdam and E. Jahani Heravi, *Guide to Convolutional Neural Networks*, DOI 10.1007/978-3-319-57550-6_2

punished because it hit the obstacle. Now, the system must figure out which actions were not correct and act accordingly.

2.1 Formulation

Supervised learning mainly breaks down into *classification* and *regression*. The main difference between them is the type of target values. While target values of a regression problem are real/discrete numbers, target values of a classification problem are categorical numbers which are called *labels*. To be more specific, assume $\mathscr{F}_r : \mathbb{R}^d \to \mathbb{R}$ is a regression model which returns a real number. Moreover, assume we have the pair $(\mathbf{x}_r, y_r)$ including a d-dimensional input vector $\mathbf{x}_r$ and real number y_r. Ideally, $\mathscr{F}_r(\mathbf{x}_r)$ must be equal to y_r. In other words, we can evaluate the accuracy of the prediction by simply computing $|\mathscr{F}_r(\mathbf{x}_r) - y_i|$.

In contrast, assume the classification model

$$\mathscr{F}_c : \mathbb{R}^d \to \{speedlimit, danger, prohibitive, mandatory\} \qquad (2.1)$$

which returns a categorical number/label. Given the pair $(\mathbf{x}_c, danger)$, $\mathscr{F}_c(\mathbf{x}_c)$ must be ideally equal to *danger*. However, it might return *mandatory* wrongly. It is not possible to simply subtract the output of $\mathscr{F}_c$ with the actual label to ascertain how much the model has deviated from actual output. The reason is that there is not a specific definition of *distance* between labels. For example, we cannot tell what is the distance between "danger" and "prohibitive" or "danger" and "mandatory". In other words, the label space is not an ordered set. Both traffic sign detection and recognition problems are formulated using a classification model. In the rest of this section, we will explain the fundamental concepts using simple examples.

Assume a set of pairs $\mathscr{X} = \{(\mathbf{x}_0, y_0), \ldots, (\mathbf{x}_n, y_n)\}$ where $\mathbf{x}_i \in \mathbb{R}^2$ is a two-dimensional input vector and $y_i \in \{0, 1\}$ is its label. Despite the fact that 0 and 1 are numbers, we treat them as categorical labels. Therefore, it is not possible to compute their distance. The target value y_i in this example can only take one of the two values. These kind of classification problems in which the target value can only take two values are called *binary classification* problems. In addition, because the input vectors are two-dimensional we can easily plot them. Figure 2.1 illustrates the scatter plot of a sample $\mathscr{X}$.

The blue squares show the points belonging to one class and the pink circles depicts the points belonging to the other class. We observe that the two classes overlap inside the green polygon. In addition, the vectors shown by the green arrows are likely to be noisy data. More importantly, these two classes are not *linearly separable*. In other words, it is not possible to perfectly separate these two classes from each other by drawing a line on the plane.

Assume we are given a $x_q \in \mathbb{R}^2$ and we are asked to tell which class x_q belongs to. This point is shown using a black arrow on the figure. Note that we do not know the target value of x_q. To answer this question, we first need to learn a model from

Fig. 2.1 A dataset of two-dimensional vectors representing two classes of objects

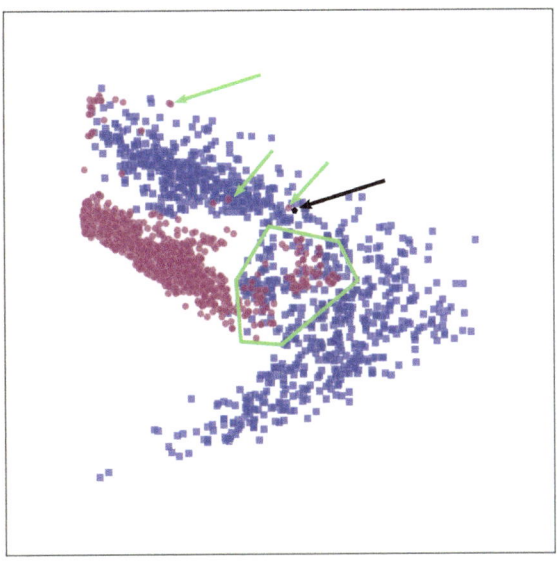

$\mathscr{X}$ which is able to discriminate the two classes. There are many ways to achieve this goal in literature. However, we are only interested in a particular technique called *linear models*. Before explaining this technique, we mention a method called *k-nearest neighbor*.

2.1.1 K-Nearest Neighbor

From one perspective, machine learning models can be categorized into *parametric* and *nonparametric* models. Roughly speaking, parametric models have some parameters which are directly learnt from data. In contrast, nonparametric models do not have any parameters to be learnt from data. K-nearest neighbor (KNN) is a nonparametric method which can be used in regression and classification problem.

Given the training set $\mathscr{X}$, KNN stores all these samples in memory. Then, given the query vector $\mathbf{x}_q$, it finds K closest samples from $\mathscr{X}$ to x_q.[1] Denoting the K closest neighbors of $\mathbf{x}_q$ with $N_K(\mathbf{x}_q; \mathscr{X})$,[2] the class of $\mathbf{x}_q$ is determined by:

$$F(\mathbf{x}_q) = \arg\max_{v \in \{0,1\}} \sum_{p \in N_K(\mathbf{x}_q)} \delta(v, f(p)) \tag{2.2}$$

---

[1]Implementations of the methods in this chapter are available at *github.com/pcnn/*.
[2]You can read this formula as "N_K of $\mathbf{x}_q$ given the dataset $\mathscr{X}$".

Fig. 2.2 K-nearest neighbor
looks for the K closets points
in the training set to the
query point

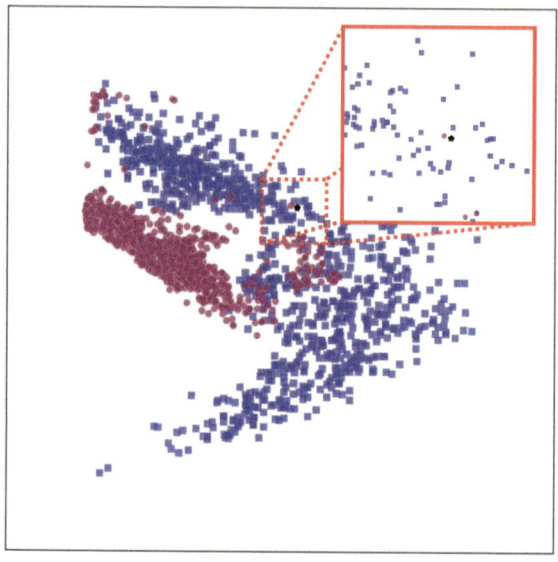

$$\delta(a, b) = \begin{cases} 1 & a = b \\ 0 & a \neq b \end{cases} \qquad (2.3)$$

where $f(p)$ returns the label of training sample $p \in \mathscr{X}$. Each of K closest neighbors
vote for $\mathbf{x}_q$ according to their label. Then, the above equation counts the votes and
returns the majority of votes as the class of $\mathbf{x}_q$. We explain the meaning of this
equation on Fig. 2.2. Assuming $K = 1$, KNN looks for the closest point to $\mathbf{x}_q$ in
the training set (shown by black polygon on the figure). According to the figure,
the red circle is the closest point. Because $K = 1$, there is no further point to vote.
Consequently, the algorithm classifies $\mathbf{x}_q$ as red.

By setting $K = 2$ the algorithm searches the two closest points which in this case
are one red circle and one blue square. Then, the algorithm counts the votes for each
label. The votes are equal in this example. Hence, the method is not confident with
its decision. For this reason, in practice, we set K to an odd number so one of the
labels always has the majority of votes. If we set $K = 3$, there will be two votes for
the blue class and one vote for the red class. As the result, $\mathbf{x}_q$ will be classified as
blue.

We classified every point on the plane using different values of K and $\mathscr{X}$.
Figure 2.3 illustrates the result. The black solid line on the plots shows the border between two regions with different class labels. This border is called *decision
boundary*. When $K = 1$ there is always a region around the noisy points, where
they are classified as the red class. However, by setting $K = 3$ those noisy regions
disappear and they become part the correct class. As the value of K increases, the
decision boundary becomes more smooth and small regions disappear.

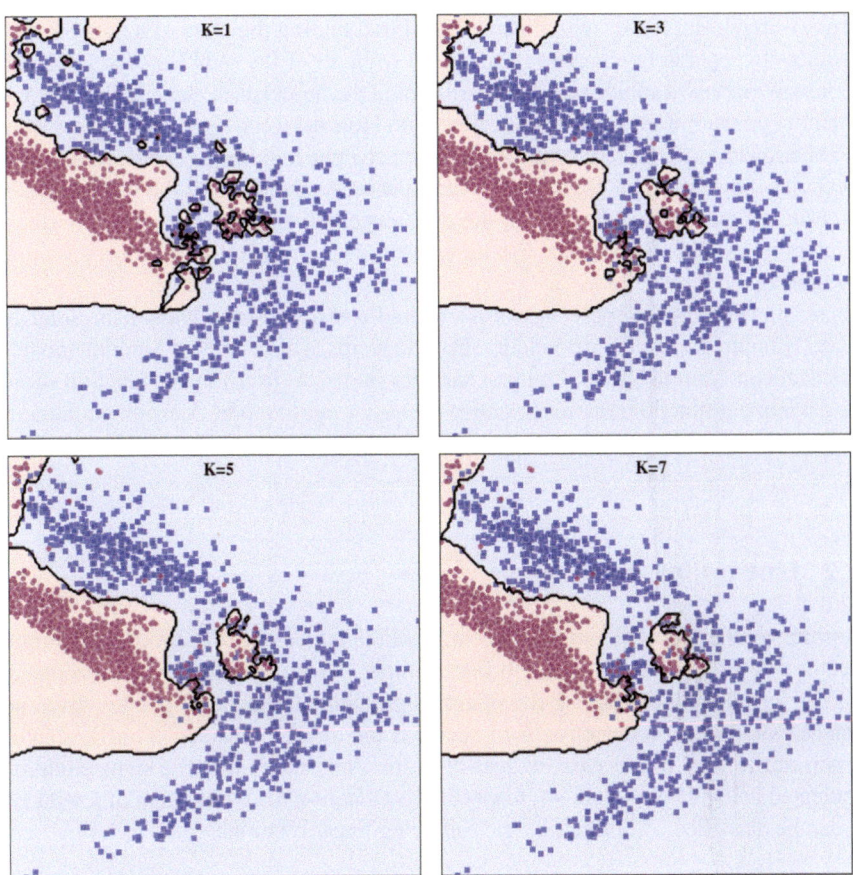

Fig. 2.3 K-nearest neighbor applied on every point on the plane for different values of K

The original KNN does not take into account the distance of its neighbor when it counts the votes. In some cases, we may want to weight the votes based on the distance from neighbors. This can be done by adding a weight term to (2.2):

$$F(\mathbf{x}_q) = \arg\max_{v \in \{0,1\}} \sum_{p \in N_K(\mathbf{x}_q)} w_i \, \delta(v, f(p)) \tag{2.4}$$

$$w_i = \frac{1}{d(\mathbf{x}_q, p)}. \tag{2.5}$$

In the above equation, $d(.)$ returns the distance between two vectors. According to this formulation, the weight of each neighbor is equal to the inverse of its distance from $\mathbf{x}_q$. Therefore, closer neighbors have higher weights. KNN can be easily extended to datasets with more than two labels without any modifications. However, there are two important issues with this method. First, finding the class of a query vector requires to separately compute the distance from all of the samples in training set. Unless we devise a solution such as partitioning the input space, this can be time and memory consuming when we are dealing with large datasets. Second, it suffers from a phenomena called *curse of dimensionality*. To put it simply, Euclidean distance becomes very similar in high-dimensional spaces. As the result, if the input of KNN is a high-dimensional vector then the difference between the closest and farthest vectors might be very similar. For this reason, it might classify the query vectors incorrectly.

To alleviate these problems, we try to find a *discriminant function* in order to directly model the decision boundary. In other words, a discriminant function models the decision boundary using training samples in $\mathscr{X}$. A discriminant function could be a nonlinear function. However, one of the easy ways to model decision boundaries is linear classifiers.

2.2 Linear Classifier

Assume a binary classification problem in which labels of the d-dimensional input vector $\mathbf{x} \in \mathbb{R}^d$ can be only 1 or -1. For example, detecting traffic signs in an image can be formulated as a binary classification problem. To be more specific, given an image patch, the aim detection is to decide if the image represents a traffic sign or a non-traffic sign. In this case, images of traffic signs and non-traffic signs might be indicated using labels 1 and -1, respectively. Denoting the ith element of $\mathbf{x}$ with x_i, it can be classified by computing the following linear relation:

$$f(\mathbf{x}) = w_1 x_1 + \cdots + w_i x_i + \cdots + w_d x_d + b \qquad (2.6)$$

where w_i is a trainable parameter associated with x_i and b is another trainable parameter which is called *intercept* or *bias*. The above equation represents a *hyperplane* in a d-dimensional Euclidean space. The set of weights $\{\forall_{i=1...d} w_i\}$ determines the orientation of the hyperplane and b indicates the distance of the hyperplane from origin. We can also write the above equation in terms of matrix multiplications:

$$f(\mathbf{x}) = \mathbf{w}\mathbf{x}^T + b \qquad (2.7)$$

where $\mathbf{w} = [w_1, \ldots, w_d]$. Likewise, it is possible to augment $\mathbf{w}$ with b and show all parameters of the above equation in a single vector $\mathbf{w}_{w|b} = [b, w_1, \ldots, w_d]$. With this formulation, we can also augment $\mathbf{x}$ with 1 to obtain $\mathbf{x}_{x|1} = [1, x_1, \ldots, x_d]$ and write the above equation using the following matrix multiplication:

$$f(\mathbf{x}) = \mathbf{w}_{w|b}\mathbf{x}_{x|1}^T. \qquad (2.8)$$

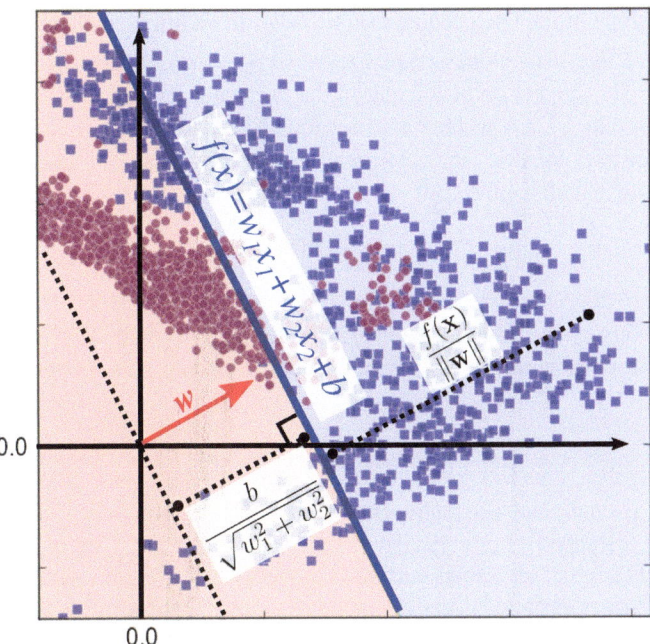

Fig. 2.4 Geometry of linear models

From now on in this chapter, when we write $\mathbf{w}$, $\mathbf{x}$ we are referring to $\mathbf{w}_{w|b}$ and $\mathbf{x}_{x|1}$, respectively. Finally, $\mathbf{x}$ is classified by applying the sign function on $f(\mathbf{x})$ as follows:

$$F(\mathbf{x}) = \begin{cases} 1 & f(\mathbf{x}) > 0 \\ NA & f(\mathbf{x}) = 0 \\ -1 & f(\mathbf{x}) < 0 \end{cases} \tag{2.9}$$

In other words, $\mathbf{x}$ is classified as 1 if $f(\mathbf{x})$ is positive and it is classified as -1 when $f(\mathbf{x})$ is negative. The special case happens when $f(\mathbf{x}) = 0$ in which $\mathbf{x}$ does not belong to any of these two classes. Although it may never happen in practice to have a $\mathbf{x}$ such that $f(\mathbf{x})$ is exactly zero, it explains an important theoretical concept which is called *decision boundary*. We shall mention this topic shortly. Before, we further analyze $\mathbf{w}$ with respect to $\mathbf{x}$. Clearly, $f(\mathbf{x})$ is zero when $\mathbf{x}$ is exactly on the hyperplane. Considering the fact that $\mathbf{w}$ and $\mathbf{x}$ are both $d+1$ dimensional vectors, (2.8) denotes the *dot product* of the two vectors. Moreover, we know from linear algebra that the dot product of two orthogonal vectors is 0. Consequently, the vector $\mathbf{w}$ is orthogonal to every point on the hyperplane.

This can be studied from another perspective. This is illustrated using a two-dimensional example on Fig. 2.4. If we rewrite (2.6) in slope-intercept form, we will obtain:

$$x_2 = -\frac{w_1}{w_2}x_1 - \frac{b}{w_2}. \tag{2.10}$$

where the slope of the line is equal to $m = -\frac{w_1}{w_2}$. In addition, a line is perpendicular to the above line if its slope is equal to $m' = \frac{-1}{m} = \frac{w_2}{w_1}$. As the result, the weight vector $\mathbf{w} = [w_1, w_2]$ is perpendicular to the every point on the above line since its slope is equal to $\frac{w_2}{w_1}$. Let us have a closer look to the geometry of the linear model. The distance of point $\mathbf{x}' = [x_1', x_2']$ from the linear model can be found by projecting $\mathbf{x} - \mathbf{x}'$ onto $\mathbf{w}$ which is given by:

$$r = \frac{|f(\mathbf{x})|}{\|\mathbf{w}\|} \tag{2.11}$$

Here, $\mathbf{w}$ refers to the weight vector before augmenting with b. Also, the signed distance can be obtained by removing the abs (absolute value) operator from the numerator:

$$r_{signed} = \frac{f(\mathbf{x})}{\|\mathbf{w}\|}. \tag{2.12}$$

When $\mathbf{x}$ is on the line (i.e., a hyperplane in N-dimensional space) then $f(\mathbf{x}) = 0$. Hence, the distance from the decision boundary will be zero. Set of all points $\{\mathbf{x} \mid \mathbf{x} \in \mathbb{R}^d \wedge f(\mathbf{x} = 0)\}$ represents the boundary between the regions with labels -1 and 1. This boundary is called decision boundary. However, if $\mathbf{x}$ is not on the decision boundary its distance will be a nonzero value. Also, the sign of the distance depends on the region that the point falls into. Intuitively, the model is more confident about its classification when a point is far from decision boundary. In contrary, as it gets closer to the decision boundary the confidence of the model decreases. This is the reason that we sometimes call $f(\mathbf{x})$ the *classification score* of $\mathbf{x}$.

2.2.1 Training a Linear Classifier

According to (2.9), output of a linear classifier could be 1 or -1. This means that labels of the training data must be also member of set $\{-1, 1\}$. Assume we are given the training set $\mathscr{X} = \{(\mathbf{x}_0, y_0), \ldots, (\mathbf{x}_n, y_n)\}$ where $\mathbf{x}_i \in \mathbb{R}^d$ is a d-dimensional vector and $y_i \in \{-1, 1\}$ showing label of the sample. In order to train a linear classifier, we need to define an *objective* function. For any $\mathbf{w}_t$, the objective function uses $\mathscr{X}$ to tell how accurate is the $f(\mathbf{x}) = \mathbf{w}_t \mathbf{x}^T$ at classification of samples in $\mathscr{X}$. The objective function may be also called *error* function or *loss* function. Without the loss function, it is not trivial to assess the goodness of a model.

Our main goal in training a classification model is to minimize the number of samples which are classified incorrectly. We can formulate this objective using the following equation:

$$\mathscr{L}_{0/1}(\mathbf{w}) = \sum_{i=1}^{n} H_{0/1}(\mathbf{w}\mathbf{x}^T, y_i) \tag{2.13}$$

$$H_{0/1}(\mathbf{w}\mathbf{x}^T, y_i) = \begin{cases} 1 & \mathbf{w}\mathbf{x}^T \times y_i < 0 \\ 0 & otherwise \end{cases} \tag{2.14}$$

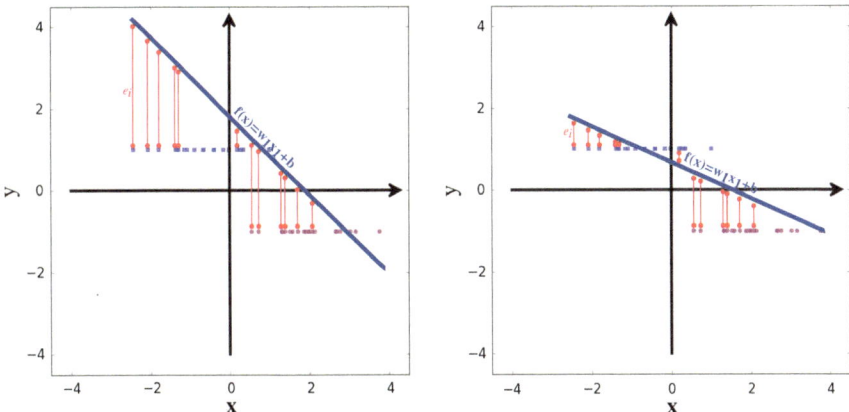

Fig. 2.5 The intuition behind squared loss function is to minimized the squared difference between the actual response and predicted value. *Left* and *right* plots show two lines with different w_1 and b. The *line* in the *right* plot is *fitted* better than the line in the left plot since its prediction error is lower in total

The above loss function is called *0/1 loss* function. A sample is classified correctly when the sign of $\mathbf{wx}^T$ and y_i are identical. If $\mathbf{x}$ is not correctly classified by the model, the signs of these two terms will not be identical. This means that one of these two terms will be negative and the other one will be positive. Therefore, their multiplication will be negative. We see that $H_{0/1}(.)$ returns 1 when the sample is classified incorrectly. Based on this explanation, the above loss function counts the number of misclassified samples. If all samples in $\mathscr{X}$ is classified correctly, the above loss function will be zero. Otherwise, it will be greater than zero. There are two problems with the above loss function which makes it impractical. First, the 0/1 loss function is nonconvex. Second, it is hard to optimize this function using gradient-based optimization methods since the function is not continuous at 0 and its gradient is zero elsewhere.

Instead of counting the number of misclassified samples, we can formulate the classification problem as a regression problem and use the *squared* loss function. This can be better described using a one-dimensional input vector $\mathbf{x} \in \mathbb{R}$ in Fig. 2.5:

In this figure, circles and squares illustrate the samples with labels -1 and 1, respectively. Since, $\mathbf{x}$ is one-dimensional (scaler), the linear model will be $f(x) = w_1 x_1 + b$ with only two trainable parameters. This model can be plotted using a line in a two-dimensional space. Assume the line shown in this figure. Given any $\mathbf{x}$ the output of the function is a real number. In the case of circles, the model should ideally return -1. Similarly, it should return 1 for all squares in this figure. Notwithstanding, because $f(\mathbf{x})$ is a linear model $f(\mathbf{x}_1) \neq f(\mathbf{x}_2)$ if $\mathbf{x}_1 \neq \mathbf{x}_2$. This means, it is impossible that our model returns 1 for every square in this figure. In contrast, it will return a unique value for each point in this figure.

For this reason, there is an *error* between the actual output of a point (circle or square) and the predicted value by the model. These errors are illustrated using red solid lines in this figure. The estimation error for $\mathbf{x}_i$ can be formulated as $e_i = (f(\mathbf{x}_i) - y_i)$ where $y_i \in \{-1, 1\}$ is the actual output of $\mathbf{x}_i$ as we defined previously in this section. Using this formulation, we can define the *squared* loss function as follows:

$$\mathcal{L}_{sq}(\mathbf{w}) = \sum_{i=1}^{n} \sqrt{(e_i)^2} = \sum_{i=1}^{n} \sqrt{(\mathbf{w}\mathbf{x}_i^T - y_i)^2}. \tag{2.15}$$

In this equation, $\mathbf{x} \in \mathbb{R}^d$ is a d-dimensional vector and $y_i \in \{-1, 1\}$ is its actual label. This loss function treat the labels as real number rather than categorical values. This makes it possible to estimate the prediction error by subtracting predicted values from actual values. Note from Fig. 2.5 that e_i can be a negative or a positive value. In order to compute the magnitude of e_i, we first compute the square of e_i and apply square root in order to compute the absolute value of e_i. It should be noted that we could define the loss function as $\sum_{i=1}^{n} |\mathbf{w}\mathbf{x}_i^T - y_i|$ instead of $\sum_{i=1}^{n} \sqrt{(\mathbf{w}\mathbf{x}_i^T - y_i)^2}$. However, as we will see shortly, the second formulation has a desirable property when we utilize a gradient-based optimization to minimize the above loss function.

We can further simplify (2.15). If we unroll the sum operator in (2.15), it will look like:

$$\mathcal{L}_{sq}(\mathbf{w}) = \sqrt{(\mathbf{w}\mathbf{x}_1^T - y_1)^2} + \cdots + \sqrt{(\mathbf{w}\mathbf{x}_n^T - y_n)^2}. \tag{2.16}$$

Taking into account the fact that square root is a monotonically increasing function and it is applied on each term individually, eliminating this operator from the above equation does not change the minimum of $\mathcal{L}(\mathbf{w})$. By applying this on the above equation, we will obtain:

$$\mathcal{L}_{sq}(\mathbf{w}) = \sum_{i=1}^{n} (\mathbf{w}\mathbf{x}_i^T - y_i)^2. \tag{2.17}$$

Our objective is to minimize the prediction error. In other words:

$$\mathbf{w} = \min_{\mathbf{w}' \in \mathbb{R}^{d+1}} \mathcal{L}(\mathbf{w}') \tag{2.18}$$

This is achievable by minimizing $\mathcal{L}_{sq}$ with respect to $\mathbf{w} \in \mathbb{R}^{d+1}$. In order to minimize the above loss function, we can use an iterative gradient-based optimization method such as *gradient descend* (Appendix A). Starting with an the initial vector $\mathbf{w}_{sol} \in \mathbb{R}^{d+1}$, this method iteratively changes $\mathbf{w}_{sol}$ proportional to the gradient vector $\nabla \mathcal{L} = [\frac{\delta \mathcal{L}}{\delta w_0}, \frac{\delta \mathcal{L}}{\delta w_1}, \ldots, \frac{\delta \mathcal{L}}{\delta w_d}]$. Here, we have shown the intercept using w_0 instead of b. Consequently, we need to calculate the partial derivative of the loss function with respect to each of parameters in $\mathbf{w}$ as follows:

$$\frac{\delta \mathcal{L}}{\delta w_i} = 2 \sum_{i=1}^{n} x_i (\mathbf{w}\mathbf{x}^T - y_i) \quad \forall i = 1 \ldots d$$

$$\frac{\delta \mathcal{L}}{\delta w_0} = 2 \sum_{i=1}^{n} (\mathbf{w}\mathbf{x}^T - y_i) \tag{2.19}$$

One problem with the above equation is that $\mathscr{L}_{sq}$ might be a large value if there are many training samples in $\mathscr{X}$. For this reason, we might need to use very small learning rate in the gradient descend method. To alleviate this problem, we can compute the *mean square error* by dividing $\mathscr{L}_{sq}$ with the total number of training samples. In addition, we can eliminate 2 in the partial derivative by multiplying $\mathscr{L}_{sq}$ by $1/2$. The final squared loss function can be defined as follows:

$$\mathscr{L}_{sq}(\mathbf{w}) = \frac{1}{2n} \sum_{i=1}^{n} (\mathbf{w}\mathbf{x}_i^T - y_i)^2 \qquad (2.20)$$

with its partial derivatives equal to:

$$\frac{\delta\mathscr{L}}{\delta w_i} = \frac{1}{n} \sum_{i=1}^{n} x_i(\mathbf{w}\mathbf{x}^T - y_i) \quad \forall i = 1\ldots d$$

$$\frac{\delta\mathscr{L}}{\delta w_0} = \frac{1}{n} \sum_{i=1}^{n} (\mathbf{w}\mathbf{x}^T - y_i) \qquad (2.21)$$

Note that the location of minimum of the (2.17) is identical to (2.20). The latter function is just multiplied by a constant value. However, adjusting the learning rate is easier when we use (2.20) to find optimal $\mathbf{w}$. One important property of the squared loss function with linear models is that it is a convex function. This means, the gradient descend method will always converge at the global minimum regardless of the initial point. It is worth mentioning this property does not hold if the classification model is nonlinear function of its parameters. We minimized the square loss function on the dataset shown in Fig. 2.1. Figure 2.6 shows the status of the gradient descend in four different iterations.

The background of the plots shows the label of each region according to sign of classification score computed for each point on the plane. The initial model is very inaccurate since most of the vectors are classified as red. However, it becomes more accurate after 400 iterations. Finally, it converges at Iteration 2000. As you can see, the amount of change in the first iterations is higher than the last iterations. By looking at the partial derivatives, we realize that the change of a parameter is directly related to the prediction error. Because the prediction error is high in the first iterations, parameters of the model changes considerably. As the error reduces, parameters also change slightly. The intuition behind the least square loss function can be studied from another perspective.

Assume the two hypothetical lines parallel to the linear model shown in Fig. 2.7. The actual distance of these lines from the linear model is equal to 1. In the case of negative region, the signed distance of the hypothetical line is -1. On the other hand, we know from our previous discussion that the normalized distance of samples $\mathbf{x}$ from the decision boundary is equal to $\frac{f(\mathbf{x})}{\|\mathbf{w}\|}$ where, here, $\mathbf{w}$ refers to the parameter vector before augmenting. If consider the projection of $\mathbf{x}$ on $\mathbf{w}$ and utilize the fact that $\mathbf{w}\mathbf{x} = \|\mathbf{w}\|\|\mathbf{x}\| \cos(\theta)$, we will see that the unnormalized distance of sample $\mathbf{x}$ from

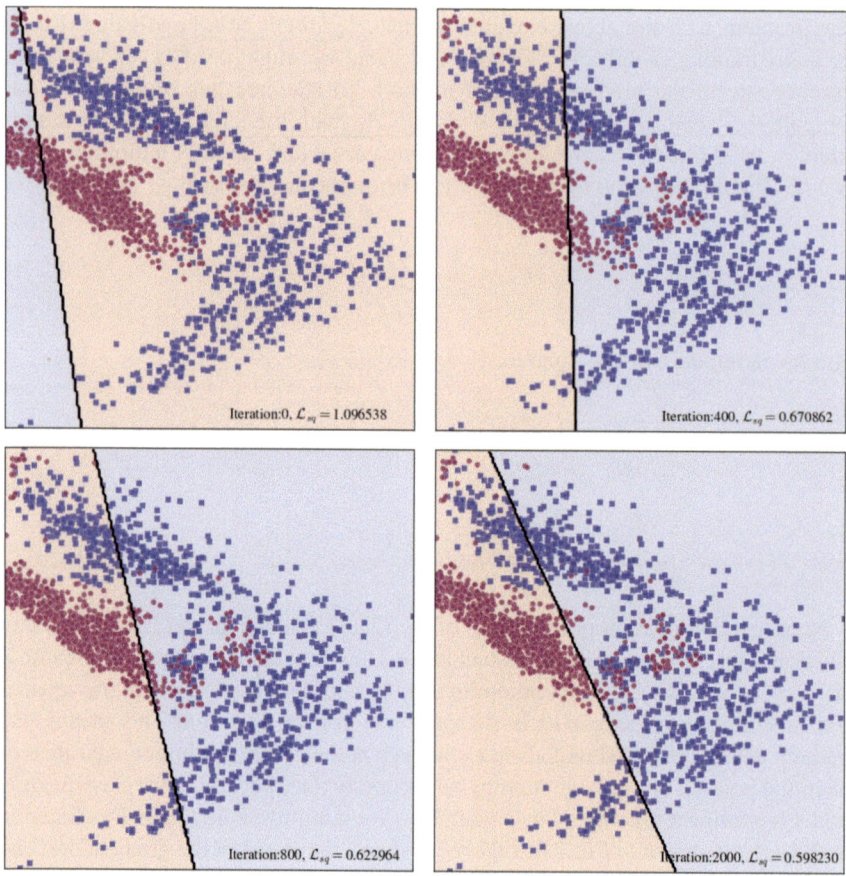

Fig. 2.6 Status of the gradient descend in four different iterations. The parameter vector **w** changes greatly in the first iterations. However, as it gets closer to the minimum of the squared loss function, it changes slightly

the linear model is equal to $f(\mathbf{x})$. Based on that, least square loss tries to minimize the sum of unnormalized distance of samples from their actual hypothetical line.

One problem with least square loss function is that it is sensitive to outliers. This is illustrated using an example on Fig. 2.8. In general, noisy samples do not come from the same distribution as clean samples. This means that they might not be close to clean samples in the d-dimensional space. On the one hand, square loss function tries to minimize the prediction error between the samples. On the other hand, because the noisy samples are located far from the clean samples, they have a large prediction error. For this reason, some of the clean samples might be sacrificed in order to reduce the error with the noisy sample. We can see in this figure that because of noisy sample, the model is not able to fit on the data accurately.

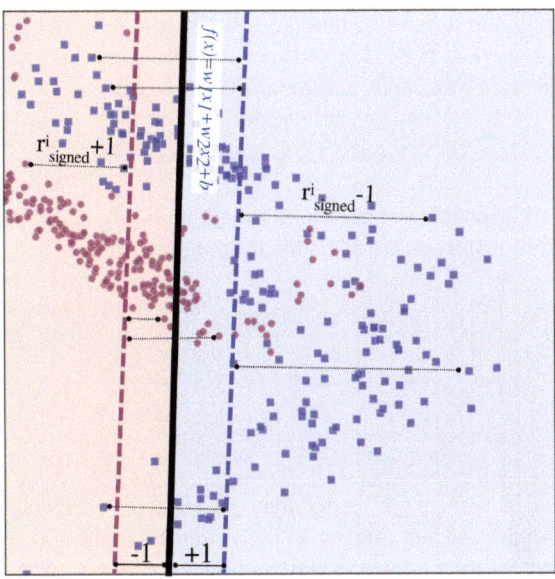

Fig. 2.7 Geometrical intuition behind least square loss function is to minimize the sum of unnormalized distances between the training samples $\mathbf{x}_i$ and their corresponding hypothetical line

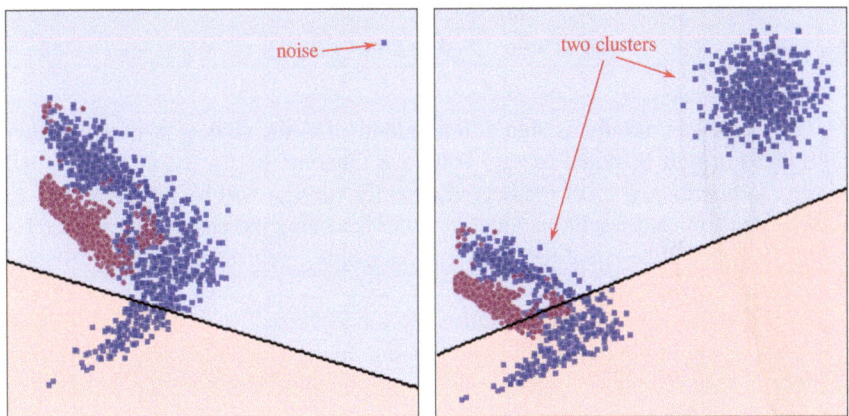

Fig. 2.8 Square loss function may fit inaccurately on training data if there are noisy samples in the dataset

It is also likely in practice that clean samples form two or more separate clusters in the d-dimensional space. Similar to the scenario of noisy samples, squared loss tries to minimize the prediction error of the samples in the far cluster as well. As we can see on the figure, the linear model might not be accurately fitted on the data if clean samples form two or more separate clusters.

This problem is due to the fact that the squared loss does not take into account the label of the prediction. Instead, it considers the classification score and computes the prediction error. For example, assume the training pairs:

$$\{(\mathbf{x}_a, 1), (\mathbf{x}_b, 1), (\mathbf{x}_c, -1), (\mathbf{x}_d, -1)\} \tag{2.22}$$

Also, suppose two different configurations $\mathbf{w}_1$ and $\mathbf{w}_2$ for the parameters of the linear model with the following responses on the training set:

$$
\begin{array}{ll}
f_{\mathbf{w}_1}(\mathbf{x}_a) = 10 & f_{\mathbf{w}_2}(\mathbf{x}_a) = 5 \\
f_{\mathbf{w}_1}(\mathbf{x}_b) = 1 & f_{\mathbf{w}_2}(\mathbf{x}_b) = 2 \\
f_{\mathbf{w}_1}(\mathbf{x}_c) = -0.5 & f_{\mathbf{w}_2}(\mathbf{x}_c) = 0.2 \\
f_{\mathbf{w}_1}(\mathbf{x}_d) = -1.1 & f_{\mathbf{w}_2}(\mathbf{x}_d) = -0.5 \\
\hline
\mathcal{L}_{sq}(w_1) = 10.15 & \mathcal{L}_{sq}(w_1) = 2.33
\end{array}
\tag{2.23}
$$

In terms of squared loss, $\mathbf{w}_2$ is better than $\mathbf{w}_1$. But, if we count the number of misclassified samples we see that $\mathbf{w}_1$ is the better configuration. In classification problems, we are mainly interested in reducing the number of incorrectly classified samples. As the result, $\mathbf{w}_1$ is favorable to $\mathbf{w}_2$ in this setting. In order to alleviate this problem of squared loss function we can define the following loss function to estimate 0/1 loss:

$$\mathcal{L}_{sg}(\mathbf{w}) = \sum_{i=1}^{n} 1 - sign(f(\mathbf{x}_i))y_i. \tag{2.24}$$

If $f(\mathbf{x})$ predicts correctly, its sign will be identical to the sign of y_i in which their multiplication will be equal to $+1$. Thus, the outcome of $1 - sign(f(\mathbf{x}_i))y_i$ will be zero. In contrary, if $f(\mathbf{x})$ predicts incorrectly, its sign will be different from y_i. So, their multiplication will be equal to -1. That being the case, the result of $1 - sign(f(\mathbf{x}_i))y_i$ will be equal to 2. For this reason, $\mathbf{w}_{sg}$ returns the twice of number of misclassified samples.

The above loss function look intuitive and it is not sensitive to far samples. However, finding the minimum of this loss function using gradient-based optimization methods is hard. The reason is because of *sign* function. One solution to solve this problem is to approximate the sign function using a differentiable function. Fortunately, *tanh* (Hyperbolic tangent) is able to accurately approximate the sign function. More specifically, $\tanh(kx) \approx sign(x)$ when $k \gg 1$. This is illustrated in Fig. 2.9. As k increases, the tanh function will be able to approximate the sign function more accurately.

By replacing the sign function with tanh in (2.24), we will obtain:

$$\mathcal{L}_{sg}(\mathbf{w}) = \sum_{i=1}^{n} 1 - \tanh(kf(\mathbf{x}_i))y_i. \tag{2.25}$$

Fig. 2.9 The sign function can be accurately approximated using tanh(kx) when $k \gg 1$

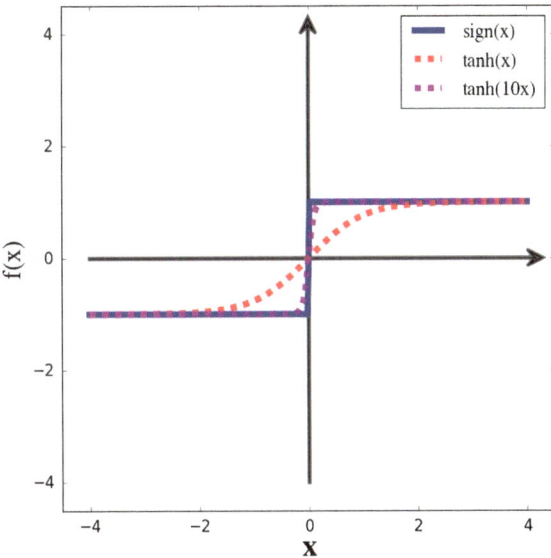

Similar to the squared loss function, the sign loss function can be minimized using the gradient descend method. To this end, we need to compute the partial derivatives of the sign loss function with respect to its parameters:

$$\frac{\delta \mathcal{L}_{sg}(\mathbf{w})}{w_i} = -kx_i y(1 - \tanh^2(kf(\mathbf{x})))$$

$$\frac{\delta \mathcal{L}_{sg}(\mathbf{w})}{w_0} = -ky(1 - \tanh^2(kf(\mathbf{x}))) \tag{2.26}$$

If we train a linear model using the sign loss function and the gradient descend method on the datasets shown in Figs. 2.1 and 2.8, we will obtain the results illustrated in Fig. 2.10. According to the results, the sign loss function is able to deal with separated clusters of samples and outliers as opposed to the squared loss function.

Even though the sign loss using the tanh approximation does a fairly good job on our sample dataset, it has one issue which makes the optimization slow. In order to explain this issue, we should study the derivative of tanh function. We know from calculus that $\frac{\delta \tanh(x)}{\delta x} = 1 - \tanh^2(x)$. Figure 2.11 shows its plot. We can see that the derivative of tanh saturates as $|x|$ increases. Also, it saturates more rapidly if we set k to a positive number greater than 1. On the other hand, we know from (2.26) that the gradient of the sign loss function directly depends on the derivative of tanh function. That means if the derivative of a sample falls into the saturated region, its magnitude is close to zero. As a consequence, parameters change very slightly. This phenomena which is called the *saturated gradients* problem slows down the convergence speed of the gradient descend method. As we shall see in the next chapters, in complex models such as neural networks with millions of parameters, the model may not

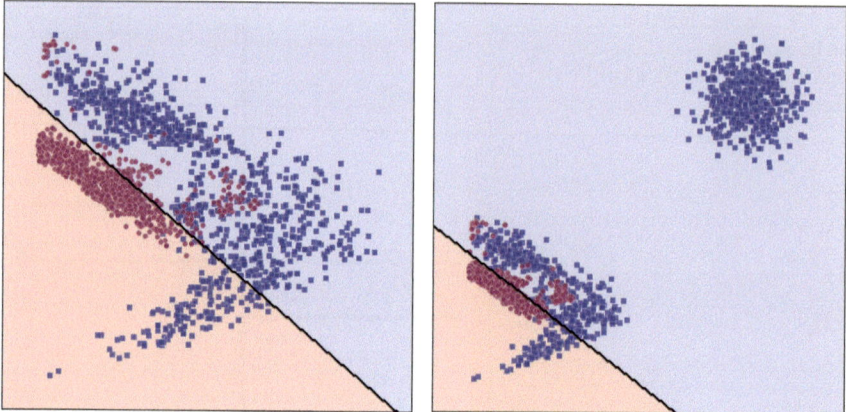

Fig. 2.10 The sign loss function is able to deal with noisy datasets and separated clusters problem mentioned previously

Fig. 2.11 Derivative of tanh(kx) function saturates as $|x|$ increases. Also, the ratio of saturation growth rapidly when $k > 1$

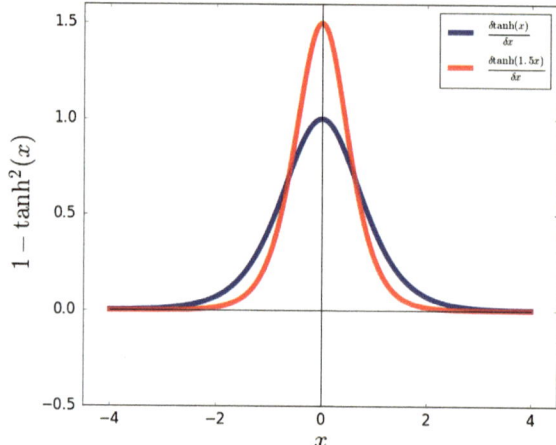

be able to adjust the parameters of initial layers since the saturated gradients are propagated from last layers back to the first layers.

2.2.2 Hinge Loss

Earlier in this chapter, we explained that the normalized distance of sample $\mathbf{x}$ from the decision boundary is equal to $\frac{|f(\mathbf{x})|}{\|\mathbf{w}\|}$. Likewise, *margin* of $\mathbf{x}$ is obtained by computing $(\mathbf{wx}^T)y$ where y is the corresponding label of $\mathbf{x}$. The margin tell us how correct is the classification of the sample. Assume that the label of $\mathbf{x}_a$ is -1. If $\mathbf{wx}_a^T$ is negative, its multiplication with $y = -1$ will be positive showing that the sample is classified correctly with a confidence analogous to $|\mathbf{wx}^T|$. Likewise, if $\mathbf{wx}_a^T$ is positive, its

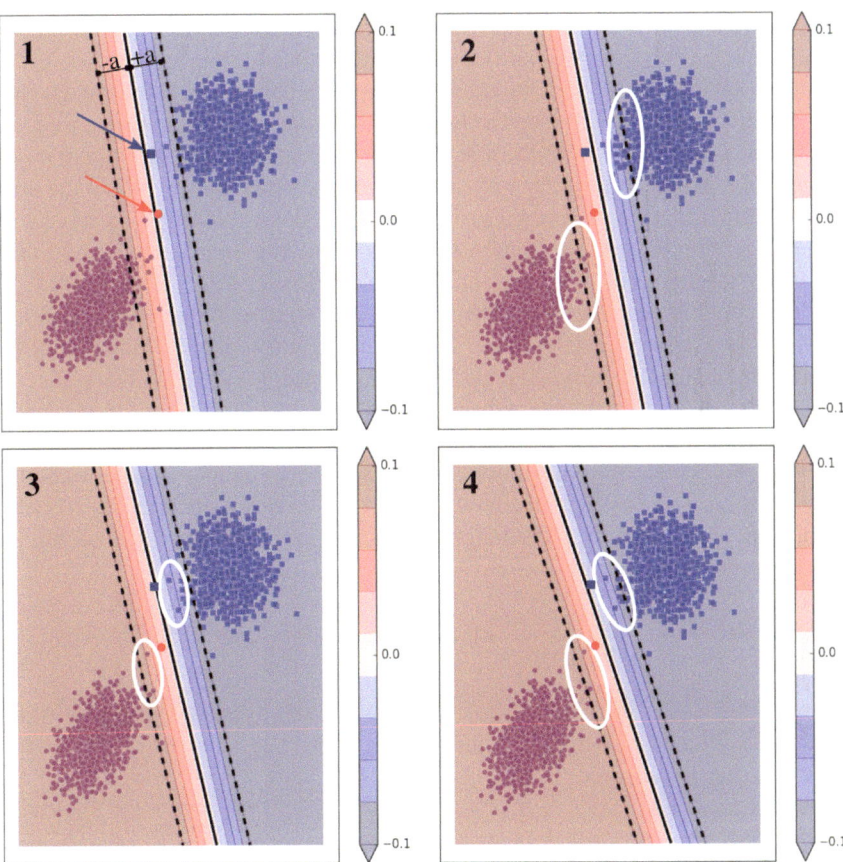

Fig. 2.12 Hinge loss increases the margin of samples while it is trying to reduce the classification error. Refer to text for more details

multiplication with $y = -1$ will be negative showing that the sample is classified incorrectly with a magnitude equal to $|\mathbf{wx}^T|$.

The basic idea behind *hinge* loss is not only to train a classifier but also to increase margin of samples. This is an important property which may increase tolerance of the classifier against noisy samples. This is illustrated in Fig. 2.12 on a synthetic dataset which are perfectly separable using a line. The solid line shows the decision boundary and the dashed lines illustrate the borders of the critical region centered at the decision boundary of this model. It means that the margin of samples in this region is less than $|a|$. In contrast, margin of samples outside this region is high which implies that the model is more confident in classification of samples outside this region. Also, the colorbar next to each plots depicts the margin corresponding to each color on the plots.

In the first plot, two *test* samples are indicated which are not used during the training phase. One of them belongs to circles and the another one belongs to squares. Although the line adjusted on the training samples is able to perfectly discriminate the training samples, it will incorrectly classify the test red sample. Comparing the model in the second plot with the first plot, we observe that fewer circles are inside the critical region but the number of squares increase inside this region. In the third plot, the overall margin of samples are better if we compare the samples marked with white ellipses on these plots. Finally, the best overall margin is found in the fourth plot where the test samples are also correctly classified.

Maximizing the margin is important since it may increase the tolerance of model against noise. The test samples in Fig. 2.12 might be noisy samples. However, if the margin of the model is large, it is likely that these samples are classified correctly. Nonetheless, it is still possible that we design a test scenario where the first plot could be more accurate than the fourth plot. But, as the number of training samples increases a classifier with maximum margin is likely to be more stable. Now, the question is how we can force the model by a loss function to increase its accuracy and margin simultaneously? The hinge loss function achieves these goals using the following relation:

$$\mathscr{L}_{hinge}(\mathbf{w}) = \frac{1}{n} \sum_{i=1}^{n} \max(0, a - \mathbf{wx}_i^T y_i) \tag{2.27}$$

where $y_i \in \{-1, 1\}$ is the label of the training sample $\mathbf{x}_i$. If signs of $\mathbf{wx}_i$ and y_i are equal, the term inside the sum operator will return 0 since the value of the second parameter in the max function will be negative. In contrast, if their sign are different, this term will be equal to $a - \mathbf{wx}_i^T y_i$ increasing the value of loss. Moreover, if $\mathbf{wx}_i^T y_i < a$ this implies that $\mathbf{x}$ is within the critical region of the model and it increases the value of loss. By minimizing the above loss function we will obtain a model with maximum margin and high accuracy at the same time. The term inside the sum operator can be written as:

$$\max(0, a - \mathbf{wx}_i^T y_i) = \begin{cases} a - \mathbf{wx}_i^T y_i & \mathbf{wx}_i^T y_i < a \\ 0 & \mathbf{wx}_i^T y_i \geq a \end{cases} \tag{2.28}$$

Using this formulation and denoting $\max(0, a - \mathbf{wx}_i^T y_i)$ with H, we can compute the partial derivatives of $\mathscr{L}_{hinge}(\mathbf{w})$ with respect to $\mathbf{w}$:

$$\begin{aligned} \frac{\delta H}{\delta w_i} &= \begin{cases} -x_i y_i & \mathbf{wx}_i^T y_i < a \\ 0 & \mathbf{wx}_i^T y_i \geq a \end{cases} \\ \frac{\delta H}{\delta w_0} &= \begin{cases} -y_i & \mathbf{wx}_i^T y_i < a \\ 0 & \mathbf{wx}_i^T y_i \geq a \end{cases} \end{aligned} \tag{2.29}$$

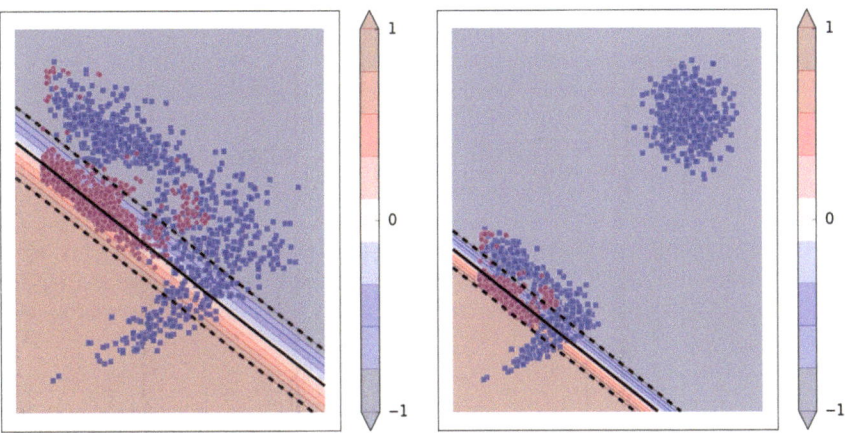

Fig. 2.13 Training a linear classifier using the hinge loss function on two different datasets

$$\frac{\delta \mathscr{L}_{hinge}(\mathbf{w})}{\delta w_i} = \frac{1}{n} \sum_{i=1}^{n} \frac{\delta H}{w_i}$$

$$\frac{\delta \mathscr{L}_{hinge}(\mathbf{w})}{\delta w_0} = \frac{1}{n} \sum_{i=1}^{n} \frac{\delta H}{w_0}$$

(2.30)

It should be noted that, $\mathscr{L}_{hinge}(\mathbf{w})$ is not continuous at $\mathbf{w}\mathbf{x}_i^T y_i = a$ and, consequently, it is not differentiable at $\mathbf{w}\mathbf{x}_i^T y_i = a$. For this reason, the better choice for optimizing the above function might be a *subgradient*-based method. However, it might never happen in a training set to have a sample in which $\mathbf{w}\mathbf{x}_i^T y_i$ is exactly equal to a. For this reason, we can still use the gradient descend method for optimizing this function.

Furthermore, the loss function does not depend on the value of a. It only affects the magnitude of $\mathbf{w}$. In other words, $\mathbf{w}$ is always adjusted such that as few training samples as possible fall into the critical region. For this reason, we always set $a = 1$ in practice. We minimized the hinge loss on the dataset shown in Figs. 2.1 and 2.8. Figure 2.13 illustrates the result. As before, the region between the two dashed lines indicates the critical region.

Based on the results, the model learned by the hinge loss function is able to deal with separated clusters problem. Also, it is able to learn an accurate model for the nonlinearly separable dataset. A variant of hinge loss called *squared hinge* loss has been also proposed which is defined as follows:

$$\mathscr{L}_{hinge}(\mathbf{w}) = \frac{1}{n} \sum_{i=1}^{n} \max(0, 1 - \mathbf{w}\mathbf{x}_i^T y_i)^2$$

(2.31)

The main difference between the hinge loss and the squared hinge loss is that the latter one is smoother and it may make the optimization easier. Another variant of the hinge loss function is called *modified Huber* and it is defined as follows:

$$\mathscr{L}_{huber}(\mathbf{w}) = \begin{cases} \max(0, 1 - y\mathbf{wx}^T)^2 & y\mathbf{wx}^T \geq -1 \\ -4y\mathbf{wx}^T & otherwise \end{cases} \quad (2.32)$$

The modified Huber loss is very close to the squared hinge and they may only differ in the convergence speed. In order to use any of these variants to train a model, we need to compute the partial derivative of the loss functions with respect to their parameters.

2.2.3 Logistic Regression

None of the previously mentioned linear models are able to compute the probability of samples $\mathbf{x}$ belonging to class $y = 1$. Formally, given a binary classification problem, we might be interested in computing $p(y = 1|\mathbf{x})$. This implies that $p(y = -1|\mathbf{x}) = 1 - p(y = 1|\mathbf{x})$. Consequently, the sample $\mathbf{x}$ belongs to class 1 if $p(y = 1|\mathbf{x}) > 0.5$. Otherwise, it belongs to class -1. In the case that $p(y = 1|\mathbf{x}) = 0.5$, the sample is exactly on the decision boundary and it does not belong to any of these two classes. The basic idea behind *logistic regression* is to learn $p(y = 1|\mathbf{x})$ using a linear model. To this end, logistic regression transforms the score of a sample into probability by passing the score through a *sigmoid* function. Formally, logistic regression computes the *posterior* probability as follows:

$$p(y = 1|\mathbf{x}; \mathbf{w}) = \sigma(\mathbf{wx}^T) = \frac{1}{1 + e^{-\mathbf{wx}^T}}. \quad (2.33)$$

In this equation, $\sigma : \mathbb{R} \rightarrow [0, 1]$ is the *logistic sigmoid* function. As it is shown in Fig. 2.14, the function has a S shape and it saturates as $|x|$ increases. In other words, derivative of function approaches to zero as $|x|$ increases.

Since range of the sigmoid function is $[0, 1]$ it satisfies requirements of a probability measure function. Note that (2.33) directly models the *posterior* probability which means by using appropriate techniques that we shall explain later, it is able to model likelihood and a priori of classes. Taking into account the fact that (2.33) returns the probability of a sample, the loss function must be also build based on probability of the whole training set given a specific $\mathbf{w}$. Formally, given a dataset of n training samples, our goal is to maximize their joint probability which is defined as:

$$\mathscr{L}_{logistic}(\mathbf{w}) = p(\mathbf{x}_1 \cap \mathbf{x}_2 \cap \cdots \cap \mathbf{x}_n) = p\left(\bigcap_{i=1}^{n} \mathbf{x}_i\right). \quad (2.34)$$

Modeling the above joint probability is not trivial. However, it is possible to decompose this probability into smaller components. To be more specific, the probability

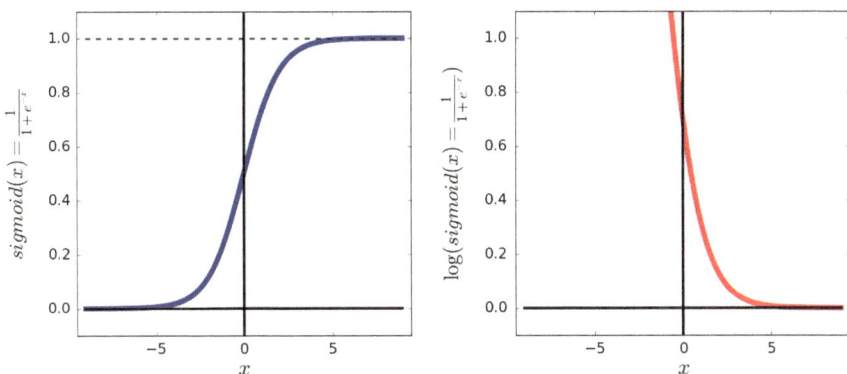

Fig. 2.14 Plot of the sigmoid function (*left*) and logarithm of the sigmoid function (*right*). The domain of the sigmoid function is real numbers and its range is [0, 1]

of $\mathbf{x}_i$ does not depend on the probability of $\mathbf{x}_j$. For this reason and taking into account the fact that $p(A, B) = p(A)p(B)$ if A and B are independent events, we can decompose the above joint probability into product of probabilities:

$$\mathcal{L}_{logistic}(\mathbf{w}) = \prod_{i=1}^{n} p(y_i | \mathbf{x}_i) \tag{2.35}$$

where $p(\mathbf{x}_i)$ is computed using:

$$p(y_i | \mathbf{x}_i) = \begin{cases} p(y = 1 | \mathbf{x}; \mathbf{w}) & y_i == 1 \\ 1 - p(y = 1 | \mathbf{x}; \mathbf{w}) & y_i == -1 \end{cases} \tag{2.36}$$

Representing the negative class with 0 rather than -1, the above equation can be written as:

$$p(\mathbf{x}_i) = p(y = 1 | \mathbf{x}; \mathbf{w})^{y_i} (1 - p(y = 1 | \mathbf{x}; \mathbf{w}))^{1-y_i}. \tag{2.37}$$

This equation which is called *Bernoulli* distribution is used to model random variables with two outcomes. Plugging (2.33) into the above equation we will obtain:

$$\mathcal{L}_{logistic}(\mathbf{w}) = \prod_{i=1}^{n} \left(\sigma(\mathbf{wx}^T)^{y_i} (1 - \sigma(\mathbf{wx}^T))^{1-y_i} \right). \tag{2.38}$$

Optimizing the above function is hard. The reason is because of $\prod$ operator which makes the derivative of the loss function intractable. However, we can apply logarithm trick to change the multiplication into summation. In other words, we can compute $\log(\mathcal{L}_{logistic}(\mathbf{w}))$:

$$\log(\mathcal{L}_{logistic}(\mathbf{w})) = \log \left(\prod_{i=1}^{n} \left(\sigma(\mathbf{wx}^T)^{y_i} (1 - \sigma(\mathbf{wx}^T))^{1-y_i} \right) \right). \tag{2.39}$$

We know from properties of logarithm that $\log(A \times B) = \log(A) + \log(B)$. As the result, the above equation can be written as:

$$\log(\mathscr{L}_{logistic}(\mathbf{w})) = \sum_{i=1}^{n} y_i \log \sigma(\mathbf{w}\mathbf{x}^T) + (1 - y_i) \log(1 - \sigma(\mathbf{w}\mathbf{x}^T)). \quad (2.40)$$

If each sample in the training set is classified correctly, $p(\mathbf{x}_i)$ will be close to 1 and if it is classified incorrectly, it will be close to zero. Therefore, the best classification will be obtained if we find the maximum of the above function. Although this can be done using gradient *ascend* methods, it is preferable to use gradient descend methods. Because gradient descend can be only applied on minimization problems, we can multiply both sides of the equation with -1 in order to change the maximum of the loss into minimum:

$$E = -\log(\mathscr{L}_{logistic}(\mathbf{w})) = -\sum_{i=1}^{n} y_i \log \sigma(\mathbf{w}\mathbf{x}^T) + (1 - y_i) \log(1 - \sigma(\mathbf{w}\mathbf{x}^T)).$$
$$(2.41)$$

Now, we can use gradient descend to find the minimum of the above loss function. This function is called *cross-entropy* loss. In general, these kind of loss functions are called *negative log-likelihood* functions. As before, we must compute the partial derivatives of the loss function with respect to its parameters in order to apply the gradient descend method. To this end, we need to compute the derivative of $\sigma(a)$ with respect to its parameter which is equal to:

$$\frac{\delta \sigma(a)}{a} = \sigma(a)(1 - \sigma(a)). \quad (2.42)$$

Then, we can utilize the chain rule to compute the partial derivative of the above loss function. Doing so, we will obtain:

$$\begin{aligned}
\frac{\delta E}{w_i} &= \left(\sigma(\mathbf{w}\mathbf{x}_i^T) - y_i\right)x_i \\
\frac{\delta E}{w_0} &= \sigma(\mathbf{w}\mathbf{x}_i^T) - y_i
\end{aligned} \quad (2.43)$$

Note that in contrast to the previous loss functions, here, $y_i \in \{0, 1\}$. In other words, the negative class is represented using 0 instead of -1. Figure 2.15 shows the result of training linear models on the two previously mentioned datasets. We see that logistic regression is find an accurate model even when the training samples are scattered in more than two clusters. Also, in contrast to the squared function, it is less sensitive to outliers.

It is possible to formulate the logistic loss with $y_i \in \{-1, 1\}$. In other words, we can represent the negative class using -1 and reformulate the logistic loss function.

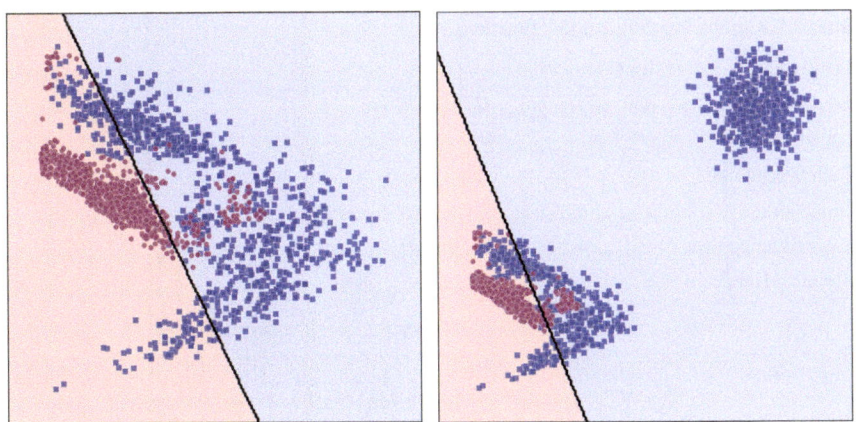

Fig. 2.15 Logistic regression is able to deal with separated clusters

More specifically, we can rewrite the logistic equations as follows:

$$p(y = 1|\mathbf{x}) = \frac{1}{1 + e^{-\mathbf{wx}^T}}$$

$$p(y = -1|\mathbf{x}) = 1 - p(y = 1|\mathbf{x}) = \frac{1}{1 + e^{+\mathbf{wx}^T}} \tag{2.44}$$

This implies that:

$$p(y_i|\mathbf{x}_i) = \frac{1}{1 + e^{-y_i\mathbf{wx}^T}} \tag{2.45}$$

Plugging this in (2.35) and taking the negative logarithm, we will obtain:

$$\mathscr{L}_{logistic}(\mathbf{w}) = \sum_{i=1}^{n} \log(1 + e^{-y_i\mathbf{wx}^T}) \tag{2.46}$$

It should be noted that (2.41) and (2.46) are identical and they can lead to the same solution. Consequently, we can use any of them to fit a linear model. As before, we only need to compute partial derivatives of the loss function and use them in the gradient descend method to minimize the loss function.

2.2.4 Comparing Loss Function

We explained 7 different loss functions for training a linear model. We also discussed some of their properties in presence of outliers and separated clusters. In this section, we compare these loss functions from different perspectives. Table 2.1 compares different loss functions. Besides, Fig. 2.16 illustrates the plot of the loss functions along with their second derivative.

Table 2.1 Comparing different loss functions

Loss function	Equation	Convex
Zero-one loss	$\mathcal{L}_{0/1}(\mathbf{w}) = \sum_{i=1}^{n} H_{0/1}(\mathbf{w}\mathbf{x}^T, y_i)$	No
Squared loss	$\mathcal{L}_{sq}(\mathbf{w}) = \sum_{i=1}^{n} (\mathbf{w}\mathbf{x}_i^T - y_i)^2$	Yes
Tanh Squared loss	$\mathbf{w}_{sg} = \sum_{i=1}^{n} 1 - \tanh(kf(\mathbf{x}_i))y_i.$	No
Hinge loss	$\mathcal{L}_{hinge}(\mathbf{w}) = \frac{1}{n}\sum_{i=1}^{n} \max(0, 1 - \mathbf{w}\mathbf{x}_i^T y_i)$	Yes
Squared hinge loss	$\mathcal{L}_{hinge}(\mathbf{w}) = \frac{1}{n}\sum_{i=1}^{n} \max(0, 1 - \mathbf{w}\mathbf{x}_i^T y_i)^2$	Yes
Modified Huber	$\mathcal{L}_{huber}(\mathbf{w}) = \begin{cases} \max(0, 1 - y\mathbf{w}\mathbf{x}^T)^2 & y\mathbf{w}\mathbf{x}^T \geq -1 \\ -4y\mathbf{w}\mathbf{x}^T & otherwise \end{cases}$	Yes
Logistic loss	$-\log(\mathcal{L}_{logistic}(\mathbf{w})) =$ $-\sum_{i=1}^{n} y_i \log \sigma(\mathbf{w}\mathbf{x}^T) + (1 - y_i)\log(1 - \sigma(\mathbf{w}\mathbf{x}^T))$	Yes

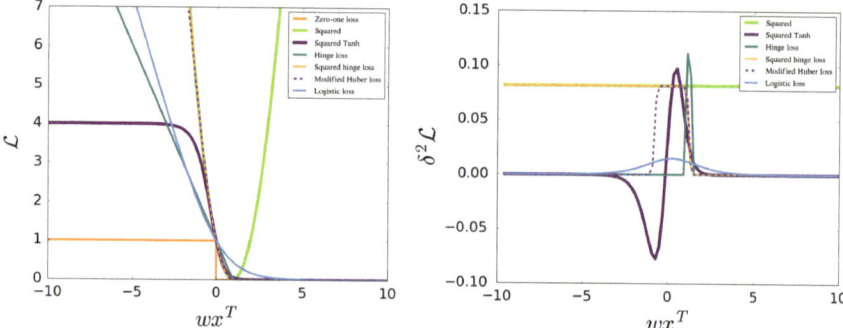

Fig. 2.16 Tanh squared loss and zero-one loss functions are not convex. In contrast, the squared loss, the hinge loss, and its variant and the logistic loss functions are convex

Informally, a one variable function is convex if for every pair of points x and y, the function falls below their connecting line. Formally, if the second derivative of a function is positive, the function is convex. Looking at the plots of each loss function and their derivatives, we realize that the Tanh squared loss and the zero-one loss functions are not convex. In contrast, hinge loss and its variants as well as the logistic loss are all convex functions. Convexity is an important property since it guarantees that the gradient descend method will find the global minimum of the function provided that the classification model is linear.

Let us have a closer look at the logistic loss function on the dataset which is linearly separable. Assume the parameter vector $\hat{\mathbf{w}}$ such that two classes are separated perfectly. This is shown by the top-left plot in Fig. 2.17. However, because the magnitude of $\hat{\mathbf{w}}$ is low $\sigma(\mathbf{w}\mathbf{x}^T)$ is smaller than 1 for the points close to the decision boundary. In order to increase the value of $\sigma(\mathbf{w}\mathbf{x}^T)$ without affecting the classification accuracy, the optimization method may increase the magnitude of $\hat{\mathbf{w}}$. As we can see in the other plots, as the magnitude increases, the logistic loss reduces. Magnitude of $\hat{\mathbf{w}}$ can increase infinitely resulting the logistic to approach zero.

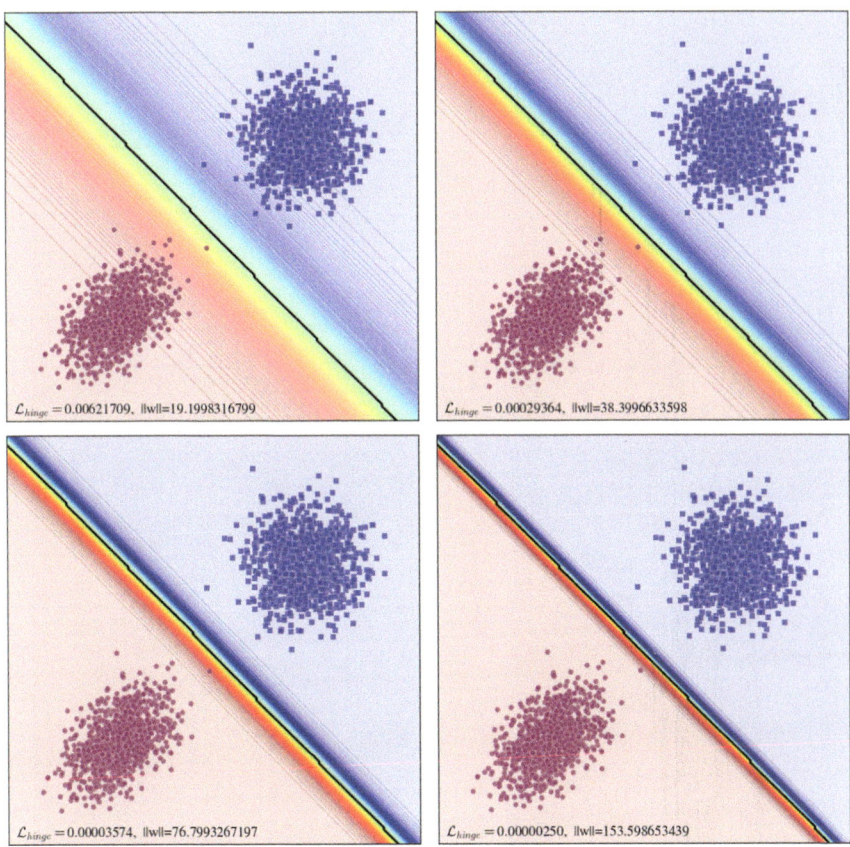

$\mathcal{L}_{hinge} = 0.00621709$, ‖w‖=19.19983 16799

$\mathcal{L}_{hinge} = 0.00029364$, ‖w‖=38.3996633598

$\mathcal{L}_{hinge} = 0.00003574$, ‖w‖=76.7993267197

$\mathcal{L}_{hinge} = 0.00000250$, ‖w‖=153.598653439

Fig. 2.17 Logistic regression tries to reduce the logistic loss even after finding a hyperplane which discriminates the classes perfectly

However, as we will explain in the next chapter, parameter vectors with high magnitude may suffer from a problem called *overfitting*. For this reason, we are usually interested in finding parameter vectors with low magnitudes. Looking at the plot of the logistic function in Fig. 2.16, we see that the function approaches to zero at infinity. This is the reason that the magnitude of model increases.

We can analyze the hinge loss function from the same perspective. Looking at the plot of the hinge loss function, we see that it becomes zero as soon as it finds a hyperplane in which all the samples are classified correctly and they are outside the critical region. We fitted a linear model using the hinge loss function on the same dataset as the previous paragraph. Figure 2.18 shows that after finding a hyperplane that classifies the samples perfectly, the magnitude of **w** increases until all the samples are outside the critical region. At this point, the error becomes zero and **w** does not change anymore. In other words, ‖**w**‖ has an upper bound when we find it using the hinge loss function.

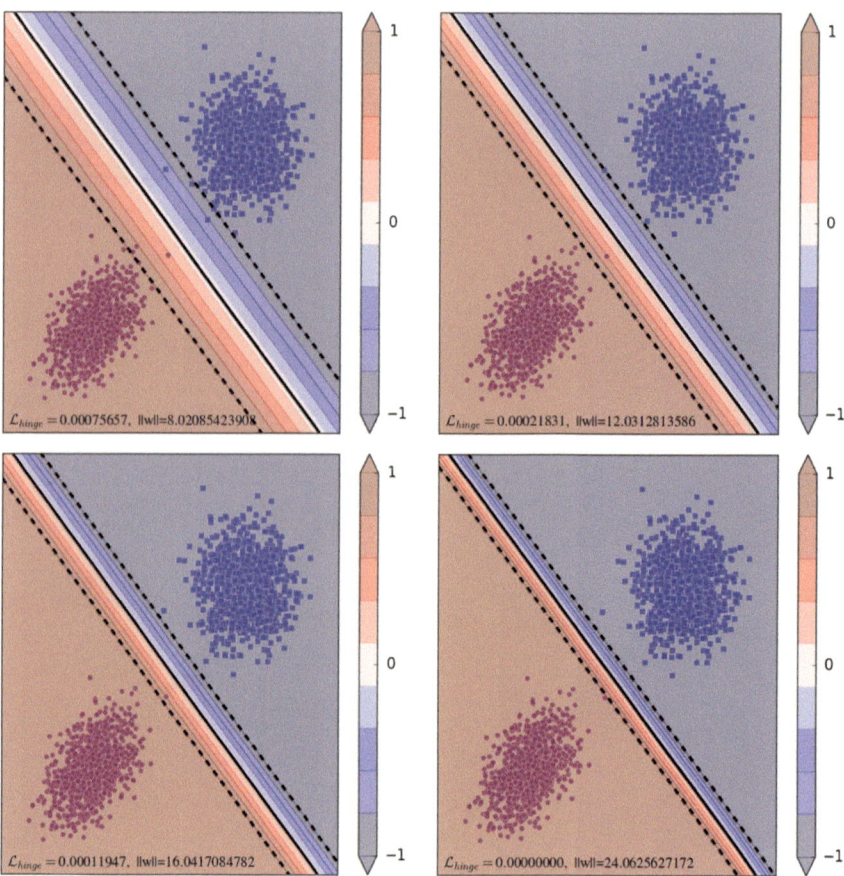

Fig. 2.18 Using the hinge loss function, the magnitude of **w** changes until all the samples are classified correctly and they do not fall into the critical region

The above argument about the logistic regression does not hold when the classes are not linearly separable. In other words, in the case that classes are nonlinearly separable, it is not possible to perfectly classify all the training samples. Consequently, some of the training samples are always classified incorrectly. In this case, as it is shown in Fig. 2.19, if $\|\mathbf{w}\|$ increases, the error of the misclassified samples also increases resulting in a higher loss. For this reason, the optimization algorithm change the value of **w** for a limited time. In other words, there could be an upper bound for $\|\mathbf{w}\|$ when the classes are not linearly separable.

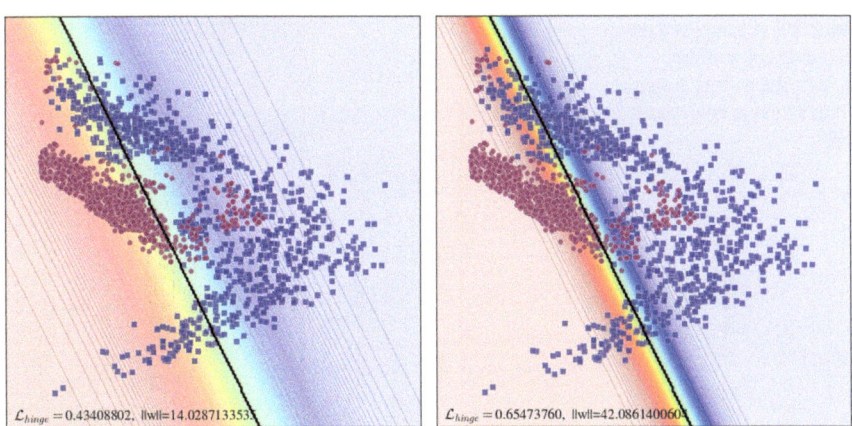

Fig. 2.19 When classes are not linearly separable, $\|\mathbf{w}\|$ may have an upper bound in logistic loss function

2.3 Multiclass Classification

In the previous section, we mentioned a few techniques for training a linear classifier on binary classification problems. Recall from the previous section that in a binary classification problem our goal is to classify the input $\mathbf{x} \in \mathbb{R}^d$ into one of two classes. A multiclass classification problem is a more generalized concept in which $\mathbf{x}$ is classified into more than two classes. For example, suppose we want to classify 10 different speed limit signs starting from 30 to 120 km/h. In this case, $\mathbf{x}$ represents the image of a speed limit sign. Then, our goal is to find the model $f : \mathbb{R}^d \to \mathcal{Y}$ where $\mathcal{Y} = \{0, 1, \ldots, 9\}$. The model $f(\mathbf{x})$ accepts a d-dimensional real vector and returns a categorical integer between 0 and 9. It is worth mentioning that $\mathcal{Y}$ is not an ordered set. It can be any set with 10 different *symbols*. However, for the sake of simplicity, we usually use integer numbers to show classes.

2.3.1 One Versus One

A multiclass classifier can be build using a group of binary classifiers. For instance, assume the 4-class classification problem illustrated in Fig. 2.20 where $\mathcal{Y} = \{0, 1, 2, 3\}$. One technique for building a multiclass classifier using a group of binary classifier is called *one-versus-one* (OVO).

Given the dataset $\mathcal{X} = \{(\mathbf{x}_0, y_0), \ldots, (\mathbf{x}_n, y_n)\}$ where $\mathbf{x}_i \in \mathbb{R}^d$ and $y_i \in \{0, 1, 2, 3\}$, we first pick the samples from $\mathcal{X}$ with label 0 or 1. Formally, we create the following dataset:

$$\mathcal{X}_{0|1} = \{\mathbf{x}_i \mid \mathbf{x}_i \in \mathcal{X} \wedge y_i \in \{0, 1\}\} \tag{2.47}$$

Fig. 2.20 A samples dataset including four different classes. Each class is shown using a unique color and shape

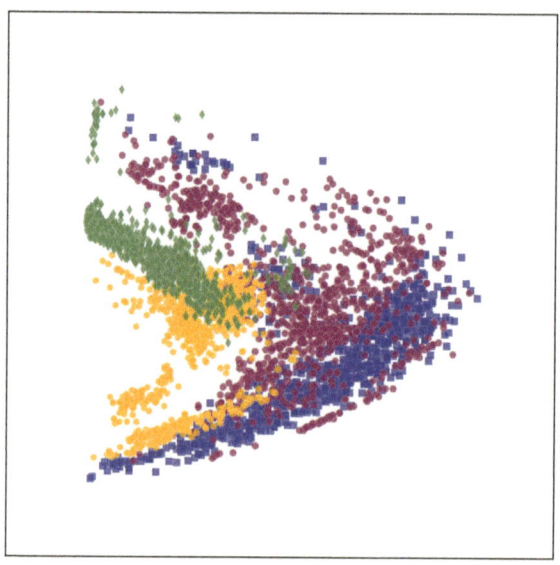

and a binary classifier is fitted on $\mathscr{X}_{0|1}$. Similarly, $\mathscr{X}_{0|2}$, $\mathscr{X}_{0|3}$, $\mathscr{X}_{1|2}$, $\mathscr{X}_{1|3}$ and $\mathscr{X}_{2|3}$ are created a separate binary classifiers are fitted on each of them. By this way, there will be six binary classifiers. In order to classify the new input $\mathbf{x}_q$ into one of four classes, it is first classified using each of these 6 classifiers. We know that each classifier will yield an integer number between 0 and 3. Since there are six classifiers, one of the integer numbers will be repeated more than others. The class of $\mathbf{x}_q$ is equal to the number with highest occurrence. From another perspective, we can think of the output of each binary classifier as a vote. Then, the winner class is the one with majority of votes. This method of classification is called *majority voting*. Figure 2.21 shows six binary classifiers trained on six pairs of classes mentioned above. Besides, it illustrates how points on the plane are classified into one of four classes using this technique.

This example can be easily extended to a multiclass classification problem with N classes. More specifically, all pairs of classes $\mathscr{X}_{a|b}$ are generated for all $a = 1 \ldots N - 1$ and $b = a + 1 \ldots N$. Then, a binary model $f_{a|b}$ is fitted on the corresponding dataset. By this way, $\frac{N(N-1)}{2}$ binary classifiers will be trained. Finally, an unseen sample $\mathbf{x}_q$ is classified by computing the majority of votes produces by all the binary classifiers.

One obvious problem of one versus one technique is that the number of binary classifiers quadratically increases with the number of classes in a dataset. This means that using this technique we need to train 31125 binary classifiers for a 250-class classification problem such as traffic sign classification. This makes the one versus one approach impractical for large values of N. In addition, sometimes ambiguous results might be generated by one versus one technique. This may happen when there are two or more classes with majority of votes. For example,

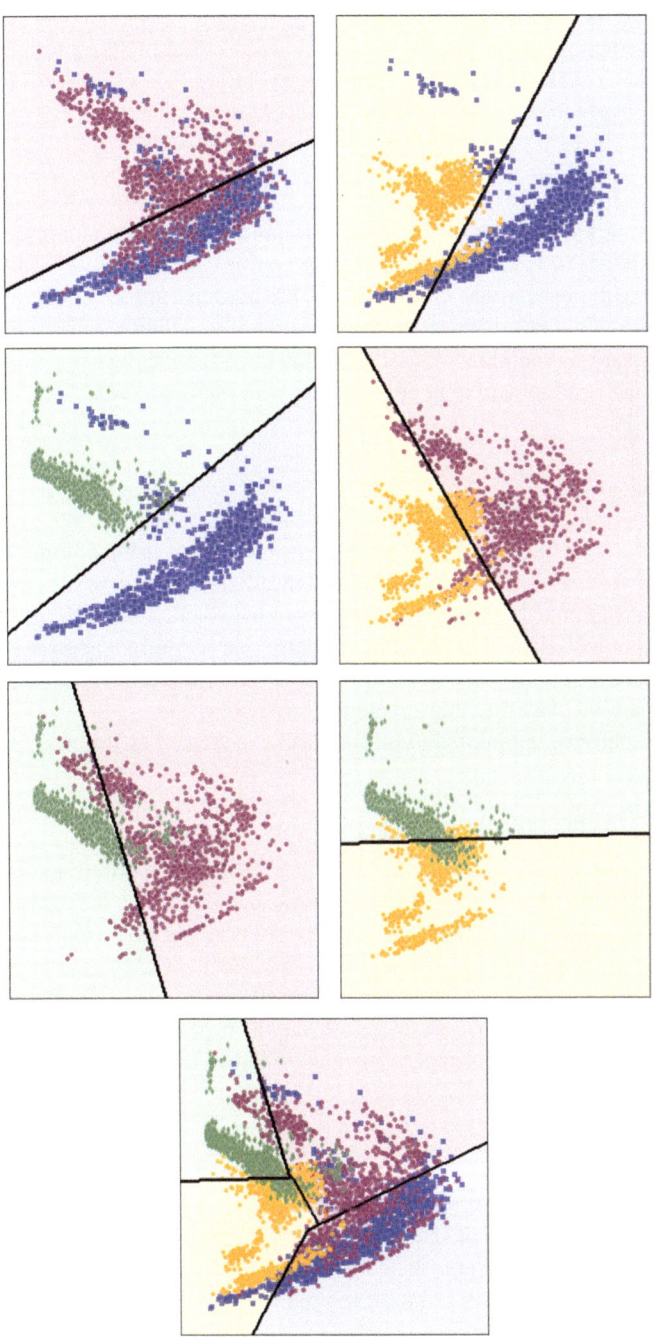

◀ **Fig. 2.21** Training six classifiers on the four class classification problem. One versus one technique considers all unordered pairs of classes in the dataset and fits a separate binary classifier on each pair. A input $\mathbf{x}$ is classified by computing the majority of votes produced by each of binary classifiers. The *bottom plot* shows the class of every point on the plane into one of four classes

assume that the votes of 6 classifiers in the above example for an unseen sample are 1, 1, 2, and 2 for classes 0, 1, 2, and 3, respectively. In this case, the Class 2 and Class 3 have equally the majority votes. Consequently, the unseen sample cannot be classified. This problem might be addressed by taking into account the classification score (i.e., $\mathbf{wx}^T$) produced by the binary classifiers. However, the fact remains that one versus one approach is not practical in applications with many classes.

2.3.2 One Versus Rest

Another popular approach for building a multiclass classifier using a group of binary classifiers is called *one versus rest* (OVR). It may also be called *one versus all* or *one against all* approach. As opposed to one versus one approach where $\frac{N(N-1)}{2}$ binary classifiers are created for a N-class classification problem, one versus rest approach trains only N binary classifiers to make predictions. The main difference between these two approaches are the way that they create the binary datasets.

In one versus rest technique, a binary dataset for class a is created as follows:

$$\mathcal{X}_{a|rest} = \{(\mathbf{x}_i, 1)|\mathbf{x}_i \in \mathcal{X} \wedge y_i = a\} \cup \{(\mathbf{x}_i, -1)|\mathbf{x}_i \in \mathcal{X} \wedge y_i \neq a\}. \quad (2.48)$$

Literally, $\mathcal{X}_{a|rest}$ is composed of all the samples in $\mathcal{X}$. The only difference is the label of samples. For creating $\mathcal{X}_{a|rest}$, we pick all the samples in $\mathcal{X}$ with label a and add them to $\mathcal{X}_{a|rest}$ after changing their label to 1. Then, the label of all the remaining samples in $\mathcal{X}$ is changed to -1 and they are added to $\mathcal{X}_{a|rest}$. For a N-class classification problem, $\mathcal{X}_{a|rest}$ is generated for all $a = 1 \ldots N$. Finally, a binary classifier $f_{a|rest}(\mathbf{x})$ is trained on each $\mathcal{X}_{a|rest}$ using the method we previously mentioned in this chapter. An unseen sample $\mathbf{x}_q$ is classified by computing:

$$\hat{y}_q = \arg\max_{a=1\ldots N} f_{a|rest}(\mathbf{x}_q). \quad (2.49)$$

In other words, the score of all the classifiers are computed. The classifier with the maximum score shows the class of the sample $\mathbf{x}_q$. We applied this technique on the dataset shown in Fig. 2.20. Figure 2.22 illustrates how the binary datasets are generated. It also shows how every point on the plane are classified using this technique.

Comparing the results from one versus one and one versus all, we observe that they are not identical. One advantage of one versus rest over one versus one approach is that the number of binary classifiers increases linearly with the number of classes.

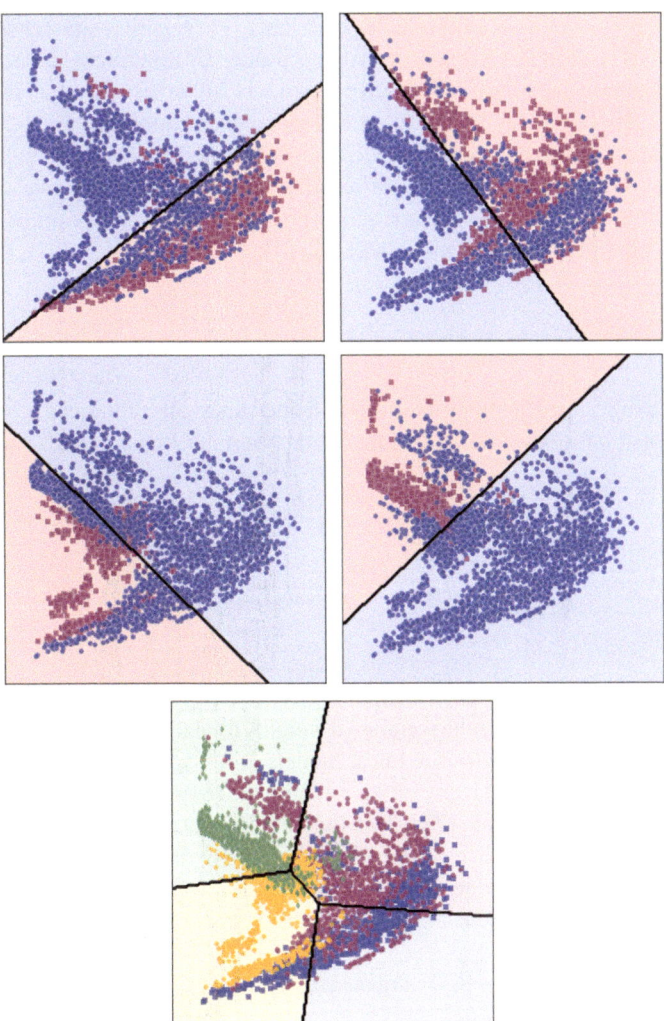

Fig. 2.22 One versus rest approach creates a binary dataset by changing the label of the class-of-interest to 1 and the label of the other classes to −1. Creating binary datasets is repeated for all classes. Then, a binary classifier is trained on each of these datasets. An unseen sample is classified based on the classification score of the binary classifiers

For this reason, one versus rest approach is practical even when the number of classes is high. However, it posses another issue which is called *imbalanced* dataset.

We will talk throughly about imbalanced datasets later in this book. But, to give an insight about this problem, consider a 250-class classification problem where each class contains 1000 training samples. This means that the training dataset contains 250,000 samples. Consequently, $\mathscr{X}_{a|rest}$ will contain 1000 samples with label 1 (positive samples) and $249,000$ samples with label -1 (negative samples). We know from previous section that a binary classifier is trained by minimizing a loss function. However, because the number of negative samples is 249 times more than the samples with label 1, the optimization algorithm will in fact try to minimize the loss occurred by the negative samples. As the result, the binary model might be highly biased toward negative samples and it might classify most of unseen positive samples as negative samples. For this reason, one versus rest approach usually requires a solution to tackle with highly imbalanced dataset $\mathscr{X}_{a|rest}$.

2.3.3 Multiclass Hinge Loss

An alternative solution to one versus one and one versus all techniques is to partition the d-dimensional space into N distinct regions using N linear models such that:

$$
\mathscr{L}_{0/1}(\mathbf{W}) = \sum_{i=1}^{n} H(\mathbf{x}, y_i)
$$

$$
H(\mathbf{x}, y_i) = \begin{cases} 0 & y_i = \arg\max_{j=1...N} f_i(\mathbf{x}_i) \\ 1 & otherwise \end{cases}
\tag{2.50}
$$

is minimum for all the samples in the training dataset. In this equation, $\mathbf{W} \in \mathbb{R}^{N \times d+1}$ is a weight matrix indicating the weights (d weights for each linear model) and biases (1 bias for each linear model) of N linear models. Also, $\mathbf{x}_i \in \mathbb{R}^d$ is defined as before and $y_i \in \{1, \ldots, N\}$ can take any of the categorical integer values between 1 and N and it indicates the class of $\mathbf{x}_i$. This loss function is in fact the generalization of the 0/1 loss function into N classes. Here also the objective of the above loss function is to minimize the number of incorrectly classified samples. After finding the optimal weight matrix $\mathbf{W}^*$, an unseen sample $\mathbf{x}_q$ is classified using:

$$
\hat{y}_q = \arg\max_{i=1...N} f_i(\mathbf{x}_q; \mathbf{W}_i^*)
\tag{2.51}
$$

where $\mathbf{W}_i^*$ depicts the ith row of the weight matrix. The weight matrix $\mathbf{W}*$ might be found by minimizing the above loss function. However, optimizing this function using iterative gradient methods is a hard task. Based on the above equation, the sample $\mathbf{x}^c$ belonging to class c is classified correctly if:

$$
\forall_{j=1...N \wedge j \neq i} \mathbf{W}_c \mathbf{x}_i > \mathbf{W}_j \mathbf{x}_i.
\tag{2.52}
$$

In other words, the score of the cth model must be greater than all other models so $\mathbf{x}^c$ is classified correctly. By rearranging the above equation, we will obtain:

$$
\forall_{j=1...N \wedge j \neq i} \mathbf{W}_j \mathbf{x}_i - \mathbf{W}_c \mathbf{x}_i \leq 0.
\tag{2.53}
$$

Assume that $\mathbf{W}_j\mathbf{x}_i$ is fixed. As $\mathbf{W}_c\mathbf{x}_i$ increases, their difference becomes more negative. In contrast, if the sample is classified incorrectly, their difference will be greater than zero. Consequently, if:

$$\max_{j=1...N \wedge j \neq i} \mathbf{W}_j\mathbf{x}_i - \mathbf{W}_c\mathbf{x}_i \qquad (2.54)$$

is negative, the sample is classified correctly. In contrary, if it is positive the sample is misclassified. In order to increase the stability of the models we can define the margin $\varepsilon \in \mathbb{R}^+$ and rewrite the above equation as follows:

$$H(\mathbf{x}_i) = \varepsilon + \max_{j=1...N \wedge j \neq i} \mathbf{W}_j\mathbf{x}_i - \mathbf{W}_c\mathbf{x}_i. \qquad (2.55)$$

The sample is classified correctly if $H(\mathbf{x}_i)$ is negative. The margin variable ε eliminates the samples which are very close to the model. Based on this equation, we can define the following loss function:

$$\mathscr{L}(\mathbf{W}) = \sum_{i=1}^{n} \max(0, \varepsilon + \max_{j \neq i} \mathbf{W}_j\mathbf{x}_i - \mathbf{W}_c\mathbf{x}_i). \qquad (2.56)$$

This loss function is called *multiclass hinge* loss. If the sample is classified correctly and it is outside the critical region, $\varepsilon + \max_{j=1...N and j \neq i} \mathbf{W}_j\mathbf{x}_i - \mathbf{W}_c\mathbf{x}_i$ will be negative. Hence, output of $\max(0, -)$ will be zero indicating that we have not made a loss on $\mathbf{x}_i$ using the current value for $\mathbf{W}$. Nonetheless, if the sample is classified in correctly or it is within the critical region $\varepsilon + \max_{j=1...N and j \neq i} \mathbf{W}_j\mathbf{x}_i - \mathbf{W}_c\mathbf{x}_i$ will be a positive number. As the result, $\max(0, +)$ will be positive indicating that we have made a loss on $\mathbf{x}_i$. By minimizing the above loss function, we will find $\mathbf{W}$ such that the number misclassified samples is minimum.

The multiclass hinge loss function is a differentiable function. For this reason, gradient-based optimization methods such as gradient descend can be used to find the minimum of this function. To achieve this goal, we have to find the partial derivatives of the loss function with respect to each of the parameters in $\mathbf{W}$. Given a sample $\mathbf{x}_i$ and its corresponding label y_i, partial derivatives of (2.56) with respect to $\mathbf{W}_{m,n}$ is calculated a follows:

$$\frac{\delta\mathscr{L}(\mathbf{W}; (\mathbf{x}_i, y_i))}{\delta\mathbf{W}_{m,n}} = \begin{cases} x_n & \varepsilon + \mathbf{W}_m\mathbf{x}_i - \mathbf{W}_{y_i}\mathbf{x}_i > 0 \ and \ m = \arg\max_{p \neq y_i} \mathbf{W}_p\mathbf{x}_i - \mathbf{W}_{y_i}\mathbf{x}_i \\ -x_n & \varepsilon + \max_{p \neq m} \mathbf{W}_p\mathbf{x}_i - \mathbf{W}_m\mathbf{x}_i > 0 \ and \ m = y_i \\ 0 & otherwise \end{cases}$$

$$\qquad (2.57)$$

$$\frac{\delta\mathscr{L}(\mathbf{W})}{\delta\mathbf{W}_{m,n}} = \sum_{i=1}^{n} \frac{\delta\mathscr{L}(\mathbf{W}; (\mathbf{x}_i, y_i))}{\delta\mathbf{W}_{m,n}} \qquad (2.58)$$

In these equations, $\mathbf{W}_{m,n}$ depicts the nth parameter of the mth model. Similar to the binary hinge loss, ε can be set to 1. In this case, the magnitude of the models will be adjusted such that the loss function is minimum. If we plug the above partial

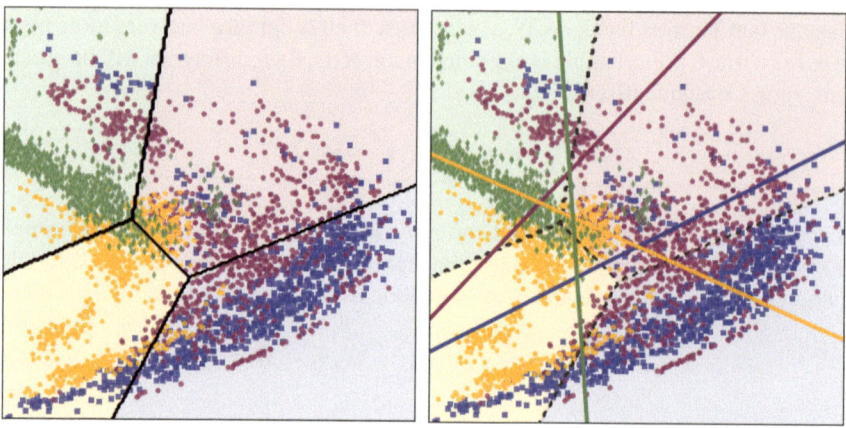

Fig. 2.23 A two-dimensional space divided into four regions using four linear models fitted using the multiclass hinge loss function. The plot on the *right* shows the linear models (lines in two-dimensional case) in the space

derivatives into the gradient descend method and apply it on the dataset illustrated in Fig. 2.20, we will obtain the result shown in Fig. 2.23.

The left plot in this figure shows how the two-dimensional space is divided into four distinct regions using the four linear models. The plot on the right also illustrates the four lines in this space. It should be noted that it is the maximum score of a sample from all the models that determined the class of the sample.

2.3.4   Multinomial Logistic Function

In the case of binary classification problems, we are able to model the probability of $\mathbf{x}$ using the logistic function in (2.33). Then, a linear model can be found by maximizing the joint probability of training samples. Alternatively, we showed in (2.46) that we can minimize the negative of logarithm of probabilities to find a linear model for a binary classification problem.

It is possible to extend the logistic function into a multiclass classification problem. We saw before that N classes can be discriminated using N different lines. In addition, we showed how to model the posterior probability of input $\mathbf{x}$ using logistic regression in (2.33). Instead of modeling $p(y = 1|\mathbf{x}; \mathbf{w})$, we can alternatively model $\ln p(y = 1|\mathbf{x}; \mathbf{w})$ given by:

$$\ln p(y = 1|\mathbf{x}; \mathbf{w}) = \mathbf{w}\mathbf{x}^T - \ln Z \qquad (2.59)$$

where $\ln Z$ is a normalization factor. This model is called *log-linear* model. Using this formulation, we can model the posterior probability of N classes using N log-linear models:

$$\ln p(y = 1|\mathbf{x}; \mathbf{w}_1) = \mathbf{w}_1\mathbf{x}^T - \ln Z$$
$$\ln p(y = 2|\mathbf{x}; \mathbf{w}_2) = \mathbf{w}_2\mathbf{x}^T - \ln Z$$
$$\cdots$$
$$\ln p(y = N|\mathbf{x}; \mathbf{w}_n) = \mathbf{w}_N\mathbf{x}^T - \ln Z$$

(2.60)

If we compute the exponential of the above equations we will obtain:

$$p(y = 1|\mathbf{x}; \mathbf{w}_1) = \frac{e^{\mathbf{w}_1\mathbf{x}^T}}{Z}$$
$$p(y = 2|\mathbf{x}; \mathbf{w}_2) = \frac{e^{\mathbf{w}_2\mathbf{x}^T}}{Z}$$
$$\cdots$$
$$p(y = N|\mathbf{x}; \mathbf{w}_N) = \frac{e^{\mathbf{w}_N\mathbf{x}^T}}{Z}$$

(2.61)

We know from probability theory that:

$$\sum_{c=1}^{N} p(y = c|\mathbf{x}; \mathbf{w}_1) = 1$$

(2.62)

Using this property, we can find the normalization factor Z that satisfies the above condition. If we set:

$$\frac{e^{\mathbf{w}_1\mathbf{x}^T}}{Z} + \frac{e^{\mathbf{w}_2\mathbf{x}^T}}{Z} + \cdots + \frac{e^{\mathbf{w}_N\mathbf{x}^T}}{Z} = 1$$

(2.63)

as solve the above equation for Z, we will obtain:

$$Z = \sum_{i=1}^{N} e^{\mathbf{w}_i\mathbf{x}^T}$$

(2.64)

Using the above normalization factor and given the sample $\mathbf{x}_i$ and its true class c, the posterior probability $p(y = c|\mathbf{x}_i)$ is computed by:

$$p(y = c|\mathbf{x}_i) = \frac{e^{\mathbf{w}_c\mathbf{x}_i^T}}{\sum_{j=1}^{N} e^{\mathbf{w}_j\mathbf{x}_i^T}}$$

(2.65)

where N is the number of classes. The denominator in the above equation is a normalization factor so $\sum_{c=1}^{N} p(y = c|\mathbf{x}_i) = 1$ holds true and, consequently, $p(y = c|\mathbf{x}_i)$ is a valid probability function. The above function which is called *softmax* function is commonly used to train convolutional neural networks. Given, a dataset of d-dimensional samples $\mathbf{x}_i$ with their corresponding labels $y_i \in \{1, \ldots N\}$ and assuming the independence relation between the samples (see Sect. 2.2.3), likelihood of all samples for a fixed $\mathbf{W}$ can be written as follows:

$$p(\mathscr{X}) = \prod_{i=1}^{n} p(y = y_i | \mathbf{x}_i). \tag{2.66}$$

As before, instead of maximizing the likelihood, we can minimize the negative of log-likelihood that is defined as follows:

$$-\log(p(\mathscr{X})) = -\sum_{i=1}^{n} \log(p(y = y_i | \mathbf{x}_i)). \tag{2.67}$$

Note that the product operator has changed to the summation operator taking into account the fact that $\log(ab) = \log(a) + \log(b)$. Now, for any $\mathbf{W}$ we can compute the following loss:

$$\mathscr{L}_{softmax}(\mathbf{W}) = -\sum_{i=1}^{n} \log(y_c) \tag{2.68}$$

where $\mathbf{W} \in \mathbb{R}^{N \times d+1}$ represents the parameters for N linear models and $y_c = p(y = y_i | \mathbf{x}_i)$. Before computing the partial derivatives of the above loss function, we explain how to show the above loss function using a *computational* graph. Assume computing $\log(y_c)$ for a sample. This can be represented using the graph in Fig. 2.24.

Computational graph is a directed acyclic graph where each non-leaf node in this graph shows a computational unit which accepts one or more inputs. Leaves also show the input of the graph. The computation starts from the leaves and follows the direction of the edges until it reaches to the final node. We can compute the gradient of each computational node with respect to its inputs. The labels next to each edge shows the gradient of its child node (top) with respect to its parent node (bottom). Assume, we want to compute $\delta\mathscr{L}/\delta\mathbf{W}_1$. To this end, we have to sum all the paths from $\mathscr{L}$ to $\mathbf{W}_1$ and multiply the gradients represented by edges along each path. This result will be equivalent to *multivariate chain rule*. According to this, $\delta\mathscr{L}/\delta\mathbf{W}_1$ will be equal to:

$$\frac{\delta\mathscr{L}}{\delta\mathbf{W}_1} = \frac{\delta\mathscr{L}}{\delta y_c} \frac{\delta y_c}{\delta z_1} \frac{\delta z_1}{\delta\mathbf{W}_1}. \tag{2.69}$$

Using this concept, we can easily compute $\delta\mathscr{L}/\delta\mathbf{W}_{i,j}$ where $\mathbf{W}_{i,j}$ refers to the jth parameter of uth model. For this purpose, we need to compute $\frac{\delta y_c}{\delta z_i}$ which is done as follows:

$$\frac{\delta y_c}{\delta z_i} = \frac{\delta \frac{e^{z_c}}{\sum_{m=1}^{N} e^{z_m}}}{\delta z_i} = \begin{cases} \frac{e^{z_c} \sum_m e^{z_m} - e^{z_c} e^{z_c}}{\left(\sum_m e^{z_m}\right)^2} = y_c(1 - y_c) & i = c \\ \frac{-e^{z_i} e^{z_c}}{\left(\sum_m e^{z_m}\right)^2} = y_i y_c & i \neq c \end{cases} \tag{2.70}$$

Fig. 2.24 Computational
graph of the softmax loss on
one sample

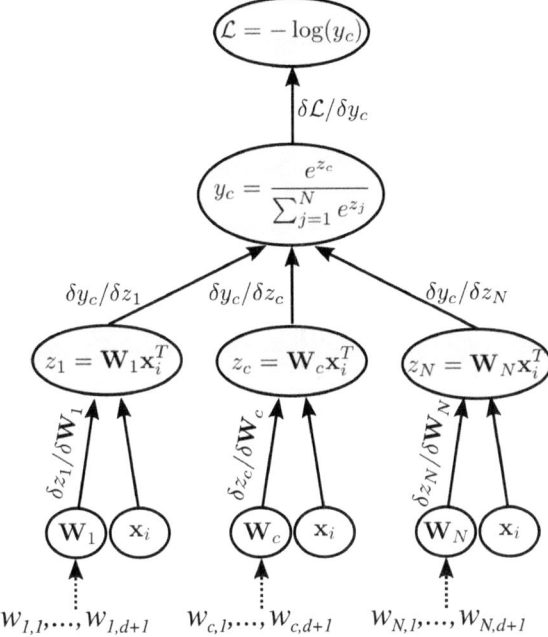

Now, we can compute $\delta\mathscr{L}/\delta\mathbf{W}_{i,j}$ by plugging the above derivative into the chain
rule obtained by the computational graph for sample $\mathbf{x}$ with label y_c.

$$\frac{\delta\mathscr{L}}{\delta\mathbf{W}_{i,j}} = \begin{cases} -(1-y_c)x_j & i = c \\ y_i x_j & i \neq c \end{cases} \tag{2.71}$$

With this formulation, the gradient of all the samples will be equal to sum of the
gradient of each sample. Now, it is possible to minimize the softmax loss function
using the gradient descend method. Figure 2.25 shows how the two-dimensional
space in our example is divided into four regions using the models trained by the
softmax loss function. Comparing the results from one versus one, one versus all,
the multiclass hinge loss and the softmax loss, we realize that their results are not
identical. However, the two former techniques is not usually used for multiclass
classification problems because of the reasons we mentioned earlier. Also, there is
not a practical rule of thumb to tell if the multiclass hinge loss better or worse than
the softmax loss function.

2.4   Feature Extraction

In practice, it is very likely that samples in the training set $\mathscr{X} = \{(\mathbf{x}_1, y_1), \dots,$
$(\mathbf{x}_n, y_n)\}$ are not linearly separable. The multiclass dataset in the previous section is

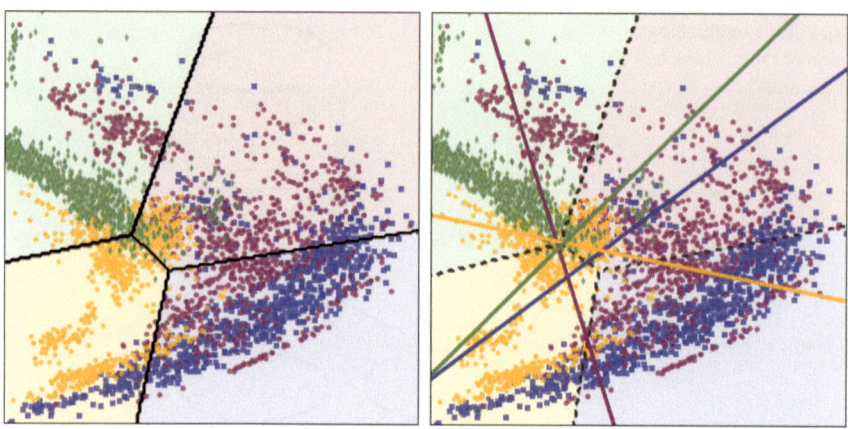

Fig. 2.25 The two-dimensional space divided into four regions using four linear models fitted using the softmax loss function. The plot on the *right* shows the linear models (lines in two-dimensional case) in the space

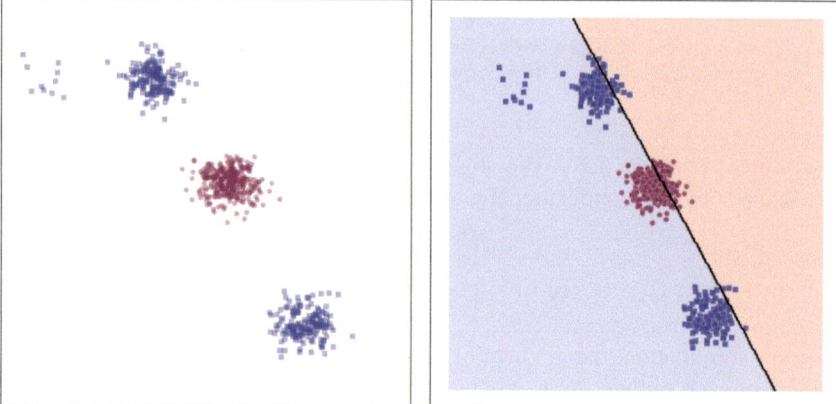

Fig. 2.26 A linear classifier is not able to accurately discriminate the samples in a nonlinear dataset

an example of such a dataset. Figure 2.26 shows a nonlinear dataset and the linear classifier fitted using logistic regression. Samples of each class are illustrated using a different marker and different color.

Clearly, it is impossible to perfectly discriminate these two classes using a line. There are mainly two solutions for solving this problem. The first solution is to train a nonlinear classifier such as random forest on the training dataset. This method is not within the scope of this book. The second method is to project the original data into another space using the transformation function $\Phi : \mathbb{R}^d \rightarrow \mathbb{R}^{\hat{d}}$ where classes are linearly separable in the transformed space. Here, $\hat{d}$ can be any arbitrary integer number. Formally, given the sample $\mathbf{x} \in \mathbb{R}^d$, it is transformed into a $\hat{d}$-dimensional space using:

$$\Phi(\mathbf{x}) = \hat{\mathbf{x}} = \begin{bmatrix} \phi_1(\mathbf{x}) \\ \phi_2(\mathbf{x}) \\ \vdots \\ \phi_{\hat{d}}(\mathbf{x}) \end{bmatrix} \qquad (2.72)$$

where $\phi_i : \mathbb{R}^d \to 1$ is a scaler function which accepts a d-dimensional input and return a scaler. Also, ϕ_i can be any function. Sometimes, an expert can design these functions based on the requirements of the problem. To transform the above nonlinear dataset, we define $\Phi(\mathbf{x})$ as follows:

$$\Phi(\mathbf{x}) = \hat{\mathbf{x}} = \begin{bmatrix} \phi_1(\mathbf{x}) = e^{-10\|\mathbf{x}-c_1\|^2} \\ \phi_2(\mathbf{x}) = e^{-20\|\mathbf{x}-c_2\|^2} \end{bmatrix} \qquad (2.73)$$

where $c_1 = (0.56, 0.67)$ and $c_2 = (0.19, 0.11)$. By applying this function on each sample, we will obtain a new two-dimensional space where the samples are non-linearly transformed. Figure 2.27 shows how samples are projected into the new two-dimensional space. It is clear that the samples in the new space become linearly separable. In other words, the dataset $\hat{\mathscr{X}} = \{(\Phi(\mathbf{x}_1), y_1), \ldots, (\Phi(\mathbf{x}_n), y_n)\}$ is linearly separable. Consequently, the samples in $\hat{\mathscr{X}}$ can be classified using a linear classifier in the previous section. Figure 2.28 shows a linear classifier fitted on the data in the new space.

The decision boundary of a linear classifier is a hyperplane (a line in this example). However, because $\Phi(\mathbf{x})$ is a nonlinear transformation, if we apply the inverse transform from the new space to the original space, the decision boundary will not be a hyperplane anymore. Instead, it will be a nonlinear decision boundary. This is illustrated in the right plot of Fig. 2.28.

Choice of $\Phi(\mathbf{x})$ is the most important step in transforming samples into a new space where they are linearly separable. In the case of high-dimensional vectors such as images, finding an appropriate $\Phi(\mathbf{x})$ becomes even harder. In some case, $\Phi(\mathbf{x})$ might be composition of multiple functions. For example, one can define $\Phi(\mathbf{x}) = \Psi(\Omega(\Gamma(\mathbf{x})))$ where $\Phi : \mathbb{R}^d \to \mathbb{R}^{\hat{d}}, \Psi : \mathbb{R}^{d_2} \to \mathbb{R}^{\hat{d}}, \Omega : \mathbb{R}^{d_1} \to \mathbb{R}^{d_2}$ and, $\Gamma : \mathbb{R}^d \to \mathbb{R}^{d_1}$. In practice, there might be infinite number of functions to make samples linearly separable.

Let us apply our discussions so far on a real world problem. Suppose the 43 classes of traffic signs shown in Fig. 2.29 that are obtained from the German traffic sign recognition benchmark (GTSRB) dataset. For the purpose of this example, we randomly picked 1500 images for each class. Assume a 50×50 RGB image. Taking into account the fact that each pixel in this image is represented by a three-dimensional vector, the flattened image will be a $50 \times 50 \times 3 = 7500$ dimensional vector. Therefore, the *training dataset* $\mathscr{X}$ is composed of 1500 training sample pair $(\mathbf{x}_i, y_i)$ where $\mathbf{x} \in \mathbb{R}^{7500}$ and $y_i \in \{0, \ldots 42\}$.

Beside the training dataset, we also randomly pick 6400 *test samples* $(\dot{\mathbf{x}}, \dot{y}_i)$ from the dataset that are not included in $\mathscr{X}$. Formally, we have another dataset $\dot{\mathscr{X}}$ of

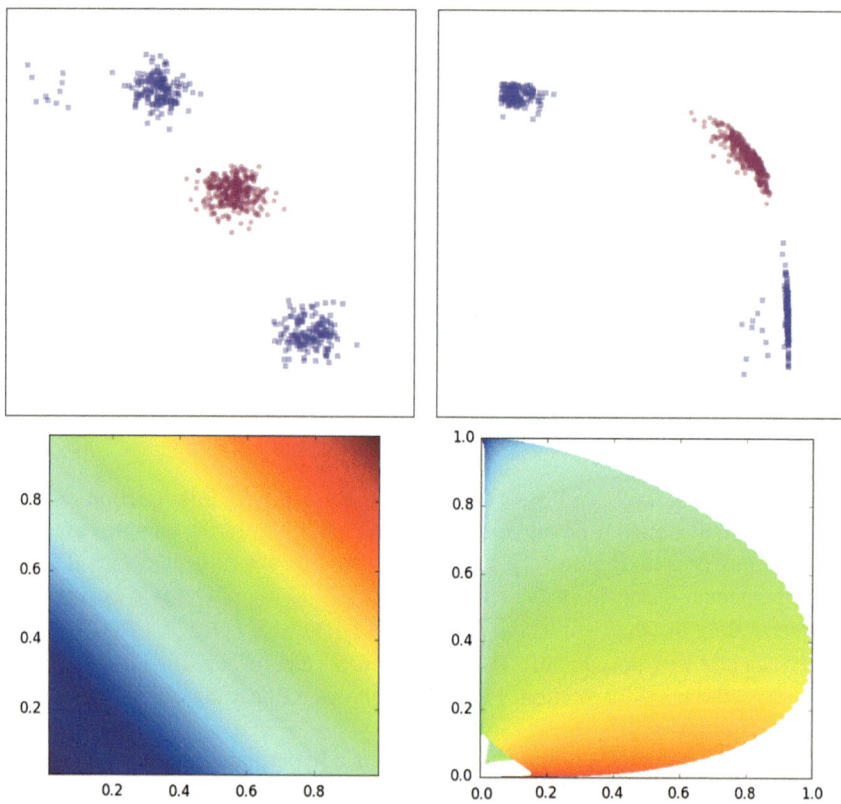

Fig. 2.27 Transforming samples from the original space (*left*) into another space (*right*) by applying $\Phi(\mathbf{x})$ on each sample. The *bottom colormaps* show how the original space is transformed using this function

traffic signs where $\dot{\mathbf{x}} \in \mathbb{R}^{7500}$ and $\dot{\mathbf{x}} \notin \mathcal{X}$ and $\dot{y}_i \in \{0, \ldots 42\}$. It is very important in testing a model to use unseen samples. We will explain this topic throughly in the next chapters. Finally, we can train a linear classifier $F(\mathbf{x})$ using $\mathcal{X}$ to discriminate the 43 classes of traffic signs. Then, $F(\mathbf{x})$ can be tested using $\dot{\mathcal{X}}$ and computing *classification accuracy*.

To be more specific, we pick every sample $\dot{\mathbf{x}}_i$ and predict its class label using $F(\dot{\mathbf{x}}_i)$. Recall from previous sections that for a softmax model with 43 linear models, the class of sample $\dot{\mathbf{x}}_i$ is computed using $F(\dot{\mathbf{x}}_i) = \arg\max_{i=1\ldots43} f_i(\dot{\mathbf{x}}_i)$ where $f_i(\dot{\mathbf{x}}_i) = \mathbf{w}\dot{\mathbf{x}}_i$ is the score computed by the ith model. With this formulation, the classification accuracy of the test samples is obtained by computing:

$$acc = \frac{1}{6400} \sum_{i=1}^{6400} \mathbf{1}[F(\dot{\mathbf{x}}_i) == \dot{y}_i] \qquad (2.74)$$

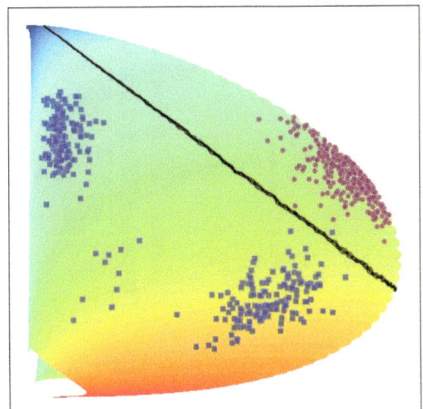
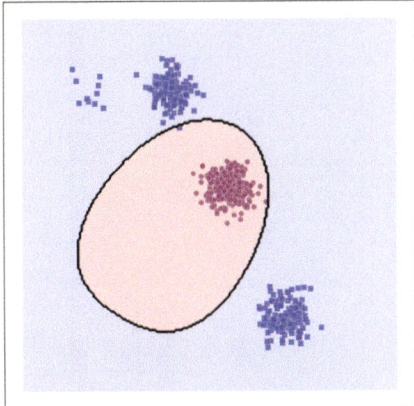

Fig. 2.28 Samples become linearly separable in the new space. As the result, a linear classifier is able to accurately discriminate these samples. If we transform the linear model from the new space into the original space, the linear decision boundary become a nonlinear boundary

Fig. 2.29 43 classes of traffic in obtained from the GTSRB dataset (Stallkamp et al. 2012)

where $\mathbf{1}[.]$ is the indicator function and it returns 1 when the input is true. The quantity *acc* is equal to 1 when all the samples are classified correctly and it is equal to 0 when all of them are misclassified. We trained a linear model on this dataset using the raw pixel values. The accuracy on the test set is equal to 73.17%. If we ignore the intercept, the parameters vector $\mathbf{w} \in \mathbb{R}^{7500}$ of the linear model $f(x) = \mathbf{w}\mathbf{x}^T$ has the same dimension as the input image. One way to visualize and study the parameter vector is to reshape $\mathbf{w}$ into a $50 \times 50 \times 3$ image. Then, we can plot each channel in this three-dimensional array using a colormap plot. Figure 2.30 shows weights of the model related to Class 1 after reshaping.

We can analyze this figure to see what a linear model trained on raw pixel intensities exactly learns. Consider the linear model $f(\mathbf{x}) = w_1 x_1 + \cdots + w_n x_n$ without the intercept term. Taking into account the fact that pixel intensities in a regular RGB image are positive values, x_i in this equation is always a positive value. Therefore, $f(\mathbf{x})$ will return a higher value if w_i is a high positive number. In contrary, $f(\mathbf{x})$ will return a smaller value if w_i is a very small negative number. From another perspective, we can interpret positive weights as "likes" and negative weights as "dislikes" of the linear model.

That being said if w_i is negative, the model does not like high values of x_i. Hence, if the intensity of pixel at x_i is higher than zero it will reduce the classification score.

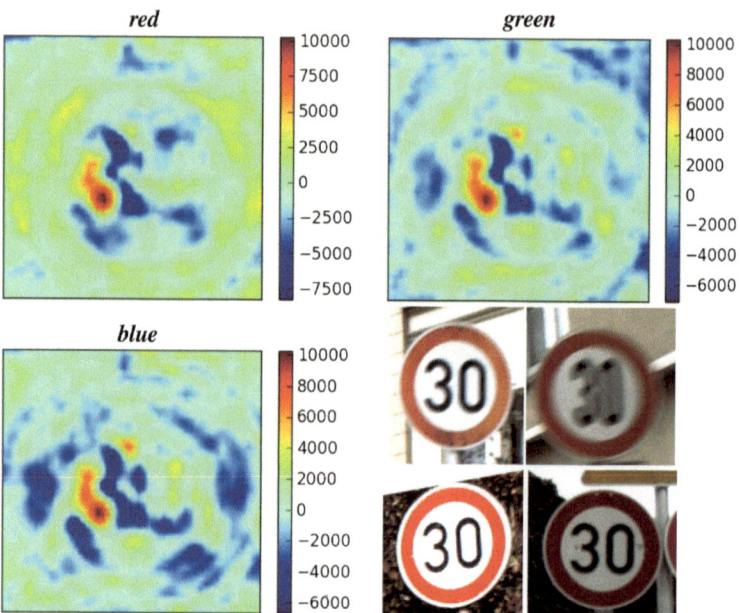

Fig. 2.30 Weights of a linear model trained directly on raw pixel intensities can be visualized by reshaping the vectors so they have the same shape as the input image. Then, each channel of the reshaped matrix can be shown using a colormap

In contrast, if w_i is positive, the model likes high values of x_i. In other words, as the intensity of x_i increases, the model becomes more confident about the classification since it increases the classification score.

Looking at this figure, we see a red region in the middle of red, green and blue channels. According to the color map next to each plot, red regions correspond to weights with high positive values. Since, the same region is red in all three channels, we can imply that the model likes to see the white color in that specific region. Then, we observe that the region analogous to the rim of the sign has high positive weight in the red channel and small negative weights in the blue channel. Also, the weights of the green channel for that region is close to zero. This means that the model likes to see high red values in that region and it dislikes blue values in that region. This choice made by the model also seems rational for a human expert. This argument can be applied on the other classes of traffic signs, as well.

Remember that the accuracy of the model trained on raw pixel intensities was equal to 73.17%. Now, the question is why the accuracy of the model is very low? To answer this question, we start with a basic concept. A two-dimensional vector (x_1, x_2) can be illustrated using a *point* in a two-dimensional space. Moreover, a three-dimensional vector (x_1, x_2, x_3) can be shown using a point in a three-dimensional space. Similarly, a d-dimensional vector $(x_1, \ldots, x_d)$ is a point in a d-dimensional space. It is trivial for a human to imagine the points in two-dimensional and three-dimensional spaces.

But, it might be difficult at first to imagine higher dimensions. For starting, it suffice to know that a d-dimensional vector is a point in a d-dimensional space.

Each RGB image in the above example will be a point in a 7500-dimensional space. We can study the above question in this space. There are mainly two possibilities that reduces the accuracy of a linear model in this space defined by raw images. First, like the dataset in Fig. 2.26 the classes of traffic signs might be completely disjoint but they might not be linearly separable. Second, similar to the dataset in Fig. 2.20, the classes might have overlap with each other. The latter problem is commonly known as *interclass similarity* meaning that samples of two or more classes are similar. In both cases, a linear model is not able to accurately discriminate the classes.

Although there might not be a quick remedy to the second problem, the first problem might be addressed by transforming the raw vectors into another space using the feature transformation function $\Phi(\mathbf{x})$. Knowing the fact that output of $\Phi(\mathbf{x})$ is a $\hat{d}$-dimensional vector, the question in designing $\Phi(\mathbf{x})$ is what should be the value of $\hat{d}$? Even if we found a way to determine the value of $\hat{d}$, the next question is what should be the transformation function $\phi_i(\mathbf{x})$, $i = 1, \ldots, \hat{d}$? There are infinite ways to define this function. For this reason, it is not trivial in practice to define $\Phi(\mathbf{x})$ for *an image* (it might not be a tedious task for other modalities with low dimensions).

To alleviate this problem, researchers came up with the idea of *feature extraction* algorithms. In general, a feature extraction algorithm processes an image and generates a more informative vector which better separates classes. Notwithstanding, a feature extraction algorithm does not guarantee that the classes will be linearly separable. Despite this, in most cases, a feature extraction is applied on an image before feeding it to a classifier. In other words, we do not classify images using raw pixel values. Instead, we always extract their feature and train a classifier on top of the feature vectors.

One of the widely used feature extraction algorithms is called *histogram of oriented gradients* (HOG). It starts by applying the gamma correction transformation on the image and computing its first derivatives. Then, the image is divided into small patches called cells. Within each cell, a histogram is computed based on the orientation of the gradient vector and its magnitude using the pixels inside that cell. Then, blocks are formed by considering neighbor cells and the histogram of the cells within that block are concatenated. Finally, the feature vector is obtained by concatenating the vectors of all blocks. The whole process of this algorithm can be easily represented in terms of mathematical equations.

Assume that $\Phi_{hog}(\mathbf{x}) : \mathbb{R}^d \to \mathbb{R}^{d_{hog}}$ denotes the HOG features. We can now apply $\Phi_{hog}(\mathbf{x})$ on each sample of the training set $\mathscr{X}$ in order to obtain $\hat{\mathscr{X}} = \{(\Phi_{hog}(\mathbf{x}_1), y_1), \ldots, (\Phi_{hog}(\mathbf{x}_n), y_n)\}$. Then, a linear classifier is trained using $\hat{\mathscr{X}}$. By doing this, the accuracy of the classification increases to 88.90%. Comparing with the accuracy of the classifier trained on raw pixel intensities (i.e., 73.17%), the accuracy increases 15.73%.

There might different reasons that the accuracy is not still very high. First, the feature extraction function $\Phi_{hog}(\mathbf{x})$ might not be able to perfectly make the classes linearly separable. This could be due to the fact that there are traffic signs such as

"left bend ahead" and "right bend ahead" with slight differences. The utilized feature
extraction function might not be able to effectively model these differences such that
these classes become linearly separable. Second, the function $\Phi_{hog}(\mathbf{x})$ may cause
some of the classes to have overlap with other classes. Both or one of these reasons
can be responsible for having a low accuracy.

Like before, it is possible to create another function whose input is $\Phi_{hog}(\mathbf{x})$ and its
output is a $\hat{d}$ dimensional vector. For example, we can define the following function:

$$\Phi(\Phi_{hog}(\mathbf{x})) = \begin{bmatrix} \phi_1(\Phi_{hog}(\mathbf{x})) \\ \phi_2(\Phi_{hog}(\mathbf{x})) \\ \vdots \\ \phi_{\hat{d}}(\Phi_{hog}(\mathbf{x})) \end{bmatrix} = \begin{bmatrix} e^{-\gamma\|\Phi_{hog}(\mathbf{x})-\mathbf{c}_1\|^2} \\ e^{-\gamma\|\Phi_{hog}(\mathbf{x})-\mathbf{c}_2\|^2} \\ \vdots \\ e^{-\gamma\|\Phi_{hog}(\mathbf{x})-\mathbf{c}_{\hat{d}}\|^2} \end{bmatrix} \tag{2.75}$$

where $\gamma \in \mathbb{R}$ is a scaling constant and $\mathbf{c}_i \in \mathbb{R}^{d_{hog}}$ is parameters which can be
defined manually or automatically. Doing so, we can generate a new dataset
$\mathscr{X} = \{\Phi((\Phi_{hog}(\mathbf{x}_1)), y_1), \ldots, (\Phi(\Phi_{hog}(\mathbf{x}_n)), y_n)\}$ and train a linear classifier on
top of this dataset. This increases the accuracy from 88.90 to 92.34%. Although the
accuracy is higher it is not still high enough to be used in practical applications.
One may add another feature transformation whose input is $\Phi(\Phi_{hog}(\mathbf{x}))$. In fact,
compositing the transformation function can be done several times. But, this does
not guarantee that the classes are going to be linearly separable. Some of the transformation function may increase the interclass overlap causing a drop in accuracy.

As it turns out, the key to accurate classification is to have a feature transformation
function $\Phi(\mathbf{x})$ which is able to make the classes linearly separable without causing
interclass overlap. But, how can we find $\Phi(\mathbf{x})$ which satisfies both these conditions?
We saw in this chapter that a classifier can be directly trained on the training dataset.
It might be also possible to learn $\Phi(\mathbf{x})$ using the same training dataset. If $\Phi(\mathbf{x})$ is
designed by a human expert (such as the HOG features), it is called a *hand-crafted*
or *hand-engineered* feature function.

2.5 Learning $\Phi(\mathbf{x})$

Despite the fairly accurate results obtained by hand-crafted features on some datasets,
as we will show in the next chapters, the best results have been achieved by learning
$\Phi(\mathbf{x})$ from a training set. In the previous section, we designed a feature function
to make the classes in Fig. 2.26 linearly separable. However, designing that feature
function by hand was a tedious task and needed many trials. Note that, the dataset
shown in that figure was composed of two-dimensional vectors. Considering the
fact that a dataset may contain high-dimensional vectors in real-world applications,
designing an accurate feature transformation function $\Phi(\mathbf{x})$ becomes even harder.

For this reason, in many cases the better approach is to learn $\Phi(\mathbf{x})$ from data.
More specifically, $\Phi(\mathbf{x}; \mathbf{w}_\phi)$ is formulated using the parameter vector $\mathbf{w}_\phi$. Then, the

linear classifier for i^{th} class is defined as:

$$f_i(\mathbf{x}) = \mathbf{w}\Phi(\mathbf{x}; \mathbf{w}_\phi)^T \tag{2.76}$$

where $\mathbf{w} \in \mathbb{R}^{\hat{d}}$ and $\mathbf{w}_\phi$ are parameter vectors that are found using training data. Depending on the formulation of $\Phi(\mathbf{x})$, $\mathbf{w}_\phi$ can be any vector with arbitrary size. The parameter vector $\mathbf{w}$ and $\mathbf{w}_\phi$ determine the weights for the linear classifier and the transformation function, respectively. The ultimate goal in a classification problem is to *jointly* learn this parameter vectors such that the classification accuracy is high.

This goal is exactly the same as learning $\mathbf{w}$ such that $\mathbf{w}\mathbf{x}^T$ accurately classifies the samples. Therefore, we can use the same loss functions in order to train both parameter vectors in (2.76). Assume that $\Phi(\mathbf{x}; \mathbf{w}_\phi)$ is defined as follows:

$$\Phi(\mathbf{x}; \mathbf{w}_\phi) = \begin{bmatrix} ln(1 + e^{(w_{11}x_1 + w_{21}x_2 + w_{01})}) \\ ln(1 + e^{(w_{12}x_1 + w_{22}x_2 + w_{02})}) \end{bmatrix} \tag{2.77}$$

In the above equation $\mathbf{w}_\phi = \{w_{11}, w_{21}, w_{01}, w_{12}, w_{22}, w_{02}\}$ is the parameter vector for the feature transformation function. Knowing the fact that the dataset in Fig. 2.26 is composed of two classes, we can minimize the binary logistic loss function for jointly finding $\mathbf{w}$ and $\mathbf{w}_\phi$. Formally, the loss function is defined as follows:

$$\mathcal{L}(\mathbf{w}, \mathbf{w}_\phi) = -\sum_{i=1}^{n} y_i log(\sigma(\mathbf{w}\Phi(\mathbf{x})^T)) + (1 - y_i)(1 - log(\sigma(\mathbf{w}\Phi(\mathbf{x})^T))) \tag{2.78}$$

The intuitive way to understanding the above loss function and computing its gradient is to build its computational graph. This is illustrated in Fig. 2.31. In the graph, $g(z) = ln(1 + e^z)$ is a nonlinear function which is responsible for nonlinearly transforming the space. First, the dot product of the input vector $\mathbf{x}$ is computed with two weigh vectors $\mathbf{w}_1^{L_0}$ and $\mathbf{w}_2^{L_0}$ in order to obtain $z_1^{L_0}$ and $z_2^{L_0}$, respectively. Then, each of these values is passed through a nonlinear function and their dot product with $\mathbf{w}^{L_2}$ is calculated. Finally, this score is passed through a sigmoid function and the loss is computed in the final node. In order to minimize the loss function (i.e., the top node in the graph), the gradient of the loss function has to be computed with respect to the nodes indicated by $\mathbf{w}$ in the figure. This can be done using the chain rule or derivatives. To this end, gradient of each node with respect to its parent must be computed. Then, for example, to compute $\delta\mathcal{L}/\delta\mathbf{w}_1^{L_0}$, we have to sum all the paths from $\mathbf{w}_1^{L_0}$ to $\mathcal{L}$ and multiply the term along each path. Since there is only from one path from $\mathbf{w}_1^{L_0}$ in this graph, the gradient will be equal to:

$$\frac{\delta\mathcal{L}}{\delta\mathbf{w}_1^{L_0}} = \frac{\delta\mathbf{z}_1^{L_0}}{\delta\mathbf{w}_1^{L_0}} \frac{\delta\mathbf{z}_1^{L_1}}{\delta\mathbf{z}_1^{L_0}} \frac{\delta\mathbf{z}^{L_2}}{\delta\mathbf{z}_1^{L_1}} \frac{\delta p}{\delta\mathbf{z}^{L_2}} \frac{\delta\mathcal{L}}{\delta p} \tag{2.79}$$

The gradient of the loss with respect to the other parameters can be obtained in a similar way. After that, we should only plug the gradient vector in the gradient descend

Fig. 2.31 Computational
graph for (2.78). Gradient of
each node with respect to its
parent is shown on the edges

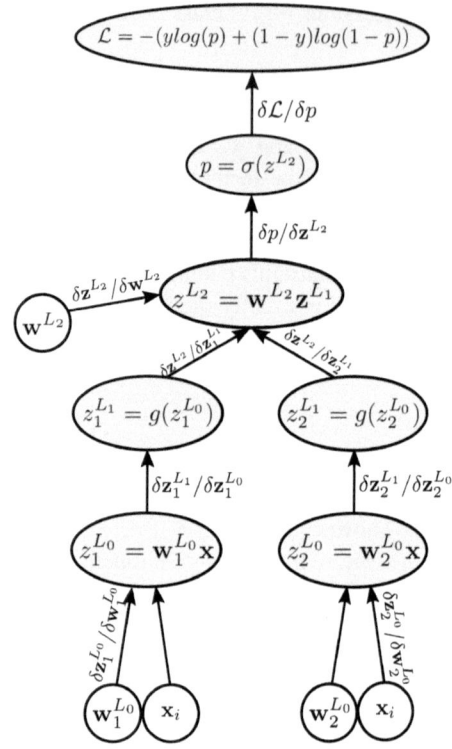

method and minimize the loss function. Figure 2.32 illustrates how the system eventually learns to transform and classify the samples. According to the plots in the
second and third rows, the model is able to find a transformation where the classes
become linearly separable. Then, classification of the samples is done in this space.
This means that the decision boundary in the transformed space is a hyperplane.
If we apply the inverse transform from the feature space to the original space, the
hyperplane is not longer a line. Instead, it is a nonlinear boundary which accurately
discriminates the classes.

In this example, the nonlinear transformation function that we used in (2.77) is
called the *softplut* function and it is defined as $g(x) = ln(1 + e^x)$. The derivative of
this function is also equal to $g'(x) = \frac{1}{1+e^{-x}}$. The softplut function can be replaced
with another function whose input is a scaler and its output is a real number. Also, we
there are many other ways to define a transformation function and find its parameters
by minimizing the loss function.

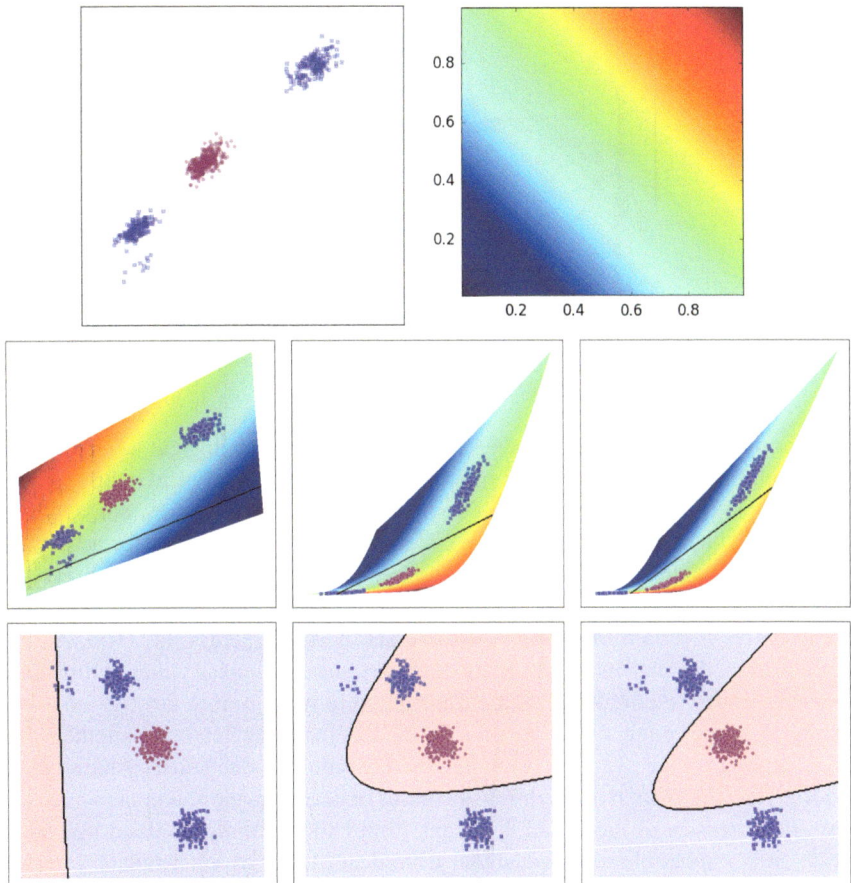

Fig. 2.32 By minimizing (2.78) the model learns to jointly transform and classify the vectors. The *first row* shows the distribution of the training samples in the two-dimensional space. The *second* and *third rows* show the status of the model in three different iterations starting from the left plots

2.6 Artificial Neural Networks

The idea of learning a feature transformation function instead of designing it by hand is very useful and it produces very accurate results in practice. However, as we pointed out above, there are infinite ways to design a trainable feature transformation function. But, not all of them might be able to make the classes linearly separable in the feature space. As the result, there might be a more general way to design a trainable feature transformation functions.

An artificial neural network (ANN) is an interconnected group of smaller computational units called neurons and it tries to mimic biological neural networks. Detailed discussion about biological neurons is not within the scope of this book. But, in order

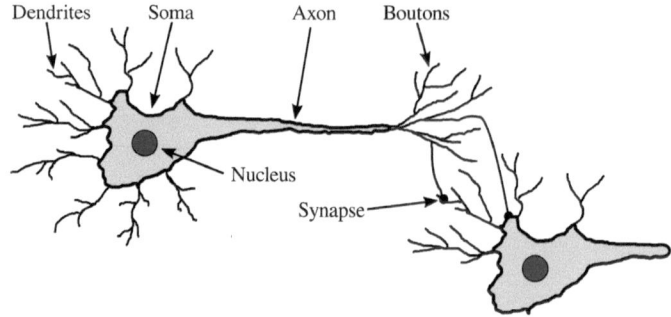

Fig. 2.33 Simplified diagram of a biological neuron

to better understand an artificial neuron we explain how a biological neuron works in general. Figure 2.33 illustrates a simplified diagram of a biological neuron.

A neuron is mainly composed of four parts including dendrites, soma, axon, nucleus and boutons. Boutons is also called axon terminals. Dendrites act as the input of the neuron. They are connected either to a sensory input (such as eye) or other neurons through synapses. Soma collects the inputs from dendrites. When the inputs passes a certain threshold it fires series of spikes across the axon. As the signal is fired, the nucleus returns to its stationary state. When it reaches to this state, the firing stops. The fired signals are transmitted to other neuron through boutons. Finally, synaptic connections transmits the signals from one neuron to another.

Depending on the synaptic strengths and the signal at one axon terminal, each dendron (i.e., one branch of dendrites) increases or decreases the potential of nucleus. Also, the direction of the signal is always from axon terminals to dendrites. That means, it is impossible to pass a signal from dendrites to axon terminals. In other words, the path from one neuron to another is always a one-way path. It is worth mentioning that each neuron might be connected to thousands of other neurons. Mathematically, a biological neuron can be formulated as follows:

$$f(\mathbf{x}) = \mathscr{G}(\mathbf{w}\mathbf{x}^T + b). \tag{2.80}$$

In this equation, $\mathbf{w} \in \mathbb{R}^d$ is the weight vector, $\mathbf{x} \in \mathbb{R}^d$ is the input and $b \in \mathbb{R}$ is the intercept term which is also called *bias*. Basically, an artificial neuron computes the weighted sum of inputs. This mimics the soma in biological neuron. The synaptic strength is modeled using $\mathbf{w}$ and inputs from other neurons or sensors are modeled using $\mathbf{x}$. In addition $\mathscr{G}(x) : \mathbb{R} \to \mathbb{R}$ is a *nonlinear* function which is called *activation function*. It accepts a real number and returns another real number after applying a nonlinear transformation on it. The activation function act as the threshold function in biological neuron. Depending on the potential of nucleus (i.e., $\mathbf{w}\mathbf{x}^T + b$), the activation function returns a real number. From computational graph perspective, a neuron is a node in the graph with the diagram illustrated in Fig. 2.34.

An artificial neural network is created by connecting one or more neurons to the input. Each pair of neurons may or may not have a connection between them. With

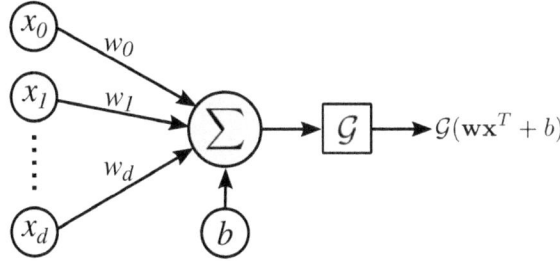

Fig. 2.34 Diagram of an artificial neuron

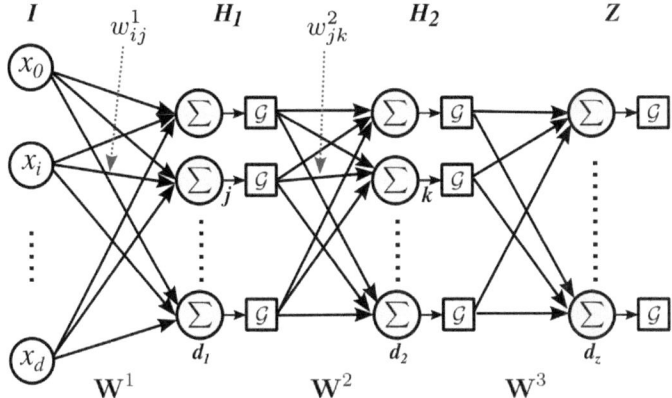

Fig. 2.35 A feedforward neural network can be seen as a directed acyclic graph where the inputs are passed through different layer until it reaches to the end

this formulation, the logistic regression model can be formulated using only one neuron where $\mathcal{G}(x)$ is the sigmoid function in (2.33). Depending on how neurons are connected, a network act differently. Among various kinds of artificial neural networks *feedforward neural network* (FNN) and *recurrent neural network* (RNN) are commonly used in computer vision community.

The main difference between these two kinds of neural networks lies in the connection between their neurons. More specifically, in a feedforward neural network the connections between neurons do not form a cycle. In contrast, in recurrent neural networks connection between neurons form a directed cycle. Convolutional neural networks are a specific type of feedforward networks. For this reason, in the remaining of this section we will only focus on feedforward networks. Figure 2.35 shows general architecture of feedforward neural networks.

A feedforward neural network includes one or more layers in which each layer contains one or more neurons. Also, number of neurons in one layer can be different from another layer. The network in the figure has one input layer and three layers with computational neurons. Any layer between the input layer and the last layer is called a *hidden layer*. The last layer is also called the *output layer*. In this chapter, the input layer is denoted by I and hidden layers are denoted by H_i where i starts from 1. Moreover, the output layer is denoted by Z. In this figure, the first hidden layer has d_1 neurons and the second hidden layer has d_2 neurons. Also, the output layer has d_z neurons.

It should be noted that every neuron in a hidden layer or the output layer is connected to all the neurons in the previous layer. That said, there is $d_1 \times d_2$ connections between H_1 and H_2 in this figure. The connection from the ith input in the input layer to the jth neuron in H_1 is denoted by w_{ij}^1. Likewise, the connection from the jth neuron in H_1 to the kth neuron in H_2 is denoted by w_{jk}^2. With this formulation, the weights connecting the input layer to H_1 can be represented using $\mathbf{W}_1 \in \mathbb{R}^{d \times d_1}$ where $W(i, j)$ shows the connection from the ith input to the jth neuron.

Finally, the activation function $\mathscr{G}$ of each neuron can be different from all other neurons. However, all the neuron in the same layer usually have the same activation function. Note that we have removed the bias connection in this figure to cut the clutter. However, each neuron in all the layers is also have a bias term beside its weights. The bias term in H_1 is represented by $\mathbf{b}^1 \in \mathbb{R}^{d_1}$. Similarly, the bias of hth layer is represented by $\mathbf{b}^h$. Using this notations, the network illustrated in this figure can be formulated as:

$$f(\mathbf{x}) = \mathscr{G}\left(\mathscr{G}\left(\mathscr{G}(\mathbf{x}\mathbf{W}^1 + \mathbf{b}^1)\mathbf{W}^2 + \mathbf{b}^2\right)\mathbf{W}^3 + \mathbf{b}^3\right). \tag{2.81}$$

In terms of feature transformation, the hidden layers act as a feature transformation function which is a composite function. Then, the output layer act as the linear classifier. In other words, the input vector $\mathbf{x}$ is transformed into a d_1-dimensional space using the first hidden layer. Then, the transformed vectors are transformed into a d_2-dimensional space using the second hidden layer. Finally, the output layer classifies the transformed d_2-dimensional vectors.

What makes a feedforward neural network very special is the fact the a feedforward network with one layer and finite number of neurons is a universal approximator. In other words, a feedforward network with one hidden layer can approximate any continuous function. This is an important property in classification problems.

Assume a multiclass classification problem where the classes are not linearly separable. Hence, we must find a transformation function which makes the classes linearly separable in the feature space. Suppose that $\Phi_{ideal}(\mathbf{x})$ is a transformation function which is able to perfectly do this job. From function perspective, $\Phi_{ideal}(\mathbf{x})$ is a vector-valued continues function. Since a feedforward neural network is a universal approximator, it is possible to design a feedforward neural network which is able to accurately approximate $\Phi_{ideal}(\mathbf{x})$. However, the beauty of feedforward networks is that we do not need to design a function. We only need to determine the number of hidden layers, number of neurons in each layer, and the type of activation functions. These are called *hyperparameters*. Among them, the first two hyperparameters is much more important than the third hyperparameter.

This implies that we do not need to design the equation of the feature transformation function by hand. Instead, we can just train a multilayer feedforward network to do both feature transformation and classification. Nonetheless, as we will see shortly, computing the gradient of loss function on a feedforward neural network using multivariate chain rule is not tractable. Fortunately, gradient of loss function can be computed using a method called *backpropagation*.

2.6.1 Backpropagation

Assume a feedforward network with a two-dimensional input layer and two hidden layers. The first hidden layer consists of four neurons and the second hidden layer consists of three neurons. Also, the output layer has three neurons. According to number of neurons in the output layer, the network is a 3-class classifier. Like multiclass logistic regression, the loss of the network is computed using a softmax function.

Also, the activation functions of the hidden layers could be any nonlinear function. But, the activation function of the output layer is the identity function $\mathcal{G}_i^3(x) = x$. The reason is that the output layer calculates the classification scores which is obtained by only computing $\mathbf{w}\mathcal{G}^2$. The classification score must be passed to the softmax function without any modifications in order to compute the multiclass logistic loss. For this reason, in practice, the activation function of the output layer is the identity function. This means that, we can ignore the activation function in the output layer. Similar to any compositional computation, a feedforward network can be illustrated using a computational graph. The computational graph analogous to this network is illustrated in Fig. 2.36.

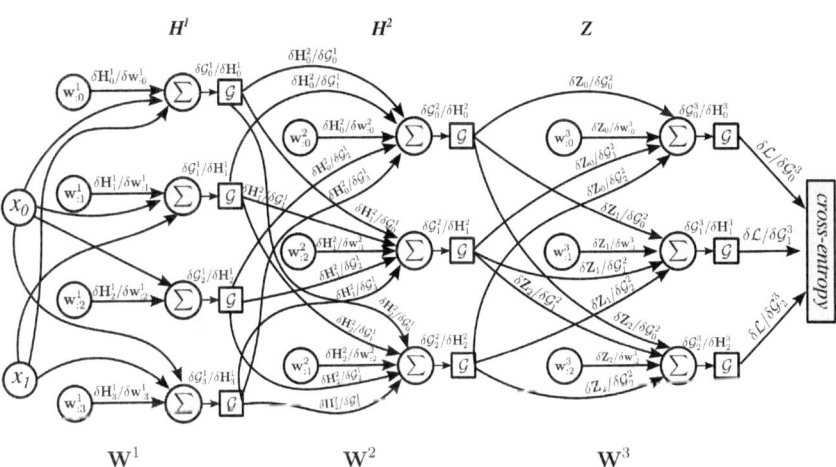

Fig. 2.36 Computational graph corresponding to a feedforward network for classification of three classes. The network accepts two-dimensional inputs and it has two hidden layers. The hidden layers consist of four and three neurons, respectively. Each neuron has two inputs including the weights and inputs from previous layer. The derivative of each node with respect to each input is shown on thee edges

Each computational node related to function of soma (the computation before applying the activation function) accepts two inputs including weights and output of the previous layer. Gradient of each node with respect to its inputs is indicated on the edges. Also note that $\mathbf{w}_b^a$ is a vector whose length is equal to the number of outputs from layer $a - 1$. Computing $\frac{\delta \mathcal{L}}{\delta \mathbf{w}_i^3}$ is straightforward and it is explained on Fig. 2.24. Assume, we want to compute $\frac{\delta \mathcal{L}}{\delta \mathbf{w}_0^1}$.

According to the multivariate chain rule, this is equal to adding all paths starting from $\mathbf{w}_0^1$ and ending at $\mathcal{L}$ in which the gradients along each path is multiplied. Based on this definition, $\frac{\delta \mathcal{L}}{\delta \mathbf{w}_0^1}$ will be equal to:

$$
\begin{aligned}
\frac{\delta \mathcal{L}}{\delta \mathbf{w}_0^1} = & \frac{\delta \mathbf{H}_0^1}{\delta \mathbf{w}_0^1} \frac{\mathcal{G}_0^1}{\delta \mathbf{H}_0^1} \frac{\delta \mathbf{H}_0^2}{\mathcal{G}_0^1} \frac{\mathcal{G}_0^2}{\delta \mathbf{H}_0^2} \frac{\delta \mathbf{Z}_0}{\mathcal{G}_0^2} \frac{\mathcal{G}_0^3}{\delta \mathbf{H}_0^3} \frac{\mathcal{L}}{\mathcal{G}_0^3} + \\
& \frac{\delta \mathbf{H}_0^1}{\delta \mathbf{w}_0^1} \frac{\mathcal{G}_0^1}{\delta \mathbf{H}_0^1} \frac{\delta \mathbf{H}_0^2}{\mathcal{G}_0^1} \frac{\mathcal{G}_0^2}{\delta \mathbf{H}_0^2} \frac{\delta \mathbf{Z}_1}{\mathcal{G}_0^2} \frac{\mathcal{G}_1^3}{\delta \mathbf{H}_1^3} \frac{\mathcal{L}}{\mathcal{G}_1^3} + \\
& \frac{\delta \mathbf{H}_0^1}{\delta \mathbf{w}_0^1} \frac{\mathcal{G}_0^1}{\delta \mathbf{H}_0^1} \frac{\delta \mathbf{H}_0^2}{\mathcal{G}_0^1} \frac{\mathcal{G}_0^2}{\delta \mathbf{H}_0^2} \frac{\delta \mathbf{Z}_2}{\mathcal{G}_0^2} \frac{\mathcal{G}_2^3}{\delta \mathbf{H}_2^3} \frac{\mathcal{L}}{\mathcal{G}_2^3} + \\
& \frac{\delta \mathbf{H}_0^1}{\delta \mathbf{w}_0^1} \frac{\mathcal{G}_0^1}{\delta \mathbf{H}_0^1} \frac{\delta \mathbf{H}_1^2}{\mathcal{G}_0^1} \frac{\mathcal{G}_1^2}{\delta \mathbf{H}_1^2} \frac{\delta \mathbf{Z}_0}{\mathcal{G}_1^2} \frac{\mathcal{G}_0^3}{\delta \mathbf{H}_0^3} \frac{\mathcal{L}}{\mathcal{G}_0^3} + \\
& \frac{\delta \mathbf{H}_0^1}{\delta \mathbf{w}_0^1} \frac{\mathcal{G}_0^1}{\delta \mathbf{H}_0^1} \frac{\delta \mathbf{H}_1^2}{\mathcal{G}_0^1} \frac{\mathcal{G}_1^2}{\delta \mathbf{H}_1^2} \frac{\delta \mathbf{Z}_1}{\mathcal{G}_1^2} \frac{\mathcal{G}_1^3}{\delta \mathbf{H}_1^3} \frac{\mathcal{L}}{\mathcal{G}_1^3} + \\
& \frac{\delta \mathbf{H}_0^1}{\delta \mathbf{w}_0^1} \frac{\mathcal{G}_0^1}{\delta \mathbf{H}_0^1} \frac{\delta \mathbf{H}_1^2}{\mathcal{G}_0^1} \frac{\mathcal{G}_1^2}{\delta \mathbf{H}_1^2} \frac{\delta \mathbf{Z}_2}{\mathcal{G}_1^2} \frac{\mathcal{G}_2^3}{\delta \mathbf{H}_2^3} \frac{\mathcal{L}}{\mathcal{G}_2^3} + \\
& \frac{\delta \mathbf{H}_0^1}{\delta \mathbf{w}_0^1} \frac{\mathcal{G}_0^1}{\delta \mathbf{H}_0^1} \frac{\delta \mathbf{H}_2^2}{\mathcal{G}_0^1} \frac{\mathcal{G}_2^2}{\delta \mathbf{H}_2^2} \frac{\delta \mathbf{Z}_0}{\mathcal{G}_2^2} \frac{\mathcal{G}_0^3}{\delta \mathbf{H}_0^3} \frac{\mathcal{L}}{\mathcal{G}_0^3} + \\
& \frac{\delta \mathbf{H}_0^1}{\delta \mathbf{w}_0^1} \frac{\mathcal{G}_0^1}{\delta \mathbf{H}_0^1} \frac{\delta \mathbf{H}_2^2}{\mathcal{G}_0^1} \frac{\mathcal{G}_2^2}{\delta \mathbf{H}_2^2} \frac{\delta \mathbf{Z}_1}{\mathcal{G}_2^2} \frac{\mathcal{G}_1^3}{\delta \mathbf{H}_1^3} \frac{\mathcal{L}}{\mathcal{G}_1^3} + \\
& \frac{\delta \mathbf{H}_0^1}{\delta \mathbf{w}_0^1} \frac{\mathcal{G}_0^1}{\delta \mathbf{H}_0^1} \frac{\delta \mathbf{H}_2^2}{\mathcal{G}_0^1} \frac{\mathcal{G}_2^2}{\delta \mathbf{H}_2^2} \frac{\delta \mathbf{Z}_2}{\mathcal{G}_2^2} \frac{\mathcal{G}_2^3}{\delta \mathbf{H}_2^3} \frac{\mathcal{L}}{\mathcal{G}_2^3}
\end{aligned}
\tag{2.82}
$$

Note that this is only for computing the gradient of the loss function with respect to the weights of one neuron in the first hidden layer. We need to repeat a similar procedure for computing the gradient of loss with respect to every node in this graph. However, although this computation is feasible for small feedforward networks, we usually need feedforward network with more layers and with thousands of neurons in each layer to classify objects in images. In this case, the simple multivariate chain rule will not be feasible to use since a single update of parameters will take a long time due do excessive number of multiplications.

It is possible to make the computation of gradients more efficient. To this end, we can factorize the above equation as follows:

$$\frac{\delta \mathcal{L}}{\delta w_0^1} = \frac{\delta H_0^1}{\delta w_0^1} \frac{\mathcal{G}_0^1}{\delta H_0^1} \left[\left(\frac{\delta H_0^2}{\mathcal{G}_0^1} \frac{\mathcal{G}_0^2}{\delta H_0^2} \left(\left(\frac{\delta Z_0}{\mathcal{G}_0^2} \left(\frac{\mathcal{G}_0^3}{\delta H_0^3} \frac{\mathcal{L}}{\mathcal{G}_0^3} \right) \right) + \left(\frac{\delta Z_1}{\mathcal{G}_0^2} \left(\frac{\mathcal{G}_1^3}{\delta H_1^3} \frac{\mathcal{L}}{\mathcal{G}_1^3} \right) \right) + \left(\frac{\delta Z_2}{\mathcal{G}_0^2} \left(\frac{\mathcal{G}_2^3}{\delta H_2^3} \frac{\mathcal{L}}{\mathcal{G}_2^3} \right) \right) \right) \right) +$$
$$\left(\frac{\delta H_1^2}{\mathcal{G}_0^1} \frac{\mathcal{G}_1^2}{\delta H_1^2} \left(\left(\frac{\delta Z_0}{\mathcal{G}_1^2} \left(\frac{\mathcal{G}_0^3}{\delta H_0^3} \frac{\mathcal{L}}{\mathcal{G}_0^3} \right) \right) + \left(\frac{\delta Z_1}{\mathcal{G}_1^2} \left(\frac{\mathcal{G}_1^3}{\delta H_1^3} \frac{\mathcal{L}}{\mathcal{G}_1^3} \right) \right) + \left(\frac{\delta Z_2}{\mathcal{G}_1^2} \left(\frac{\mathcal{G}_2^3}{\delta H_2^3} \frac{\mathcal{L}}{\mathcal{G}_2^3} \right) \right) \right) \right) +$$
$$\left(\frac{\delta H_2^2}{\mathcal{G}_0^1} \frac{\mathcal{G}_2^2}{\delta H_2^2} \left(\left(\frac{\delta Z_0}{\mathcal{G}_2^2} \left(\frac{\mathcal{G}_0^3}{\delta H_0^3} \frac{\mathcal{L}}{\mathcal{G}_0^3} \right) \right) + \left(\frac{\delta Z_1}{\mathcal{G}_2^2} \left(\frac{\mathcal{G}_1^3}{\delta H_1^3} \frac{\mathcal{L}}{\mathcal{G}_1^3} \right) \right) + \left(\frac{\delta Z_2}{\mathcal{G}_2^2} \left(\frac{\mathcal{G}_2^3}{\delta H_2^3} \frac{\mathcal{L}}{\mathcal{G}_2^3} \right) \right) \right) \right) \right]$$
$$(2.83)$$

Compared with (2.82), the above equation requires much less multiplications which makes it more efficient in practice. The computations starts with the most inner parenthesizes and moves to the most outer terms. The above factorization has a very nice property. If we carefully study the above factorization it looks like that the direction of the edges are hypothetically reversed and instead of moving from w_0^1 to $\mathcal{L}$ the gradient computations moves in the reverse direction. Figure 2.37 shows the nodes analogous to each inner computation in the above equation.

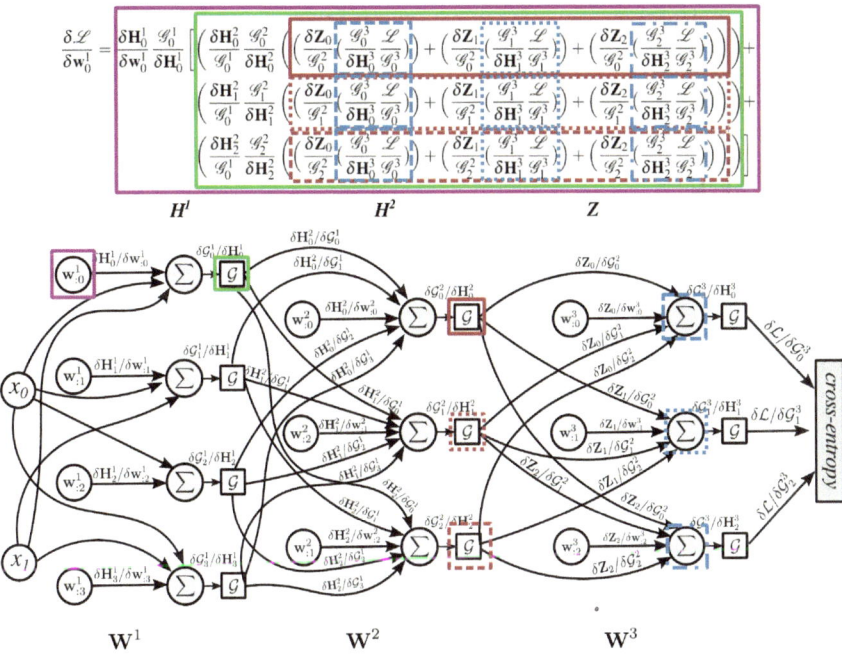

Fig. 2.37 Forward mode differentiation starts from the end node to the starting node. At each node, it sums the output edges of the node where the value of each edge is computed by multiplying the edge with the derivative of the child node. Each *rectangle* with different color and line style shows which part of the partial derivative is computed until that point

More precisely, assume the blue rectangles with dashed lines. These rectangles denote $\frac{\mathscr{G}_0^3}{\delta \mathbf{H}_0^3} \frac{\mathscr{L}}{\mathscr{G}_0^3}$ which corresponds to the node $\delta \mathbf{Z}_0$ on the graph. Furthermore, these rectangles in fact are equal to $\frac{\mathscr{L}}{\mathscr{Z}_0}$. Likewise, the blue rectangles with dotted lines and dashed-dotted lines denote $\frac{\mathscr{L}}{\mathscr{Z}_1} = \frac{\mathscr{G}_1^3}{\delta \mathbf{H}_1^3} \frac{\mathscr{L}}{\mathscr{G}_1^3}$ and $\frac{\mathscr{L}}{\mathscr{Z}_2} = \frac{\mathscr{G}_2^3}{\delta \mathbf{H}_2^3} \frac{\mathscr{L}}{\mathscr{G}_2^3}$ respectively.

The rectangles with solid red lines denote $\left(\frac{\delta \mathbf{Z}_0}{\mathscr{G}_0^2} \left(\frac{\mathscr{G}_0^3}{\delta \mathbf{H}_0^3} \frac{\mathscr{L}}{\mathscr{G}_0^3} \right) \right) + \left(\frac{\delta \mathbf{Z}_1}{\mathscr{G}_1^2} \left(\frac{\mathscr{G}_1^3}{\delta \mathbf{H}_1^3} \frac{\mathscr{L}}{\mathscr{G}_1^3} \right) \right) + \left(\frac{\delta \mathbf{Z}_2}{\mathscr{G}_2^2} \left(\frac{\mathscr{G}_2^3}{\delta \mathbf{H}_2^3} \frac{\mathscr{L}}{\mathscr{G}_2^3} \right) \right)$ which is analogous the derivative of the loss function with respect to $\delta \mathbf{H}_0^2$. In other words, before computing this rectangle, we have in fact computed $\frac{\mathscr{L}}{\mathbf{H}_0^2}$. Similarly, the dotted and dashed red rectangles illustrate $\frac{\mathscr{L}}{\mathbf{H}_1^2}$ and $\frac{\mathscr{L}}{\mathbf{H}_2^2}$ respectively. The same argument holds true with the green and purple rectangles.

Assume we want to compute $\frac{\delta \mathscr{L}}{\delta \mathbf{w}_1^1}$ afterwards. In that case, we do not need to compute none of the terms inside the red and blue rectangles since they have been computed once for $\frac{\delta \mathscr{L}}{\delta \mathbf{w}_0^1}$. This saves a great amount of computations especially when the network has many layers and neurons.

The *backpropagation* algorithm has been developed based on this factorization. It is a method for efficiently computing the gradient of leaf nodes with respect to each node on the graph using only one backward pass from the leaf nodes to input nodes. This algorithm can be applied on any computational graph. Formally, let $G = \langle \mathbf{V}, \mathbf{E} \rangle$ denotes a *directed acyclic graph* where $\mathbf{V} = \{v_1, \ldots, v_K\}$ is set of nodes in the computational graph and $\mathbf{E} = (v_i, v_j) | v_i, v_j \in V$ is the set of ordered pairs (v_i, v_j) showing a directed edge from v_i to v_j. Number of edges going into a node is called *indegree* and the number of edges coming out of a node is called *outdegree*.

Formally, if $in(v_a) = \{(v_i, v_j) | (v_i, v_j) \in E \wedge v_j = v_a\}$ returns set of input edges to v_a, indegree of v_a will be equal to $|in(v_a)|$ where $|.|$ returns the cardinality of a set. Likewise, $out(v_a) = \{(v_i, v_j) | (v_i, v_j) \in E \wedge v_i = v_a\}$ shows the set of output edges from v_a and $|out(v_a)|$ is equal to the outdegree of v_a. The computational node v_a is called an input if $in(v_a) = 0$ and $out(v_a) > 0$. Also, the computational node v_a is called a leaf if $out(v_a) = 0$ and $in(v_a) > 0$. Note that there must be only one leaf node in a computational graph which is typically the loss. This is due to the fact the we are always interested in computing the derivative of one node with respect to all other nodes in the graph. If there are more than one leaf node in the graph, the gradient of the leaf node of interest with respect to all other leaf nodes will be equal to zero.

Suppose that the leaf node of the graph is denoted by v_{leaf}. In addition, let $child(v_a) = \{v_j | (v_i, v_j) \in E \wedge v_i = v_a\}$ and $parent(v_a) = \{v_i | (v_i, v_j) \in E \wedge v_j = v_a\}$ returns the child nodes and parent nodes of v_a. Finally, depth of v_a is equal to number of edges on the *longest* path from input nodes to v_a. We denote the depth of v_a by $dep(v_a)$. It is noteworthy that for any node v_i in the graph that $dep(v_i) \geq dep(v_{leaf})$ the gradient of v_{leaf} with respect to v_i will be equal to zero. Based on the above discussion, the backpropagation algorithm is defined as follows:

Algorithm 1 The backpropagation algorithm

$G :< \mathbf{V}, \mathbf{E} >$ is a directed graph.
$\mathbf{V}$ is set of vertices
$\mathbf{E}$ is set of edges
v_{leaf} is the leaf node in $\mathbf{V}$
$d_{leaf} \leftarrow dep(v_{leaf})$
$v_{leaf}.d = 1$
for $d = d_{leaf} - 1$ to 0 **do**
    for $v_a \in \{v_i | v_i \in \mathbf{V} \wedge dep(v_i) == d\}$ **do**
        $v_a.d \leftarrow 0$
        for $v_c \in child(v_a)$ **do**
            $v_a.d \leftarrow v_a.d + \frac{\delta v_c}{\delta v_a} \times v_c.d$

The above algorithm can be applied on any computational graph. Generally, the it computes gradient of a loss function (leaf node) with respect to all other nodes in the graph using only one backward pass from the loss node to the input node. In the above algorithm, each node is a data structure which stores information related to the computational unit including their derivative. Specifically, the derivative of v_a is stored in $v_a.d$. We execute the above algorithm on the computational graph shown in Fig. 2.38.

Based on the above discussion, $loss$ is the leaf node. Also, the longest path from input nodes to the leaf node is equal to $d_{leaf} = dep(loss) = 4$. According to the algorithm, $v_{leaf}.d$ must be set to 1 before executing the loop. In the figure, $v_{leaf}.d$ is illustrated using d_8. Then, the loop start with $d = d_{leaf} - 1 = 3$. The first inner loop, iterates over all nodes in which their depth is equal to 3. This is equivalent to $\mathbf{Z}_0$ and $\mathbf{Z}_1$ on this graph. Therefore, v_a is set to $\mathbf{Z}_0$ in the first iteration. The most inner loop also iterates over children of v_a. This is analogous to $child(\mathbf{Z}_0) = \{loss\}$ which only has one child. Then, the derivative of v_a ($\mathbf{Z}_0$) is set to $d_6 = v_a.d = 0 + r \times 1$.

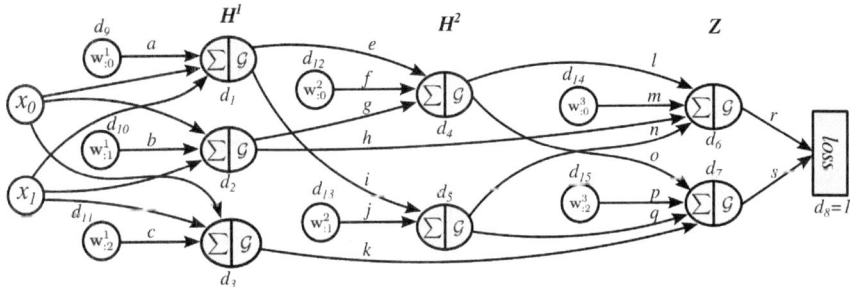

Fig. 2.38 A sample computational graph with a loss function. To cut the clutter, activations functions have been fused with the soma function of the neuron. Also, the derivatives on edges are illustrated using small letters. For example, g denotes $\frac{\delta \mathbf{H}_0^2}{\delta \mathbf{H}_1^1}$

Table 2.2 Trace of the backpropagation algorithm applied on Fig. 2.38

Depth	Node	Derivative
3	$\mathbf{Z}_0$	$d_6 = r \times 1$
3	$\mathbf{Z}_1$	$d_7 = s \times 1$
2	$\mathbf{H}_0^2$	$d_4 = l \times d_6 + o \times d_7$
2	$\mathbf{H}_1^2$	$d_5 = n \times d_6 + q \times d_7$
1	$\mathbf{H}_0^1$	$d_1 = e \times d_4 + i \times d_5$
1	$\mathbf{H}_1^1$	$d_2 = g \times d_4 + h \times d_6$
1	$\mathbf{H}_1^1$	$d_3 = k \times d_7$
0	$\mathbf{w}_{:0}^3$	$d_{14} = m \times d_6$
0	$\mathbf{w}_{:1}^3$	$d_{15} = p \times d_7$
0	$\mathbf{w}_{:0}^2$	$d_{12} = f \times d_4$
0	$\mathbf{w}_{:1}^2$	$d_{13} = j \times d_5$
0	$\mathbf{w}_{:0}^1$	$d_9 = a \times d_1$
0	$\mathbf{w}_{:1}^1$	$d_{10} = b \times d_2$
0	$\mathbf{w}_{:2}^1$	$d_{11} = c \times d_3$
0	x_0	$d_{16} = t \times d_1 + w \times d_2 + x \times d_3$
0	x_1	$d_{17} = y \times d_1 + z \times d_2 + zz \times d_3$

After that, the inner loop goes to $\mathbf{Z}_1$ and the most inner loop sets derivative of $\mathbf{Z}_1$ to $d_7 = v_a.d = 0 + s \times 1$.

At this point the inner loop finishes and the next iteration of the main loop start by setting d to 2. Then, the inner loop iterates over $\mathbf{H}_0^2$ and $\mathbf{H}_1^2$. In the first iteration of the inner loop, $\mathbf{H}_0^2$ is selected and its derivative d_4 is set to 0. Next, the most inner loop iterates over children of $\mathbf{H}_0^2$ which are $\mathbf{Z}_0$ and $\mathbf{Z}_1$. In the first iteration of the most inner loop d_4 is set to $d_4 = 0 + l \times d_6$ and in the second iteration it is set to $d_4 = l \times d_6 + o \times d_7$. At this point, the most inner loop is terminated and the algorithm proceeds with $\mathbf{H}_1^2$. After finishing the most inner loop, the d_5 will be equal to $d_5 = n \times d_6 + q \times d_7$. Likewise, derivative of other nodes are updated. Table 2.2 shows how derivative of nodes in different depths are calculated using the backpropagation algorithm.

We encourage the reader to carefully study the backpropagation algorithm since it is a very efficient way for computing gradients in complex computational graphs. Since we are able to compute the gradient of loss function with respect to every parameter in a feedforward neural network, we can train a feedforward network using the gradient descend method (Appendix A).

Given an input $\mathbf{x}$, the data is forwarded throughout the network until it reaches to the leaf node. Then, the backpropagation algorithm is executed and the gradient of loss with respect to every node given the input $\mathbf{x}$ is computed. Using this gradient, the parameters vectors are updated.

2.6.2 Activation Functions

There are different kinds of activation functions that can be used in neural networks. However, we are mainly interested in activation functions that are nonlinear and continuously differentiable. A nonlinear activation function makes it possible that a neural network learns any nonlinear functions provided that the network has enough neurons and layers. In fact, a feedforward network with linear activations in all neurons is just a linear function. Consequently, it is important that to have at least one neuron with a nonlinear activation function to make a neural network nonlinear.

Differentiability property is also important since we mainly train a neural network using gradient descend method. Although non-gradient-based optimization methods such as *genetic algorithms* and *particle swarm optimization* are used for optimizing simple functions, but gradient-based methods are the most commonly used methods for training neural networks. However, using non-gradient-based methods for training a neural network is an active research area.

Beside the above factors, it is also desirable that the activation function approximates the identity mapping near origin. To explain this, we should consider the activation of a neuron. Formally, the activation of a neuron is given by $\mathscr{G}(\mathbf{w}\mathbf{x}^T + b)$ where $\mathscr{G}$ is the activation function. Usually, the weight vector $\mathbf{w}$ and bias b are initialized with values close to zero by the gradient descend method. Consequently, $\mathbf{w}\mathbf{x}^T + b$ will be close to zero. If $\mathscr{G}$ approximates the identity function near zero, its gradient will be approximately equal to its input. In other words, $\delta\mathscr{G} \approx \mathbf{w}\mathbf{x}^T + b \iff \mathbf{w}\mathbf{x}^T + b \approx 0$. In terms of the gradient descend, it is a strong gradient which helps the training algorithm to converge faster.

2.6.2.1 Sigmoid
The sigmoid activation function and its derivative are given by the following equations. Figure 2.39 shows their plots.

$$\mathscr{G}_{sigmoid}(x) = \frac{1}{1 + e^{-x}} \tag{2.84}$$

and

$$\mathscr{G}'_{sigmoid}(x) = \mathscr{G}(x)(1 - \mathscr{G}(x)). \tag{2.85}$$

The sigmoid activation $\mathscr{G}_{sigmoid}(x) : \mathbb{R} \to [0, 1]$ is smooth and it is differentiable everywhere. In addition, it is a biologically inspired activation function. In the past, sigmoid was very popular activation function in feedforward neural networks. However, it has two problems. First, it does not approximate the identity function near zero. This is dues to the fact that $\mathscr{G}_{sigmoid}(0)$ is not close to zero and $\mathscr{G}'_{sigmoid}(x)$ is not close to 1. More importantly, sigmoid is a *squashing* function meaning that it saturates as $|x|$ increases. In other words, its gradient becomes very small if x is not close to origin.

This causes a serious problem in backpropagation which is known as *vanishing gradients* problem. The backpropagation algorithm multiplies the gradient of the

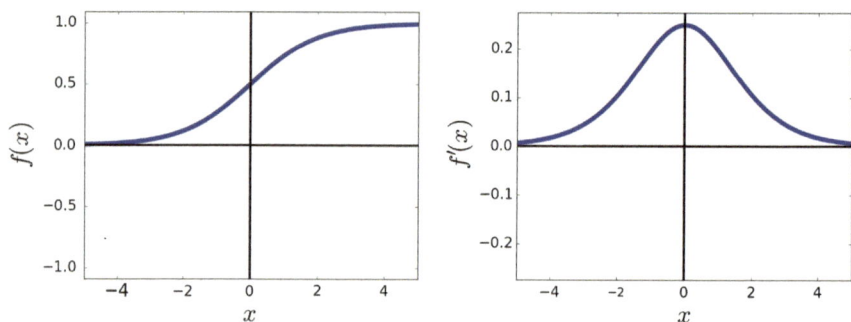

Fig. 2.39 Sigmoid activation function and its derivative

activation function with its children in order to compute the gradient of the loss function with respect to the current node. If x is far from origin, $\mathscr{G}_{sigmoid}$ will be very small. When it is multiplied by its children, the gradient of the loss with respect to that node will become smaller. If there are many layers with sigmoid activation, the gradient starts to become approximately zero (i.e., gradient vanishes) in the first layers. For this reason, the weight changes will be very small or even negligible. This cause the network to stuck in the current configuration of parameters and do not learn anymore. For these reasons, sigmoid activation function is not used in deep architectures since training the network become nearly impossible.

2.6.2.2 Hyperbolic Tangent

The *hyperbolic tangent* activation function is in fact a rescaled version of the sigmoid function. Its defined by the following equations. Figure 2.40 illustrates the plot of the function and its derivative.

$$\mathscr{G}_{tanh}(x) = \frac{e^x + e^{-x}}{e^x + e^{-x}} = \frac{2}{1 + e^{-2x}} - 1 \tag{2.86}$$

$$\mathscr{G}'_{tanh}(x) = 1 - \mathscr{G}_{tanh}(x)^2 \tag{2.87}$$

The hyperbolic tangent function $\mathscr{G}_{tanh}(x) : \mathbb{R} \to [-1, 1]$ is a smooth function which is differentiable everywhere. Its range is $[-1, 1]$ as opposed to range of the sigmoid function which is $[0, 1]$. More importantly, the hyperbolic tangent function approximates the identity function close to origin. This is easily observable from the plots where $\mathscr{G}_{tanh}(0) \approx 0$ and $\mathscr{G}'_{tanh}(0) \approx 1$. This is a desirable property which increases the convergence speed of the gradient descend algorithm. However, similar to the sigmoid activation function, it saturates as $|x|$ increases. Therefore, it may suffer from vanishing gradient problems in feedforward neural networks with many layers. Nonetheless, the hyperbolic activation function is preferred over the sigmoid function because it approximates the identity function near origin.

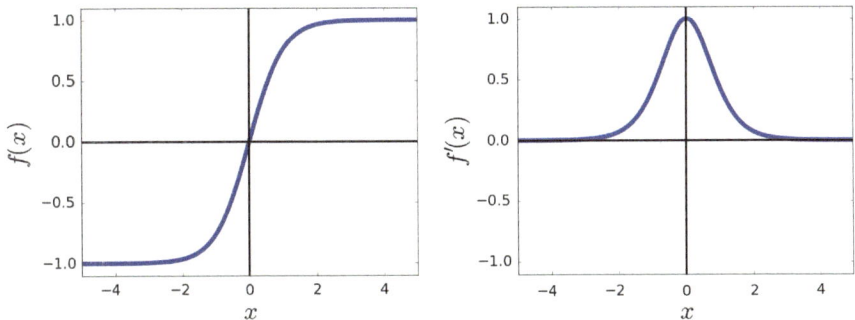

Fig. 2.40   Tangent hyperbolic activation function and its derivative

2.6.2.3   Softsign

The softsign activation function is closely related to the hyperbolic tangent function. However, it has more desirable properties. Formally, the softsign activation function and its derivative are defined as follows:

$$\mathscr{G}_{softsign}(x) = \frac{x}{1 + |x|} \tag{2.88}$$

$$\mathscr{G}'_{softsign}(x) = \frac{1}{(1 + |x|)^2} \tag{2.89}$$

Similar to the hyperbolic tangent function, the range of the softsign function is $[-1, 1]$. Also, the function is equal to zero at origin and its derivative at origin is equal to 1. Therefore, is approximates the identity function at origin. Comparing the function and its derivative with hyperbolic tangent, we observe that it also saturates as $|x|$ increases. However, the saturation ratio of the softsign function is less than the hyperbolic tangent function which is a desirable property. In addition, gradient of the softsign function near origin drops with a greater ratio compared with the hyperbolic tangent. In terms of computational complexity, softsign requires less computation than the hyperbolic tangent function. The softsign activation function can be used as an alternative to the hyperbolic tangent activation function (Fig. 2.41).

2.6.2.4   Rectified Linear Unit

Using the sigmoid, hyperbolic tangent and softsign activation functions is mainly limited to neural networks with a few layers. When a feedforward network has few hidden layers it is called a *shallow* neural network. In contrast, a network with many hidden layers is called a *deep* neural network. The main reason is that in deep neural networks, gradient of these three activation functions vanishes during the backpropagation which causes the network to stop learning in deep networks.

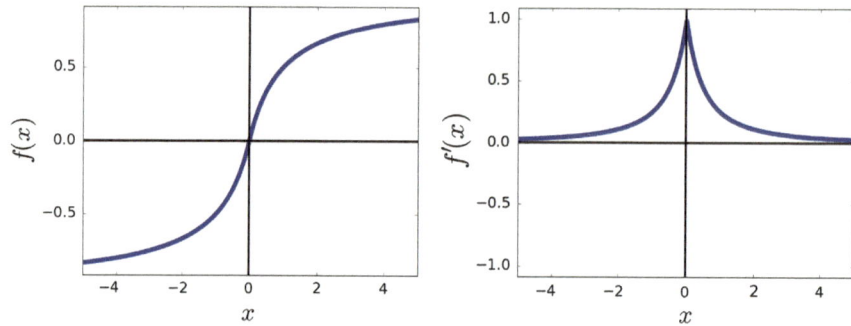

Fig. 2.41 The softsign activation function and its derivative

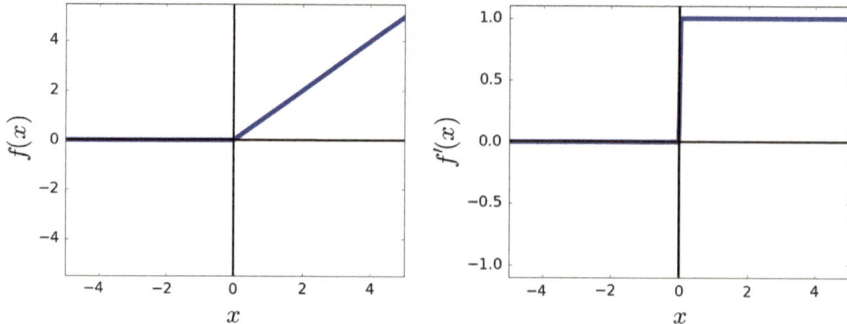

Fig. 2.42 The rectified linear unit activation function and its derivative

A *rectified linear unit* (ReLU) is an activation function which is computationally very efficient and it is defined as follows:

$$\mathscr{G}_{relu}(x) = \max(0, x) \tag{2.90}$$

$$\mathscr{G}'_{relu}(x) = \begin{cases} 0 & x < 0 \\ 1 & x \geq 0 \end{cases} \tag{2.91}$$

ReLU is a very simple nonlinear activation function which actually works very well in practice. Its derivative in $\mathbb{R}^+$ is always 1 and it does not saturate in $\mathbb{R}^+$. In other words, the range of this function is $[0, \infty)$. However, this function does not approximate the identity function near origin. But because it does not saturate in $\mathbb{R}^+$ it always produce a strong gradient in this region. Consequently, it does not suffer from the vanishing gradient problem. For this reason, it is a good choice for deep neural networks (Fig. 2.42).

One property of the ReLU activation is that it may produce dead neurons during the training. A dead neuron always return 0 for every sample in the dataset. This may happen because the weight of a dead neuron have been adjusted such that **wx** for the neuron is always negative. As the result, when it is passed to the ReLU activation function, it always return zero. The advantage of this property is that, the output of a layer may have entries which are always zero. This outputs can be removed from the network to make it computationally more efficient. The negative side of this property is that dead neuron may affect the overall accuracy of the network. So, it is always a good practice to check the network during training for dead neurons.

2.6.2.5   Leaky Rectified Linear Unit

The basic idea behind *Leaky ReLU* (Maas et al. 2013) is to solve the problem of dead neuron which is inherent in ReLU function. The leaky ReLU is defined as follows:

$$\mathscr{G}_{rrelu}(x) = \begin{cases} \alpha x & x < 0 \\ x & x \geq 0 \end{cases} \tag{2.92}$$

$$\mathscr{G}'_{rrelu}(x) = \begin{cases} \alpha & x < 0 \\ 1 & x \geq 0 \end{cases} \tag{2.93}$$

One interesting property of leaky ReLU is that its gradient does not vanish in negative region as opposed to ReLU function. Rather, it returns the constant value α. The hyperparameter α usually takes a value between [0, 1]. Common value is to set α to 0.01. But, on some datasets it works better with higher values as it is proposed in Xu et al. (2015). In practice, leaky ReLU and ReLU may produce similar results. This might be due to the fact that the positive region of these function is identical (Fig. 2.43).

2.6.2.6   Parameterized Rectified Linear Unit

Parameterized rectified linear unit is in fact (PReLU) the leaky ReLU (He et al. 2015). The difference is that α is treated as a parameter of the neural network so it

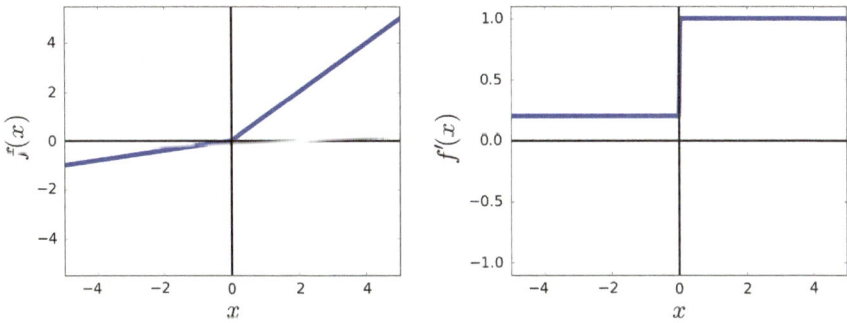

Fig. 2.43 The leaky rectified linear unit activation function and its derivative

can be learned from data. The only thing that needs to be done is to compute the gradient of the leaky ReLU function with respect to α which is given by:

$$\frac{\delta \mathscr{G}_{prelu}(x)}{\delta \alpha} = \begin{cases} x & x < 0 \\ \alpha & x \geq 0 \end{cases} \tag{2.94}$$

Then, the gradient of the loss function with respect to α is obtained using the back-propagation algorithm and it is updated similar to other parameters of the neural network.

2.6.2.7 Randomized Leaky Rectified Linear Unit

The main idea behind *randomized rectified linear unit* (RReLU) is to add randomness to the activations during training of a neural network. To achieve this goal, the RReLU activation draws the value of α from the uniform distribution $\mathscr{U}(a, b)$ where $a, b \in [0, 1)$ during training of the network. Drawing the value of α can be done once for all the network or it can be done for each layer separately. To increase the randomness, one may draw different α from the uniform distribution for each neuron in the network. Figure 2.44 illustrates how the function and its derivative vary using this method.

In the test time, the parameter α is set to the constant value $\bar{\alpha}$. This value is obtained by computing the mean value of α for each neuron that is assigned during training. Since the value of alpha is drawn from $\mathscr{U}(a, b)$, then value of $\bar{\alpha}$ can be easily obtained by computing the expected value of $\mathscr{U}(a, b)$ which is equal to $\bar{\alpha} = \frac{a+b}{2}$.

2.6.2.8 Exponential Linear Unit

Exponential linear units (ELU) (Clevert et al. 2015) can be seen as a smoothed version of the shifted ReLU activation function. By shifted ReLU we mean to change the original ReLU from $\max(0, x)$ to $\max(-1, x)$. Using this shift, the activation passes a negative number near origin. The exponential linear unit approximates the shifted

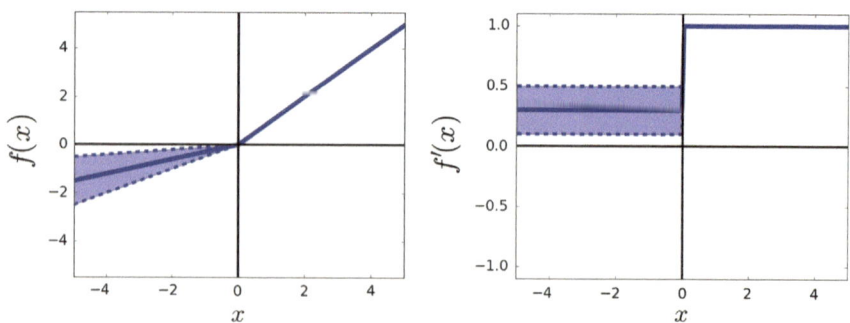

Fig. 2.44 The softplus activation function and its derivative

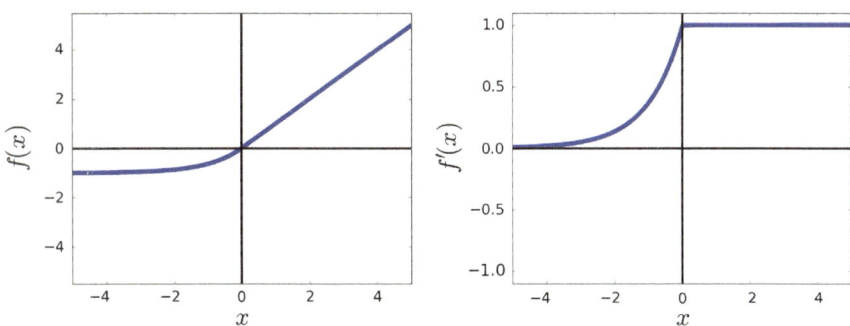

Fig. 2.45 The exponential linear unit activation function and its derivative

ReLU using a smooth function which is given by:

$$\mathcal{G}_{elu}(x) = \begin{cases} \alpha(e^x - 1) & x < 0 \\ x & x \geq 0 \end{cases} \tag{2.95}$$

$$\mathcal{G}'_{elu}(x) = \begin{cases} \mathcal{G}(x) + \alpha & x < 0 \\ 1 & x \geq 0 \end{cases} \tag{2.96}$$

The ELU activation usually speeds up the learning. Also, as it is illustrated in the plot, its derivative does not drop immediately in the negative region. Instead, the gradient of the negative region saturates nonlinearly (Fig. 2.45).

2.6.2.9 Softplus
The last activation function that we explain in this book is called *Softplus*. Broadly speaking, we can think of the softplus activation function as a smooth version of the ReLU function. In contrast to the ReLU which is not differentiable at origin, the softplus function is differentiable everywhere. In addition, similar to the ReLU activation, its range is $[0, \infty)$. The function and its derivative are defined as follows:

$$\mathcal{G}_{softplus} = \ln(1 + e^x) \tag{2.97}$$

$$\mathcal{G}'_{softplus} = \frac{1}{1 + e^{-x}} \tag{2.98}$$

The derivative of the softplus function is the sigmoid function which means the range of derivative is $[0, 1]$. The difference with ReLU is the fact that the derivative of softplus is also a smooth function which saturates as $|x|$ increases (Fig. 2.46).

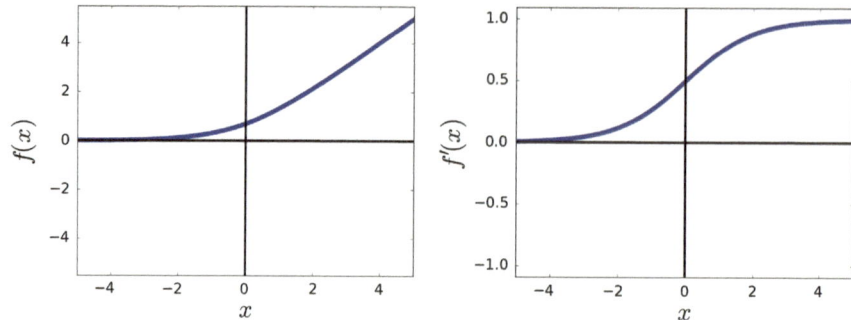

Fig. 2.46 The softplus activation function and its derivative

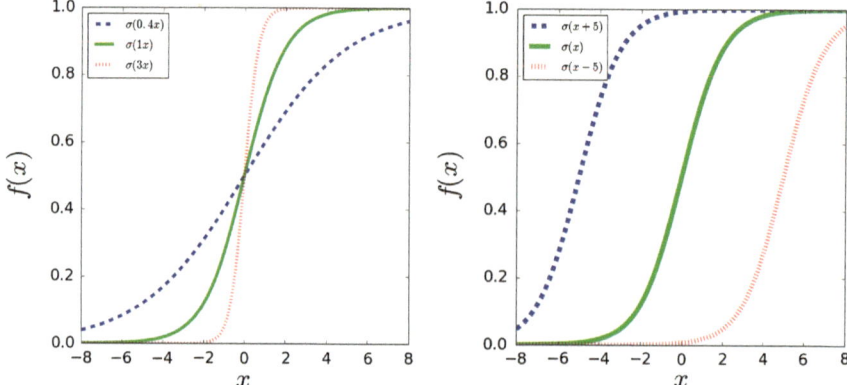

Fig. 2.47 The weights affect the magnitude of the function for a fixed value of bias and **x** (*left*). The bias term shifts the function to left or right for a fixed value of **w** and **x** (*right*)

2.6.3 Role of Bias

Basically, the input to an activation function is $\mathbf{w}\mathbf{x}^T + b$. The first term in this equation, computes the dot product between **w** and **x**. Assume that x is a one-dimensional vector (scaler). To see the effect of w, we can set $b = 0$ and keep the value of x fixed. Then, the effect of **w** can be illustrated by plotting the activation function for different values of **w**. This is shown in left plot in Fig. 2.47.

We observe that changing the weights affects the magnitude of activation function. For example, assume a neural network without a hidden layer where the output layer has only one neuron with sigmoid activation function. The output of the neural network for inputs $\mathbf{x}_1 = 6$ and $\mathbf{x}_2 = -6$ are equal to $\sigma(6w + b) = 0.997$ and $\sigma(-6w + b) = 0.002$ when $w = 1$ and $b = 0$. Suppose we want to find **w** and keep $b = 0$ such that $\sigma(6w + b) = 0.999$ and $\sigma(-6w + b) = 0.269$. There is no **w** which perfectly satisfies these two conditions. But, it is possible to find **w** that approximates the above values as accurate as possible. To this end, we only need to minimize the squared error loss of the neuron. If we do this, the approximation error will high indicating that it is not possible to approximate these values accurately.

However, it is possible to find b where $\sigma(6w + b) = 0.999$ and $\sigma(-6w + b) = 0.269$ when $\mathbf{w} = 1$. To see the effect of b, we can keep $\mathbf{w}$ and $\mathbf{x}$ fixed and change the value of b. The right plot in Fig. 2.47 shows the result. It is clear that the bias term shifts the activation function to left or right. It gives a neuron more freedom to be fitted on data.

According to the above discussion, using bias term in a neuron seems necessary. However, bias term might be omitted in very deep neural networks. Assume the final goal of a neural network is to estimated $(x = 6, f(x) = 0.999)$ and $(x = -6, f(x) = 0.269)$. If we are forced to only use a single layer neural network with only one neuron in the layer, the estimation error will be high without a bias term. But, if we are allowed to use more layers and neurons, then it is possible to design a neural network that accurately approximates these pairs of data.

In deep neural networks, even if the bias term is omitted, the network might be able to shift the input across different layers if it reduces the loss. Though, it is a common practice to keep the bias term and train it using data. Omitting the bias term may only increase the computational efficiency of a neural network. If the computational resources are not limited, it is not necessary to remove this term from neurons.

2.6.4 Initialization

The gradient descend algorithm starts by setting an initial value for parameters. A feedforward neural network has mainly two kind of parameters including weights and biases. All biases are usually initialized to zero. There are different algorithms for initializing the weights. To common approach is to initialize them using a uniform or a normal distribution. We will explain initialization methods in the next chapter.

The most important thing to keep in mind is that, weights of the neurons must be different. If they all have the same value. Neurons in the same layer will have identical gradients leading to the same update rule. For this reason, weights must be initialized with different values. Also, they are commonly initialized very close to zero.

2.6.5 How to Apply on Images

Assume the dataset $\mathscr{X} = \{(\mathbf{x}_1, y_1), \ldots, (\mathbf{x}_n, y_n)\}$ where the input vector $\mathbf{x}_i \in \mathbb{R}^{1000}$ is a 1000-dimensional vector and $y_i = [0, \ldots, c]$ is an integer number indicating the class of the vector. A rule of thumb in designing a neural network for classification of these vectors is to have more neurons in the first hidden layer and start to decrease the number of neurons in the subsequent layers. For instance, we can design a neural network with three hidden layers where the first hidden layer has 5000 neurons, the second hidden layer has 2000 neurons and the third hidden layer hast 500 neurons. Finally, the output layer also will contain c neurons.

One important step in designing a neural network is to count the total number of parameters in the network. For example, there are $5000 \times 1000 = 5,000,000$ weights between the input layer and the first hidden layer. Also, the first hidden layer has 5000 biases. Similarly, the number of parameters between the first hidden layer and second hidden layer is equal to $5000 \times 2000 = 10,000,000$ plus 2000 biases. The number of parameters between the second hidden layer and the third hidden layer is also equal to $2000 \times 500 = 1,000,000$ plus 500 biases. Finally, the number of weights and biases between the third hidden layer and the output layer is equal to $500 \times c + c$. Overall, this neural network is formulated using $16,007,200 + 500c + c$ parameters. Even for this shallow neural network, the number of parameters is very high. Training this neural network requires a dataset with many training samples. Collecting this dataset might not be practical.

Now, suppose our aim is to classify traffic signs. The input of the classifier might be $50 \times 50 \times 3$ images. Our aim is to classify 100 classes of traffic signs. We mentioned before that training a classifier directly on pixel intensities does not produce accurate results. Better results were obtained by extracting features using the histogram of oriented gradients. We also mentioned that neural networks learn the feature transformation function automatically from data.

Consequently, we can design a neural network where the input of the network is raw images and its output is the classification scores of the image per each class of traffic sign. The neural network learns to extract features from the image so that they become linearly separable in the last hidden layer. A $50 \times 50 \times 3$ image can be stored in a three-dimensional matrix. If we flatten this matrix, the results will be a 7500-dimensional vector.

Suppose a neural network containing three hidden layers with 10000-8000-3000 neurons in these layers. This network is parameterized using $179,312,100$ parameters. A dramatically smaller neural network with three hidden layers such as 500-300-250 will also have $4,001,150$ parameters. Although the number of parameters in the latter neural network is still hight, it may not produce accurate results. In addition, the number of parameters in the former network is very high which makes it impossible to train this network with the current algorithms, hardware and datasets.

Besides, classification of objects is a complex problem. The reason is that some of traffic signs differ only slightly. Also, their illumination changes during day. There are also other factors that we will discuss in the later chapters. For these reasons, accurately learning a feature transformation function that traffic signs linearly separable in the feature space requires a deeper architecture. As the depth of neural network increases, the number of parameters may also increase. The reason that a deeper model is preferable over a shallower model is described on Fig. 2.48.

The wide black line on this figure shows the function that must be approximated using a neural network. The red line illustrates the output of a neural network including four hidden layers with 10-10-9-6 architecture using the hyperbolic tangent activation functions. In addition, the white line shows the output of a neural network consisting of five layers with 8-6-4-3-2 architecture using the hyperbolic tangent activation function. Comparing the number of parameters in these two networks, the shallower network has 296 parameters and the deeper network has 124 parameters. In

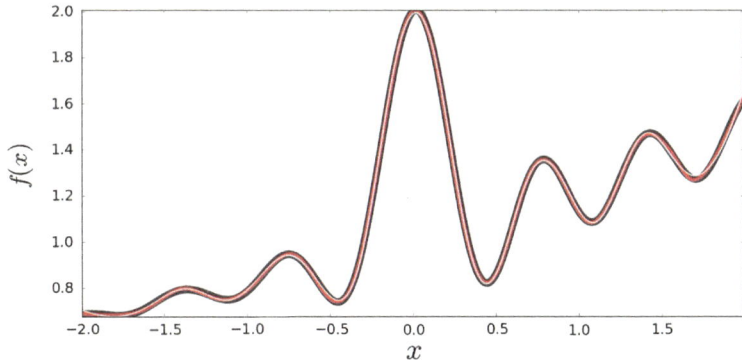

Fig. 2.48 A deeper network requires less neurons to approximate a function

general, deeper models require less parameters for modeling a complex function. It is obvious from figure that the deeper model is approximated the function accurately despite the fact that it has much less parameters.

Feedforward neural networks that we have explained in this section are called *fully connected* feedforward neural networks. The reason is that every neuron in one layer is connected to *all* neurons in the previous layer. As we explained above, modeling complex functions such as extracting features from an image may require deep neural networks. Training deep fully connected networks on dataset of images is not tractable due to very high number of parameters. In the next chapter, we will explain a way to dramatically reduce the number of parameters in a neural network and train them on images.

2.7 Summary

In this chapter, we first explained what are classification problems and what is a decision boundary. Then, we showed how to model a decision boundary using linear models. In order to better understand the intuition behind a linear model, they were also studied from geometrical perspective. A linear model needs to be trained on a training dataset. To this end, there must be a way to assess how good is a linear model in classification of training samples. For this purpose, we thoroughly explained different loss functions including 0/1 loss, squared loss, hinge loss, and logistic loss, Then, methods for extending binary models to multiclass models including one-versus-one and one-versus-rest were reviewed. It is possible to generalize a binary linear model directly into a multiclass model. This requires loss functions that can be applied on multiclass dataset. We showed how to extend hinge loss and logistic loss into multiclass datasets.

The big issue with linear models is that they perform poorly on datasets in which classes are not linearly separable. To overcome this problem, we introduced the idea of feature transformation function and applied it on a toy example. Designing a feature transformation function by hand could be a tedious task especially when they have to be applied on high-dimensional datasets. A better solution is to learn a feature transformation function directly from training data and training a linear classifier on top of it.

We developed the idea of feature transformation from simple functions to compositional functions and explained how neural networks can be used for simultaneously learning a feature transformation function together with a linear classifier. Training a complex model such as neural network requires computing gradient of loss function with respect to every parameter in the model. Computing gradients using conventional chain rule might not be tractable. We explained how to factorize a multivariate chain rule and reduce the number of arithmetic operations. Using this formulation, we explained the backpropagation algorithm for computing gradients on any computational graph.

Next, we explained different activation functions that can be used in designing neural networks. We mentioned why ReLU activations are preferable over traditional activations such as hyperbolic tangent. Role of bias in neural networks is also discussed in detail. Finally, we finished the chapter by mentioning how an image can be used as the input of a neural network.

2.8 Exercises

2.1 Find an equation to compute the distance of point $\mathbf{p}$ from a line.

2.2 Given the convex set $X \subset \mathbb{R}^d$, we know that function $f(x) : X \to \mathbb{R}$ is convex if:

$$\forall_{x_1, x_2 \in X, \alpha \in [0,1]} f(\alpha \mathbf{x}_1 + (1 - \alpha)\mathbf{x}_2) \leq \alpha f(\mathbf{x}_1) + (1 - \alpha) f(\mathbf{x}_2). \qquad (2.99)$$

Using the above definition, show why 0/1 loss function is nonconvex?

2.3 Prove that square loss is a convex function.

2.4 Why setting a in the hinge loss to different values does not affect the classification accuracy of the learn model?

2.5 Compute the partial derivative of the squared hinge loss and modified Huber loss functions.

2.6 Apply $\log(A \times B) = \log(A) \log(B)$ on (2.39) to obtain (2.39).

2.7 Show that:

$$\frac{\delta\sigma(a)}{a} = \sigma(a)(1 - \sigma(a)). \tag{2.100}$$

2.8 Find the partial derivative of (2.41) with respect to w_i using the chain rule of derivatives.

2.9 Show how we obtained (2.46).

2.10 Compute the partial derivatives of (2.46) and use them in the gradient descend method for minimizing the loss represented by this equation.

2.11 Compute the partial derivatives of (2.56) and obtain (2.57).

2.12 Draw an arbitrary computation graph with three leaf nodes and call them A, B and C. Show that $\delta C/\delta A = 0$ and $\delta C/\delta B = 0$

2.13 Show that a feedforward neural network with linear activation functions in all layers is in fact just a linear function.

2.14 Show that it is impossible to find a **w** such that:

$$\sigma(6w) = \frac{1}{1 + e^{-6w}} = 0.999$$
$$\sigma(-6w) = \frac{1}{1 + e^{6w}} = 0.269 \tag{2.101}$$

References

Clevert DA, Unterthiner T, Hochreiter S (2015) Fast and accurate deep network learning by exponential linear units (ELUs). 1997, pp 1–13. arXiv:1511.07289

He K, Zhang X, Ren S, Sun J (2015) Delving deep into rectifiers: surpassing human-level performance on ImageNet classification. arXiv:1502.01852

Maas AL, Hannun AY, Ng AY (2013) Rectifier nonlinearities improve neural network acoustic models. In: ICML workshop on deep learning for audio, speech and language processing, vol 28. http://www.stanford.edu/~awni/papers/relu_hybrid_icml2013_final.pdf

Stallkamp J, Schlipsing M, Salmen J, Igel C (2012) Man vs. computer: benchmarking machine learning algorithms for traffic sign recognition. Neural Netw 32:323–332. doi:10.1016/j.neunet.2012.02.016

Xu B, Wang N, Chen T (2015) Empirical evaluation of rectified activations in convolutional network. arXiv:1505.00853v2

Convolutional Neural Networks

3

In the previous chapter, we explained how to train a linear classifier using loss functions. The main problem of linear classifiers is that the classification accuracy drops if the classes are not separable using a hyperplane. To overcome this problem, the data can be transformed to a new space where classes in this new space are linearly separable. Clearly, the transformation must be nonlinear.

There are two common approaches to designing a transformation function. In the first approach, an expert designs a function manually. This method could be tedious especially when dimensionality of the input vector is high. Also, it may not produce accurate results and it may require many trials and errors for creating an accurate feature transformation function. In the second approach, the feature transformation function is learned from data.

Fully connected feedforward neural networks are commonly used for simultaneously learning features and classifying data. The main problem with using a fully connected feedforward neural network on images is that the number of neurons could be very high even for shallow architectures which makes them impractical for applying on images. The basic idea behind *convolutional neural networks* (ConvNets) is to devise a solution for reducing the number of parameters allowing a network to be deeper with much less parameters.

In this chapter, we will explain principals of ConvNets and we will describe a few examples where ConvNets with different architectures have been used for classification of objects.

3.1 Deriving Convolution from a Fully Connected Layer

Recall from Sect. 2.6 that in a fully connected layer, all neurons are connected to every neuron in the previous layer. In the case of grayscale images, input of first hidden layer is a $W \times H$ matrix which is denoted by $\mathbf{x} \in [0, 1]^{W \times H}$. Here, we have

© Springer International Publishing AG 2017
H. Habibi Aghdam and E. Jahani Heravi, *Guide to Convolutional Neural Networks*, DOI 10.1007/978-3-319-57550-6_3

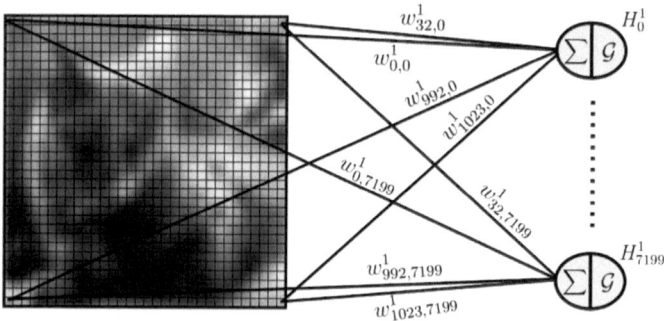

Fig. 3.1 Every neuron in a fully connected layers is connected to every pixel in a grayscale image

indicated intensity of pixels by a real number between 0 and 1. But, the following argument holds true for intensities within any range. Assuming that there are K neurons in the first hidden layer, each neuron H_i^1, $i = 0 \ldots K$ in the hidden layer is connected to all pixels in the grayscale image leading to $W \times H$ connections only for H_i^1.

Assume the 16×16 grayscale image illustrated in Fig. 3.1 which is connected to a hidden layer consisting of 7200 neurons. As it is indicated in this figure, the image can be thought as a $16 \times 16 = 1024$ dimensional vector. The first neuron in the hidden layer is connected to 1024 elements in the input. Similarly, other neurons are also connected on every element of the input image. Consequently, this fully connected layer is formulated using $1024 \times 7200 = 7,372,800$ distinct parameters.

One way to reduce the number of parameters is to reduce the number of neurons in the hidden layer. However, this may adversely affect the classification performance. For this reason, we usually need to keep the number of neurons in the first hidden layer high. In order to reduce the number of parameters, we first hypothetically rearrange the 7200 neurons into 50 blocks of 12×12 neurons. This is illustrated in Fig. 3.2. Here, f_i, $i = 0 \ldots 49$ shows the number of the block. Each block is formed using 12×12 neurons.

The number of required parameters is still $1024 \times 50 \times 12 \times = 7,372,800$. We can dramatically reduce the number of parameters by considering the geometry of pixels in an image. Concretely, the pixel (m, n) in an image is highly correlated with its close neighbors than its far neighbors. Assume that neuron $(0, 0)$ in each block is intended to extract information from a region around pixel $(2, 2)$ in the image. Likewise, neuron $(11, 11)$ in all blocks is intended to extract information from pixel $(14, 14)$ in the image.

Since the correlation between far pixels is very low, neuron $(0, 0)$ needs only information from pixel $(2, 2)$ and its neighbors in order to extract information from this region. For example, in Fig. 3.3, we have connected each neuron in each block to a 5×5 region in the image. Neurons in a block cover all the input image and extract information for each 5×5 patch in the input image.

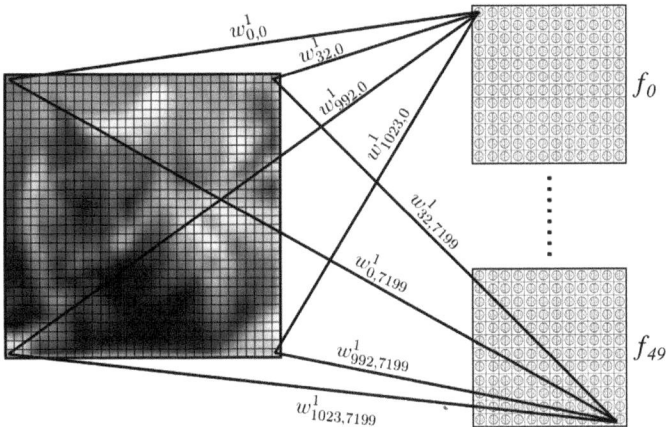

Fig. 3.2 We can hypothetically arrange the neurons in blocks. Here, the neurons in the hidden layer have been arranged into 50 blocks of size 12×12

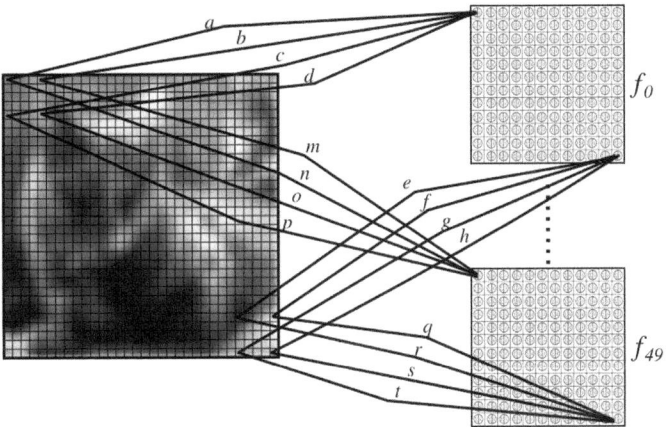

Fig. 3.3 Neurons in each block can be connected locally to the input image. In this figure, each neuron is connected to a 5×5 region in the image

By this way, the number of parameters is reduced to $(5 \times 5) \times 50 \times 12 \times 12 = 180,000$ which 97.5% reduction in number of parameters compared with the fully connected approach. But this number can be reduced further. The weights in Fig. 3.3 have been illustrated using small letters. We observe that each neuron in a block has a different weight compared with other neurons in the same block. To further reduce the number of parameters, we can assume that all neurons in one block share the same weights. This is shown in Fig. 3.4. This means that, each block is formulated using only 25 weights. Consequently, there are only $5 \times 5 \times 50 = 1250$ weights between the image and the hidden layer leading to 99.98% reduction in the number of parameters compared with the fully connected layer.

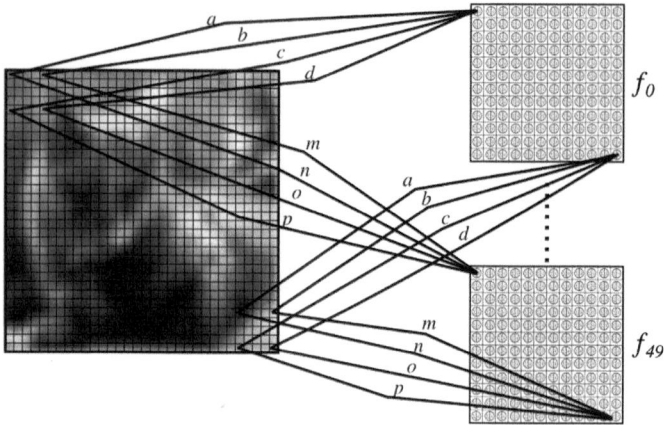

Fig. 3.4 Neurons in one block can share the same set of weights leading to reduction in the number of parameters

This great amount of reduction was achieved using a technique called *weight sharing* between neurons. Denoting the neuron (p, q) in block l in the above figure by $f_{p,q}^l$, the output of this neuron is given by

$$f_{p,q}^l = (\mathscr{G})\left(\sum_{i=0}^{4}\sum_{j=0}^{4} im(p+i, q+j)w_{i,j}^l\right) \tag{3.1}$$

where $w_{a,b}^l$ shows the weight (a, b) in block l and $p, q = 0, \ldots, 11$. In the above example, a and b varies between 0 and 4 since each neuron is connected to a 5×5 region. The region to which a neuron is connected is called *receptive field* of the neuron. In this example, the receptive field of the neuron is a 5×5 region. The output of each block will have the same size as its block. Hence, in this example, the output of each block will be a 12×12 matrix. With this formulation and denoting the output matrix of lth with $\mathbf{f}^l$, this matrix can be obtained by computing

$$\mathbf{f}^l(p, q) = (\mathscr{G})\left(\sum_{i=0}^{4}\sum_{j=0}^{4} im(p+i, q+j)w_{i,j}^l\right)\forall p, q \in 0, 11. \tag{3.2}$$

The above equation is exactly analogous to *convolving* the 5×5 filter w with the input image.[1] As the result, output of the lth block is obtained by convolving the filter w on the input image. The convolution operator is usually denoted by $*$ in literature. Based on the above discussion, the layer in Fig. 3.4 can be represented using a filter and convolution operator that is illustrated in Fig. 3.5.

---

[1]Readers that are not familiar with convolution can refer to textbooks of image processing for detailed information.

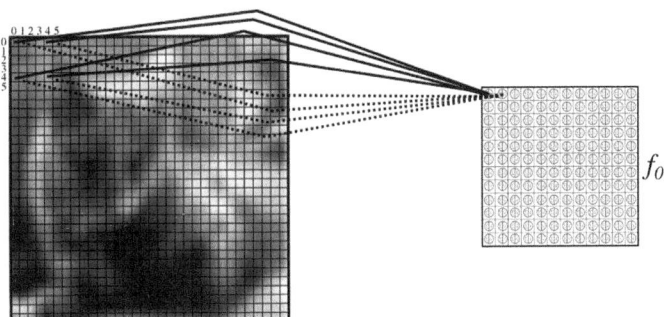

Fig. 3.5 The above convolution layer is composed of 49 filters of size $5\times$. The output of the layer is obtained by convolving each filter on the image

The output of a convolution layer is obtained by convolving each filter on the input image. The output of the convolution layer will be series of images where the number of images is analogous to the number of filters. Then, the activation function is applied on each image separately in element-wise fashion. In general if the size of the image is $W \times H$ and the convolution layer is composed of L filters of size $M \times N$, the output of the convolution layer will be L images of size $W - M + 1 \times H - N + 1$ where each image is obtained by convolving the corresponding filter with the input image.

A deep convolutional neural network normally contains more than one convolution layer. From image processing perspective, a convolution filter is a two-dimensional array (i.e. matrix) which is applied on a grayscale image. In the case of multichannel images such as RGB images, the convolution filter might be still a two-dimensional array which is separately applied on each channel.

However, the main idea behind convolution filters in ConvNets is that the result of convolving a filter with a multichannel input is always *single channel*. In other words, if convolution filter f is applied on the RGB image X with three channels, $X * f$ must be a single-channel image. A multichannel image can be seen as a three-dimensional array where the first two dimensions show the spatial coordinate of pixels and the third dimension shows the channel. For example, a 800×600 RGB image is stored in a $600 \times 800 \times 3$ array. In the same way, a 640×480 multispectral image which is taken in 7 different spectrum is stored in a $480 \times 600 \times 7$ array.

Assuming that $X \in \mathbb{R}^{H \times W \times C}$ is a multichannel image with C channels, our aim is to design the filter f such that $X * f \in \mathbb{R}^{H' \times W' \times 1}$ where H' and W' depends on the height and width of the filter respectively. To this end, f must be a three-dimensional filter where the third dimensional is alway equal to the number of input channels. Formally, if $f \in \mathbb{R}^{h \times w \times C}$ then $X * f \in \mathbb{R}^{H-h+1 \times W-w+1 \times 1}$. Based on this definition, we can easily design multiple convolution layers. As example is illustrated in Fig. 3.6.

In this example, the input of the network is a single-channel image (i.e., a grayscale image). The first convolution layers contains L_1 filters of size $M_1 \times N_1 \times 1$. The third dimension of filters is 1 because the input of this layer is a single-channel image.

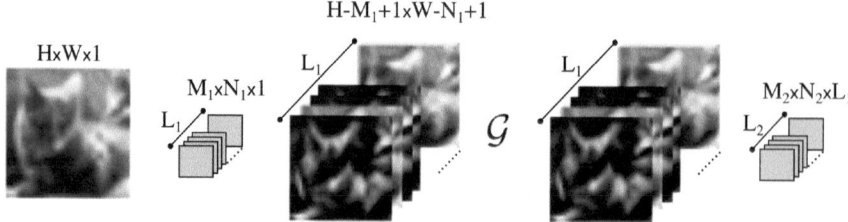

Fig. 3.6 Normally, convolution filters in a ConvNet are three-dimensional array where the first two dimensions are arbitrary numbers and the third dimension is always equal to the number out channels in the previous layer

Applying these filters on the input will produce L_1 images of size $H - M_1 + 1 \times W - N_1 + 1$. From another perspective, the output of the first convolution layer can be seen as a multichannel image with L_1 channels. Then, the activation function $\mathcal{G}$ is applied on every element of this multichannel image, separately.

Based on the above discussion, the filter of the second convolution layer must be $M_2 \times N_2 \times L_1$ so that convolving a filter with the L_1-channel input will have always single channel. In addition, M_2 and N_2 could be any arbitrary numbers. Similarly, output of the second convolution layer will be a L_2 channel image. In terms of ConvNets, output of convolution layers is called *feature maps* where a feature map is the result of convolving a filter with the input of layer. In sum, it is important to keep this in mind that the convolution filter in ConvNets are mainly three-dimensional filters where the third dimension is always equal to the number of channels in the input.[2]

3.1.1 Role of Convolution

Bank of Gabor filters is one of powerful methods for extracting features. The core of this method is to create a bank of N Gabor filters. Then, each filter is convolved with the input image. This way, N different images will be produced. Then, pixels of each image are pooled to extract information from each image. We shall discuss about pooling later in this chapter. There are mainly two steps in this method including convolution and pooling.

In a similar approach, ConvNets extract features based on this fundamental idea. Specifically, ConvNets apply series of convolution and pooling in order to extract features. it is noteworthy to study the role of convolution in ConvNets. For this purpose, we generated two consecutive convolution layers where the input of the first layer is an RGB image. Also, the first layer has six filters of size $7 \times 7 \times 3$ and

---

[2]In the case of video, convolution filters could be four-dimensional. But, for the scope of this book we only mention usual filters which are applied on images.

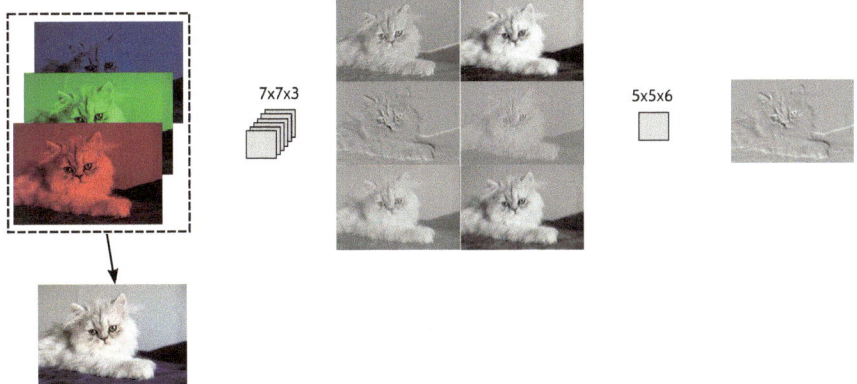

Fig. 3.7 From ConvNet point of view, an RGB image is a three-channel input. The image is taken from www.flickr.com

the second layer has one filter of size $5 \times 5 \times 6$. This simple ConvNet is illustrated in Fig. 3.7.

The input has three channels. Therefore, the third dimension of the first convolution layers has to be equal to 3. Applying each filter in the first layer on the image will produce a single-channel image. Hence, output of the first convolution layer will be a six-channel image. Then, because there is not activation function after the first convolution layer, it is directly fed to the second convolution layer. Based on our previous discussion, the third dimension of the filter in the second convolution layer has to be 6. Since there is only one filter in the second convolution layer, the output of this layer will be a single-channel image.

For the purpose of this example, we generated random filters for both first and second convolution layers. Looking at the results of the first convolution layer, we observe that two filters have acted as low-pass filters (i.e., smoothing filters) and rest of the filters have acted as high-pass filters (i.e., edge detection filter). Then, the second convolution layer has generated a single-channel image where the value at location (m, n) is obtained by linearly combining all six channels of the first convolution layer in the 5×5 neighborhood at location (m, n). Comparing the result of the second layer with results of the first layer, we see that the second layer has intensified the strong edges around eyes of cat and her nose. In addition, although edges generated by fur of cat are stronger in the output of the first layer, they have diminished by the second layer.

Note that filters in the above example are just random filters. In practice, a ConvNet learns to adjust the weights of filters such that different classes become linearly separable in the last layer of the network. This is done during training procedure.

3.1.2 Backpropagation of Convolution Layers

In Sect. 2.6.1, we explained how to compute the gradient of a leaf node in a computational graph with respect to every node in the graph using a method called backpropagation. Training a convolution layer also requires the gradient of convolution layer with respect to its parameters and to its inputs. To simplify the problem, we study backpropagation on a one-dimensional convolutional layer. Figure 3.8 shows two layers from a ConvNet where the neurons of the layer in right share the same weight and they are also locally connected. This integrally shows that the output of the second layer is obtained by convolving the weights $\mathbf{W}^2$ with $\mathbf{H}^1$.

In this graph, $W = \{w_0, w_1, w_2\}$, $w_i \in \mathbb{R}$ is the weight vector. Moreover, assume that we already know the gradient of loss function (i.e., the leaf node in computational graph) with respect to the computational node in $\mathbf{H}^2$. This is illustrated using δ_i, $i = 0, \ldots, 3$ on figure. According to backpropagation algorithm, $\frac{\delta \mathcal{L}}{\delta w_i}$ is given by

$$
\begin{aligned}
\frac{\delta \mathcal{L}}{\delta w_i} &= \frac{\delta \mathbf{H}_0^2}{\delta w_i} \frac{\delta \mathcal{L}}{\delta \mathbf{H}_0^2} + \frac{\delta \mathbf{H}_1^2}{\delta w_i} \frac{\delta \mathcal{L}}{\delta \mathbf{H}_1^2} + \frac{\delta \mathbf{H}_2^2}{\delta w_i} \frac{\delta \mathcal{L}}{\delta \mathbf{H}_2^2} + \frac{\delta \mathbf{H}_3^2}{\delta w_i} \frac{\delta \mathcal{L}}{\delta \mathbf{H}_3^2} \\
&= \frac{\delta \mathbf{H}_0^2}{\delta w_i} \delta_0 + \frac{\delta \mathbf{H}_1^2}{\delta w_i} \delta_1 + \frac{\delta \mathbf{H}_2^2}{\delta w_i} \delta_2 + \frac{\delta \mathbf{H}_3^2}{\delta w_i} \delta_3
\end{aligned}
\tag{3.3}
$$

Fig. 3.8 Two layers from middle of a neural network indicating the one-dimensional convolution. The weight $\mathbf{W}^2$ is shared among the neurons of $\mathbf{H}^2$. Also, δ_i shows the gradient of loss functions with respect to H_i^2

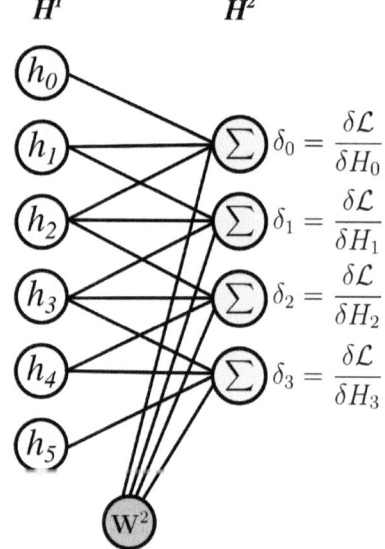

By computing the above equation for each w_i in the graph we will obtain

$$\frac{\delta \mathscr{L}}{\delta w_0} = h_0 \delta_0 + h_1 \delta_1 + h_2 \delta_2 + h_3 \delta_3$$

$$\frac{\delta \mathscr{L}}{\delta w_1} = h_1 \delta_0 + h_2 \delta_1 + h_3 \delta_2 + h_4 \delta_3 \qquad (3.4)$$

$$\frac{\delta \mathscr{L}}{\delta w_2} = h_0 \delta_0 + h_3 \delta_1 + h_4 \delta_2 + h_5 \delta_3$$

Let $\delta^2 = [\delta_0, \delta_1, \delta_2, \delta_3]$ denotes the vector of gradients in of $\mathbf{H}^2$ and $\mathbf{h}^1 = [h_0, h_1, h_2, h_3, h_4, h_5]$ denotes the output of neurons in $\mathbf{H}^1$. If we carefully study the above equation, we will realize that computing

$$\mathbf{h}^1 * \delta^2 \qquad (3.5)$$

will return $\frac{\mathscr{L}}{W} = [\frac{\mathscr{L}}{w_0}, \frac{\mathscr{L}}{w_1}, \frac{\mathscr{L}}{w_2}]$. As before, the operator $*$ denotes the *valid* convolution operation. In general, gradient of loss function with respect to the convolution filters is obtained by convolving δ of current layer with the inputs of the layer.

Beside $\frac{\delta \mathscr{L}}{\delta w_i}$, we also need to compute $\frac{\delta \mathscr{L}}{\delta h_i}$ in order to pass the error to previous layer. According to the backpropagation algorithm, we can compute these gradients as follows:

$$\frac{\delta \mathscr{L}}{\delta h_0} = \frac{\mathbf{H}_0^2}{h_0} \delta_0$$

$$\frac{\delta \mathscr{L}}{\delta h_1} = \frac{\mathbf{H}_0^2}{h_1} \delta_0 + \frac{\mathbf{H}_1^2}{h_1} \delta_1$$

$$\frac{\delta \mathscr{L}}{\delta h_2} = \frac{\mathbf{H}_0^2}{h_2} \delta_0 + \frac{\mathbf{H}_1^2}{h_2} \delta_1 + \frac{\mathbf{H}_2^2}{h_2} \delta_2$$

$$\frac{\delta \mathscr{L}}{\delta h_3} = \frac{\mathbf{H}_1^2}{h_3} \delta_1 + \frac{\mathbf{H}_2^2}{h_3} \delta_2 + \frac{\mathbf{H}_3^2}{h_3} \delta_3 \qquad (3.6)$$

$$\frac{\delta \mathscr{L}}{\delta h_4} = \frac{\mathbf{H}_2^2}{h_4} \delta_2 + \frac{\mathbf{H}_3^2}{h_4} \delta_3$$

$$\frac{\delta \mathscr{L}}{\delta h_4} = \frac{\mathbf{H}_3^2}{h_5} \delta_3$$

By computing $\frac{\mathbf{H}_i^2}{h_j}$ and plugging it in the above equation, we will obtain

$$\frac{\delta \mathcal{L}}{\delta h_0} = w_0 \delta_0$$

$$\frac{\delta \mathcal{L}}{\delta h_1} = w_1 \delta_0 + w_0 \delta_1$$

$$\frac{\delta \mathcal{L}}{\delta h_2} = w_2 \delta_0 + w_1 \delta_1 + w_0 \delta_2$$

$$\frac{\delta \mathcal{L}}{\delta h_3} = w_2 \delta_1 + w_1 \delta_2 + w_0 \delta_3 \qquad (3.7)$$

$$\frac{\delta \mathcal{L}}{\delta h_4} = w_2 \delta_2 + w_1 \delta_3$$

$$\frac{\delta \mathcal{L}}{\delta h_4} = w_2 \delta_3$$

If we carefully study the above equation, we will realize that computing

$$\delta^2 * flip(W) \qquad (3.8)$$

will give us the gradient of loss with respect to every node in the previous layer. Note that $*$ here refers to the *full* convolution and *flip* is a function that reverses the direction of W. In general, in a convolution layer, the gradient of current layer with respect to the nodes in previous layer is obtained by convolving δ of current layer with the reverse of convolution filters.

3.1.3 Stride in Convolution

Given the image $\mathbf{X} \in \mathbb{R}^{W \times H}$, convolution of kernel $f \in \mathbb{R}^{P \times Q}$ with the image is given by

$$(\mathbf{X} * f)(m, n) = \sum_{i=0}^{P-1} \sum_{j=0}^{Q-1} \mathbf{X}(m+i, n+j) f(i, j) \qquad m = 0, \dots, H-1, n = 0, \dots, W-1 \quad (3.9)$$

The output of the above equation is a $W - P + 1 \times H - Q + 1$ image where value of each element is computed using the above equation. Technically, we say that the *stride* of the convolution is equal to one meaning that the above equation is computed for every m and n in $\mathbf{X}$.

As we will discuss shortly, in some cases, we might be interested in computing the convolution with a larger stride. For example, we want to compute the convolution of alternate pixels. In this case, we say that the stride of convolution is equal to two, leading to the equation below

$$(\mathbf{X} * f)(m, n) = \sum_{i=0}^{P-1} \sum_{j=0}^{Q-1} \mathbf{X}(m+i, n+j) f(i, j) \quad m = 0, 2, 4 \ldots, H-1, n = 0, 2, 4, \ldots, W-1$$

$$(3.10)$$

The result of the above convolution will be a $\frac{W-P}{2} + 1 \times \frac{H-Q}{2} + 1$ image. Common values for stride are 1 and 2 and you may rarely find a convolution layer with a stride greater than 3. In general, denoting the stride with s, size of the output matrix will be equal to $\frac{W-P}{s} + 1 \times \frac{H-Q}{s} + 1$. Note that the value of stride and filter size must be chosen such that $\frac{W-P}{2} + 1$ and $\frac{H-Q}{2} + 1$ become integer number. Otherwise, $\mathbf{X}$ has to be cropped so they become integer numbers.

3.2 Pooling

In Sect. 3.1.1 we explained that the feature extraction based on bank of Gabor filters is done in two steps. After convolving input image with many filters in the first step, the second step in this method locally pools pixels to extract information.

A similar approach is also used in ConvNets. Specifically, assume a 190×190 image which is connected to a convolution layer containing 50 filters of size 7×7. Output of the convolution layer will contain 50 feature maps of size 184×184 which collectively represent a 50-channel image. From another perspective, output of this layer can be seen as a $184 \times 184 \times 50 = 1,692,800$ dimensional vector. Clearly, this number of dimensions is very high.

The major goal of a pooling layer is to reduce the dimensionality of feature maps. For this reason, it is also called *downsampling*. The factor to which the downsampling will be done is called *stride* or *downsampling factor*. We denote the pooling stride by s. For example, assume the 12-dimensional vector $x = [1, 10, 8, 2, 3, 6, 7, 0, 5, 4, 9, 2]$. Downsampling x with stride $s = 2$ means we have to pick every alternate pixel starting from the element at index 0 which will generate the vector $[1, 8, 3, 7, 5, 9]$. By doing this, dimensionality of x is divided by $s = 2$ and it becomes a six-dimensional vector.

Suppose that x in the above example shows the response of a model to an input. When the dimensionality of x is reduced using downsampling, it ignores the effect of other value between alternate pixels. For example, downsampling vectors $x_1 = [1, 10, 8, 2, 3, 6, 7, 0, 5, 4, 9, 2]$ and $x_2 = [1, 100, 8, 20, 3, 60, 7, 0, 5, 40, 9, 20]$ with stride $s = 2$ will both produce $[1, 8, 3, 7, 5, 9]$. However, x_1 and x_2 may represent two different states of input. As the result, important information might be discarded using simple downsampling approach.

Pooling generalizes downsampling by considering the elements between alternate pixels as well. For instance, a *max pooling* with stride $s = 2$ and size $d = 2$ will downsample x as

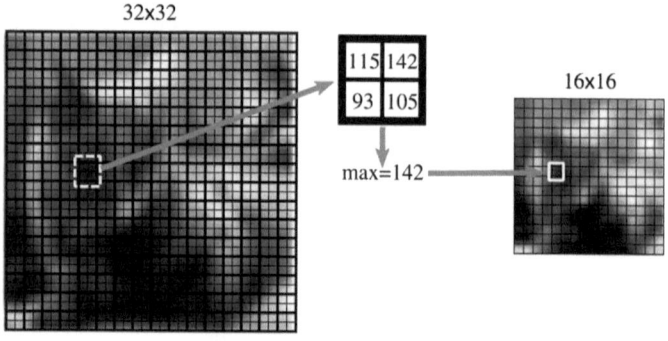

Fig. 3.9 A pooling layer reduces the dimensionality of each feature map separately

$$x_{max-pool} = [\max(1, 10), \max(8, 2), \max(3, 6), \max(7, 0), \max(5, 4), \max(9, 2)] \tag{3.11}$$

that is equal to $x_{max-pool} = [10, 8, 6, 7, 5, 9]$. Likewise, max pooling x_1 and x_2 will produce $[10, 8, 6, 7, 5, 9]$ and $[100, 20, 60, 7, 50, 20]$, respectively. Contrary to the simple downsampling, we observe that max pooling does not ignore any element. Instead, it intelligently reduces the dimension of the vector taking into account the values in local neighborhood of the current element, the size of local neighborhood is determined by d.

Pooling feature maps of a convolution layer is done in a similar way. This is illustrated in Fig. 3.9 where the 32×32 image is max-pooled with stride $s = 2$ and $d = 2$. To be more specific, the image is divided into a $d \times d$ region every s pixels row-wise and column-wise. Each region corresponds to a pixel in the output feature map. The value at each location in the output feature map is obtained by computing the maximum value in the corresponding $d \times d$ region in the input feature map. In this figure, it is shown how the value at location $(7, 4)$ has been computed using its corresponding $d \times d$ region.

It is worth mentioning that pooling is applied on each feature map separately. That means if the output of a convolution layer has 50 feature maps (i.e., the layer has 50 filters), the pooling operation is applied on each of these feature maps separately and produce another 50-channel feature maps. However, dimensionality of feature maps are *spatially* reduced by factor of s. For example, if the output of a convolution layer is a $184 \times 184 \times 50$ image, it will be a $92 \times 92 \times 50$ image after max-pooling with stride $s = 2$.

Regions of pooling may overlap with each other. For example, there is not overlap in the $d \times d$ regions when the stride is set to $s = d$. However, by setting $s = a, a < d$ the region will overlap with surrounding regions. Pooling stride is usually set to 2 and the size pooling region is commonly set to 2 or 3.

As we mentioned earlier, the major goal of pooling is to reduce the dimensionality of feature maps. As we will see shortly, this makes it possible to design a ConvNet where the dimensionality of the feature vector in the last layer is very low. However, the need for pooling layer has been studied by researchers such as Springenberg et al. (2015). In this work, the authors show that a pooling layer can be replaced by a convolution layer with convolution stride $s = 2$. Some ConvNets such as Aghdam et al. (2016) and Dong et al. (2014) do not use pooling layers since their aim is to generate a new image for a given input image.

Average pooling is an alternative max pooling in which instead of computing the maximum value in a region, the average value of the region is calculated. However, Scherer et al. (2010) showed that max pooling produces superior results than average-pooling layer. In practice, average pooling is rarely used in middle layers. Another pooling method is called stochastic pooling (Zeiler and Fergus 2013). In this approach, a value from the region is randomly picked where elements with higher values are more likely to be picked by the algorithm.

3.2.1 Backpropagation in Pooling Layer

Pooling layers are also part of the computational graph. However, in contrast to convolution layers which are formulated using some parameters, the pooling layers that we mentioned in the previous section do not have trainable parameters. For this reason, we only need to compute their gradient with respect to the previous layer. Assume the one-dimensional layer in Fig. 3.10. Each neuron in the right layer computes the maximum of its inputs.

We need to compute gradient of each neuron with respect to it inputs. This can be easily computed as follows:

Fig. 3.10 A one-dimensional max-pooling layer where the neurons in $\mathbf{H}^2$ compute the maximum of their inputs

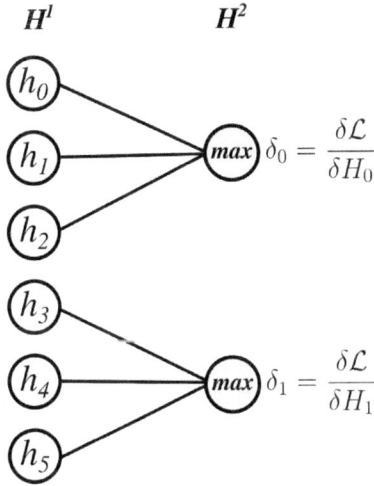

$$\frac{\delta \mathbf{H}_0^2}{\delta h_0} = \begin{cases} 1 & max(h_0, h_1, h_2) == h_0 \\ 0 & otherwise \end{cases}$$

$$\frac{\delta \mathbf{H}_0^2}{\delta h_1} = \begin{cases} 1 & max(h_0, h_1, h_2) == h_1 \\ 0 & otherwise \end{cases}$$

$$\frac{\delta \mathbf{H}_0^2}{\delta h_2} = \begin{cases} 1 & max(h_0, h_1, h_2) == h_2 \\ 0 & otherwise \end{cases}$$

$$\frac{\delta \mathbf{H}_1^2}{\delta h_3} = \begin{cases} 1 & max(h_3, h_4, h_5) == h_3 \\ 0 & otherwise \end{cases} \tag{3.12}$$

$$\frac{\delta \mathbf{H}_1^2}{\delta h_4} = \begin{cases} 1 & max(h_3, h_4, h_5) == h_4 \\ 0 & otherwise \end{cases}$$

$$\frac{\delta \mathbf{H}_1^2}{\delta h_5} = \begin{cases} 1 & max(h_3, h_4, h_5) == h_5 \\ 0 & otherwise \end{cases}$$

According to the above equation, if neuron h_i is selected during the max-pooling operation, the gradient from next layer will be passed to h_i. Otherwise, the gradient will not be passed to h_i. In other words, if h_i is not selected during max pooling, $\frac{\mathscr{L}}{h_i} = 0$. Gradient of the stochastic pooling is also computed in a similar way. Concretely, the gradient is passed to the selected neurons and it is blocked to other neurons. In the case of average pooling, gradient to all input neurons are equal to $1/n$ where n denotes the number of inputs of the pooling neuron.

3.3 LeNet

The basic concept of ConvNets dates back to 1979 when Kunihiko Fukushima proposed an artificial neural network including simple and complex cells which were very similar to convolution and pooling layers in modern ConvNets (Schmidhuber 2015). In 1989, LeCun et al. (1998) proposed the weight sharing paradigm and derived convolution and pooling layers. Then, they designed a ConvNet which is called LeNet-5. The architecture of this ConvNet is illustrated in Fig. 3.11.

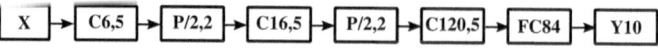

Fig. 3.11 Representing LeNet-5 using a DAG

In this DAG, Ca, b shows a convolution layer with a filters of size $b \times b$ and the phrase $/a$ in any nodes shows the stride of that operation. Moreover, $P/a, b$ denotes a pooling operation with stride a and size b, FCa shows a fully connected layer with a neurons, Ya shows the output layer with a neurons.

This ConvNet that is originally proposed for recognizing handwritten digits consists of four convolution-pooling layers. The input of the ConvNet is a single-channel 32×32 image. Also, the last pooling layer (S4) is connected to the fully connected layer C5. The convolution layer C1 contains six filters of size 5×5. Convolving a 32×32 image with these filters produces six feature maps of size 28×28 (recall from previous discussion that $32(width) - 5(filterwidth) + 1 = 28$). Since the input of the network is a single-channel image, the convolution filter in C1 are actually $5 \times 5 \times 1$ filters.

The convolution layer C1 is followed by the pooling layer S2 with stride 2. Thus, the output of S2 is six feature maps of size 14×14 which collectively show a six-channel input. Then, 16 filters of size 5×5 are applied on the six-channel image in the convolution layer C3. In fact, the size of convolution filters in C3 is $5 \times 5 \times 6$. As the result, the output of C3 will be 16 images of size 10×10 which, together, show a 16-channel input. Next, the layer S4 applies a pooling operation with stride 2 and produces 16 images of size 5×5. The layer C5 is a fully connected layer in which every neuron in C5 is connected to all the neuron in S4. In other words, every neuron in C5 is connected to $16 \times 56 \times 5 = 400$ neurons in S4. From another perspective, C5 can be seen as a convolution layer with 120 filters of size 5×5. Likewise, S6 is also a fully connected layer that is connected to S5. Finally, the classification layer is a radial basis function layer where the inputs of the radial basis function are 84 dimensional vectors. However, for the purpose of this book, we consider the classification layer a fully connected layer composed of 10 neurons (one neuron for each digit).

The pooling operation in this particular ConvNet is not the max-pooling or average-pooling operations. Instead, it sums the four inputs and divides them by the trainable parameter a and adds the trainable bias b to this result. Also, the activation functions are applied after the pooling layer and there is no activation function after the convolution layers. In this ConvNet, the sigmoid activation functions are used.

One important question that we have to always ask is that how many parameters are there in the ConvNet that we have designed. Let us compute this quantity for LeNet-5. The first layer consists of six filters of size $5 \times 5 \times 1$. Assuming that each filter has also a *bias* term, C1 is formulated using $6 \times 5 \times 5 \times 1 + 6 = 156$ trainable parameters. Then, in this particular ConvNet, each pooling unit is formulated using two parameters. Hence, S2 contains 12 trainable parameters. Then, taking into account the fact that C3 is composed of 16 filters of size $5 \times 5 \times 6$, it will contain $16 \times 5 \times 5 \times 6 + 16 = 2416$ trainable parameters. S4 will also contain 32 parameters since each pooling unit is formulated using two parameters. In the case of C5, it consists of $120 \times 5 \times 5 \times 16 + 120 = 48120$ parameters. Similarly, F6 contains $84 \times 120 + 84 = 10164$ trainable parameters and the output includes $10 \times 84 + 10 = 850$ trainable parameters. Therefore, the LeNet 5 ConvNet requires training $156 + 12 + 2416 + 32 + 48120 + 10164 + 850 = 61750$ parameters.

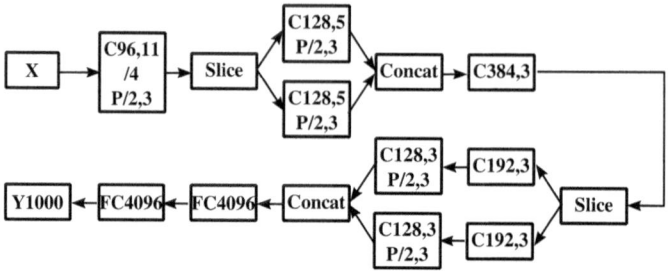

Fig. 3.12 Representing AlexNet using a DAG

3.4 AlexNet

In 2012, Krizhevsky et al. (2012) trained a large ConvNet on the ImageNet dataset (Deng et al. 2009) and won the image classification competition on this dataset. The challenge is to classify 1000 classes of natural objects. Afterwards, this ConvNet became popular and it was called AlexNet.[3] The architecture of this ConvNet is shown in Fig. 3.12.

In this diagram, *Slice* illustrates a node that slices a feature maps through its depth and *Concat* shows a node that concatenates feature maps coming from different nodes across their depth. The first convolution layer in this ConvNet contains 96 filters of size 11×11 which are applied with stride $s = 4$ on 224×3 images. Then, the ReLU activation is applied on the 96 feature maps. After that a 3×3 max pooling with stride $s = 2$ is applied on activation maps. Before applying the second convolution layer, the 96 activation maps is divided into two 48 channel maps. The second convolution layer consists of 256 filters of size $5 \times 5 \times 48$ in which the first 128 filters are applied on the first 48-channel map from the first layer and the second 128 filters are applied on the second 48-channel map from the first layer. The feature maps of the second convolution layer are passed through the ReLU activation and a max-pooling layer. The third convolution layer has 384 filters of size $3 \times 3 \times 256$.

It turns out that each filter in the third convolution layer is connected to both 128-channel maps from the second layer. The ReLU activation is applied on the third convolution layer but there is not pooling after the third convolution layer. At this point, the third convolution layer is divided into two 192-channel maps. The fourth convolution layer in this ConvNet has 384 filters of size $3 \times 3 \times 192$. As before, the first 192 filters are connected to the first 192-channel map from the third convolution layer and the second 192 filters are connected to the second 192-channel map from the third convolution layer. The output of the fourth convolution layer is passed through a ReLU activation and it directly goes into the fifth convolution layer

---

[3] Alex is the first name of the first author.

without passing through a pooling layer. The fifth convolution layer has 256 filters of size $3 \times 3 \times 192$ where each of 128 filters is connected to one 192-channel feature map from the fourth layer. Here, output of this layer goes into a ReLU activation and it is passed through a max-pooling layer. Finally, there are two consecutive fully connected layers each containing 4096 neurons and ReLU activation after the fifth convolution layer. The output of the ConvNet is also a fully connected layer with 1000 neurons.

AlexNet has 60,965,224 trainable parameters. Also, it is worth mentioning that there are *local response normalization* (LRN) layers after some of these layers. We will explain this layer in this chapter. In short, a LRN layer does not have any trainable parameter and it applies a nonlinear transformation on feature maps. Also, it does not change any of the dimensions of feature maps.

3.5 Designing a ConvNet

In general, one of the difficulties in neural networks is finding a good architecture which produces accurate results and it is computationally efficient. In fact, there is no golden rule in finding such an architecture. Even people with years of experience in neural networks may require many trials to find a good architecture.

Arguably, the practical way is to immediately start with an architecture, implement, and train it on the training set. Then, the ConvNet is evaluated on the validation set. If the results are not satisfactory, we change the architecture or hyperparameters of the ConvNet and repeat the aforementioned procedure. This approach is illustrated in Fig. 3.13.

In the rest of this section, we thoroughly explain each of these steps.

Fig. 3.13 Designing a ConvNet is an iterative process. Finding a good architecture may require several iterations of design–implement–evaluate

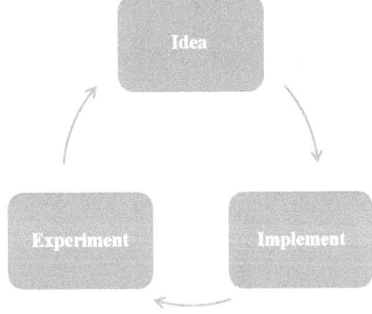

3.5.1 ConvNet Architecture

Although there is no golden rule in designing a ConvNet, there are a few rule-of-thumbs that can be found in many successful architectures. A ConvNet typically consists of several convolution-pooling layers followed by a few fully connected layers. Also, the last layer is always the output layer. From another perspective, a ConvNet is a directed acyclic graph (DAG) with one leaf node. In this DAG, each node represents a layer and edges show the connection between layers.

A convenient way of designing a ConvNet is to use a DAG diagram like the one illustrated in Fig. 3.12. One can define other nodes or combine several nodes into one node. You can design any DAG to represent a ConvNet. However, two rules have to be followed. First, there is always one leaf node in a ConvNet which represents the classification layer or the loss function. That does not make sense to have more than one classification layer in a ConvNet. Second, inputs of a node must have the same spatial dimension. The exception could be the concatenation node where you can also concatenate the inputs spatially. As long as these two rules are observed in the DAG, the architecture is valid and correct. Also, all operations represented by nodes in the DAG must be differentiable so that the backpropagation algorithm can be applied on the graph.

As the first rule of thumb, remember to always compute the size of feature maps for each node in the DAG. Usually, nodes that are connected to fully connected layers have spatial size less than 8×8. Common sizes are 2×2, 3×3, and 4×4. However, the channels (third dimension) of the nodes connecting to fully connected layer could be any arbitrary size. The second rule of thumb is that the number of feature maps, usually, has a direct relation with depth of each node in the DAG. That means we start with small number of feature maps in early layers and the number of feature maps increases as they the depth of nodes increases. However, some flat architectures have been also proposed in literature where all layers have the same number of feature maps or they have a repetitive pattern.

The third rule of thumb is that state-of-the-art ConvNets commonly use convolution filters of size 3×3, 5×5, and 7×7. Among them, AlexNet is the only one that has utilized 11×11 convolution filters. The fourth rule of thumb is activation functions usually come immediately after a convolution layer. However, there are a few works that put the activation function after the pooling layer. As the fifth rule of thumb remember that while putting several convolution layers consecutively makes sense, it is not common to add two or more consecutive activation function layers. The sixth rule of thumb is to use an activation function from the family of ReLU function (ReLU, Leaky ReLU, PReLU, ELU or Noisy ReLU). Also, always compute the number of trainable parameters of your ConvNet. If you do not have plenty of data and you design a ConvNet with millions of parameters, the ConvNet might not generalize well on the test set.

In the simplest scenario, the *idea* in Fig. 3.13 might be just designing a ConvNet. However, there are many other points that must be considered. For example, we may need to preprocess the data. Also, we have to split the data into several parts. We shall discuss about this later in this chapter. For now, let just assume that the idea refers to designing a ConvNet. Having the idea defined clearly, the next step is to implement the ideas.

3.5.2 Software Libraries

The main scope of this book is ConvNets. For this reason, we will only discuss how to efficiently implement a ConvNet in a practical application. Other ideas such as preprocessing data or splitting data might be done in any programming languages.

There are several commonly used libraries for implementing ConvNets which are actively updated as new methods and ideas are developed in this field. There have been other libraries such as *cudaconvnet* which are not active anymore. Also, there are many other libraries in addition to the following list. But, the following list is widely used in academia as well as industry:

- Theano (deeplearning.net/software/theano/)
- Lasagne (lasagne.readthedocs.io/en/latest/)
- TensorFlow (www.tensorflow.org/)
- Keras (keras.io/)
- Torch (torch.ch/)
- cuDNN (developer.nvidia.com/cudnn)
- mxnet (mxnet.io)
- Caffe (caffe.berkeleyvision.org/)

3.5.2.1 Theano

Theano is a library which can be used in Python for symbolic numerical computations. In this library, a computational DAG such as ConvNet is defined using symbolic expressions. Then, the symbolic expressions are compiled using its built-in compiler into executable functions. These functions can be called similar to other functions in Python. There are two important features in Theano which are very important. First, based on the user configuration, compiling the functions can be done either on CPUs or a GPU. Even a user with little knowledge about GPU programming can easily use this library for running heavy expressions on GPU.

Second, Theano represents any expression in terms of computational graphs. For this reason, it is also able to compute the gradient of a leaf node with respect to all other nodes in the graph automatically. For this reason, user can easily implement gradient based optimization algorithms to train a ConvNet. Gradient of convolution and pooling layers are also computed efficiently using Theano. These features make Theano a good choice for doing research. However, it might not be easily utilized in commercial products.

3.5.2.2 Lasagne

Despite its great power, Theano is a low-level library. For example, every time that you need to design a convolution layer followed by the ReLU activation, you must write codes for each part separately. *Lasagne* has been built on top of Theano and it has developed the common patterns in ConvNets so you do not need to implement them every time. In fact, using Lasagne, you can only design neural networks including ConvNets. Nonetheless, to use Lasagne one must have the basic knowledge about Theano as well.

3.5.2.3 TensorFlow

TensorFlow is another library for numerical computations. It has interfaces for both Python and C++. Similar to Theano it expresses mathematical equations in terms of a DAG. It supports automatic differentiation as well. Also, it can compile the DAG on CPUs or GPUs.

3.5.2.4 Keras

Keras is a high-level library which is written in Python. Keras is able to run either of TensorFlow or Theano depending on the user configuration. Using Keras, it is possible to rapidly develop your idea and train it. Note that it is also possible to rapidly develop the ideas in Theano and TensorFlow.

3.5.2.5 Torch

Torch is also a library for scientific computing and it supports ConvNets. It is based on Lua programming language and uses the scripting language LuaJIT. Similar to other libraries it supports computations on CPUs and GPUs.

3.5.2.6 cuDNN

cuDNN has been developed by NVIDIA and it can be used only for implementing deep neural networks on GPUs created by NVIDIA. Also, it supports forward and backward propagation. Hence, not only it can be used in developing products but it can be also used for training ConvNets. cuDNN is a great choice for commercial products. In fact, all other libraries in our list use cuDNN for compiling their code on GPUs.

3.5.2.7 mxnet

Another commonly used library is called *mxnet*.[4] Similar to Theano, Tensorflow and Torch it supports auto differentiation and symbolic expressions. It also supports

---

[4]mxnet.io.

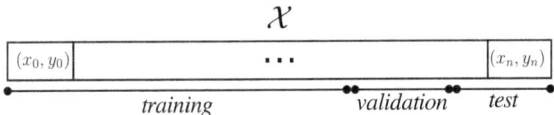

Fig. 3.14 A dataset is usually partitioned into three different parts namely *training set*, *development set* and *test set*

distributed computing which is very useful in the case that you want to train a model on several GPUs.

3.5.2.8 Caffe

Caffe is the last library in our list. It is written in C++ and it has interfaces for MATLAB and Python. It can be only used for developing deep neural networks. Also, it supports all state-of-the-art methods proposed in community for ConvNets. It can be used both for research and commercial products. However, developing new layers in Caffe is not as easy as Theano, TensorFlow, mxnet, or Torch. But, creating a ConvNet to solve a problem can be done quickly and effectively. More importantly, the trained ConvNet can be easily ported on embedded systems. We will develop all our ConvNets using the Caffe library. In the next chapter, we will mention how to design, train, and test a ConvNet using Caffe library.

There are also other libraries such as Deeplearning4j,[5] Microsoft Cognitive Toolkit,[6] Pylearn2[7] and MatConvNet.[8] But, the above list is more common in academia.

3.5.3 Evaluating a ConvNet

After implementing your idea using one of the libraries in the previous section, it is time to evaluate how good is the idea for solving the problem. Concretely, evaluation must be done empirically using a dataset. In practice, evaluation is done using three different partitions of data. Assume the dataset $\mathcal{X} = \{(\mathbf{x}_0, y_0), \ldots, (\mathbf{x}_n, y_n)\}$ containing n samples where $\mathbf{x}_i \in \mathbb{R}^{W \times H \times 3}$ is a color image and $y_i \in \{1, \ldots, c\}$ is its corresponding class label. This dataset must be partitioned into three *disjoint sets* namely *training set*, *development set* and *test set* as it is illustrated in Fig. 3.14.

Formally, the dataset $\mathcal{X}$ is partitioned into $\mathcal{X}_{train}$, $\mathcal{X}_{dev}$ and $\mathcal{X}_{test}$ such that

$$\mathcal{X} = \mathcal{X}_{train} \bigcup \mathcal{X}_{dev} \bigcup \mathcal{X}_{test} \tag{3.13}$$

---

[5]deeplearning4j.org.
[6]www.microsoft.com/en-us/research/product/cognitive-toolkit/
[7]github.com/lisa-lab/pylearn2.
[8]www.vlfeat.org/matconvnet.

and

$$\mathcal{X}_{train} \bigcap \mathcal{X}_{dev} = \mathcal{X}_{train} \bigcap \mathcal{X}_{test} = \mathcal{X}_{dev} \bigcap \mathcal{X}_{test} = \emptyset. \qquad (3.14)$$

The training set will be only and only used during training (i.e., minimizing loss function) the ConvNet. During training the ConvNet, its performance is regularly evaluated on the development set. If the performance is not acceptable, we go back to the idea and refine the idea or design a new idea from scratch. Then, the new idea will be implemented and trained on the same training set. Then, it is evaluated on the development set. This procedure will be repeated until we are happy with the performance of model on the development set. After that, we carry out a final evaluation using the test set. The performance on the test set will tell us how good our model will be in real world. It is worth mentioning that the development set is commonly called *validation set*. In this book, we use validation set and development set interchangeably.

Splitting data into three partitions is very important step toward developing a good and reliable model. We should note that evaluation on the test set is done only once. We never try to refine our model based on the performance of the test set. Instead, if we see that the performance on the test set is not acceptable and we need to develop a new idea, the new idea will be refined and evaluated only on the training and development sets. The test set will be only used to ascertain whether or not the model is good for the real-world application. If we refine the idea based on performance of test set rather than the development set we may end up with a model which might not yield accurate results in practice.

3.5.3.1 Classification Metrics

Evaluating a model on the development set or the test set can be done using *classification metric functions* or simply a metric function. On the one hand, the output of a ConvNet which is trained for a classification task is the label (class) of its input. We call the label produced by a ConvNet predicted label. On the other hand, we also know the actual label of each sample in $\mathcal{X}_{dev}$ and $\mathcal{X}_{test}$. Therefore, we can use the predicted labels and actual labels of samples in these sets to assess our ConvNet. Mathematically, a classification metric function usually accepts the actual labels and predicted labels and returns a score or set of scores. The following metric functions can be applied on any classification dataset regardless if it is a training set, development set or test set.

3.5.3.2 Classification Accuracy

The simplest metric function for the task of classification is the *classification accuracy*. It calculates fraction of samples that are classified correctly. Given the set $\mathcal{X}' = \{(\mathbf{x}_1, y_1), \ldots, (\mathbf{x}_N, y_N)\}$ containing N pair of samples, the classification score is computed as follows:

$$accuracy = \frac{1}{N} \sum_{i=1}^{N} \mathbf{1}[y_i == \hat{y}_i] \qquad (3.15)$$

where y_i and $\hat{y}_i$ are the actual label and the predicted label of the ith sample in $\mathscr{X}'$. Also, $\mathbf{1}[.]$ returns 1 when the value of its argument evaluates to True and 0 otherwise. Clearly, *accuracy* will take a value in $[0, 1]$. If the accuracy is equal to 1 that means all the samples in $\mathscr{X}'$ are classified correctly. In contrast, if accuracy is equal to 0 that means none of the samples in $\mathscr{X}'$ is classified correctly.

Computing accuracy is straightforward and it is commonly used for assessing classification models. However, accuracy posses one serious limitation. We explain this limitation using an example. Assume the set $\mathscr{X}' = \{(\mathbf{x}_1, y_1), \ldots, (\mathbf{x}_{3000}, y_{3000})\}$ with 3000 samples where $y_i \in 1, 2, 3$ show that samples in this dataset belongs to one of three classes. Suppose 1500 samples within $\mathscr{X}'$ belong to class 1, 1400 samples belong to class 2 and 100 samples belong to class 3. Further assume that all samples belonging to class 1 and 2 are classified correctly but all samples belonging to class 3 are classified incorrectly. In this case, the accuracy will be equal to $\frac{2900}{3000} = 0.9666$ showing that 96.66% of samples in $\mathscr{X}'$ are classified correctly. If we only look at the accuracy, we might think that 96.66% is very accurate for our application and we decide that our ConvNet is finalized.

However, the accuracy in the above example is high because the number of samples belonging to class 3 is much less than the number of samples belonging to class 1 or 2. In other words, the set $\mathscr{X}'$ is *imbalanced*. To alleviate this problem, we can set a weight for each sample where the weight of a sample in class A is proportional to the number of samples in class A and total number of samples in $\mathscr{X}'$. Based on this formulation, the weighted accuracy is given by

$$accuracy = \frac{1}{N} \sum_{i=1}^{N} w_i \times \mathbf{1}[y_i == \hat{y}_i] \qquad (3.16)$$

where w_i denotes the weight of ith sample. If there are C classes in $\mathscr{X}'$, the weight of a sample belonging to class A is usually equal to

$$\frac{1}{C \times number\ of\ samples\ in\ class\ A}. \qquad (3.17)$$

In the above example, weights of samples of class 1 will be equal to $\frac{1}{3 \times 1500} = 0.00022$ and weights of samples of class 2 will be equal to $\frac{1}{3 \times 1400} = 0.00024$. Similarly, the weight of samples of class 3 will be equal to $\frac{1}{3 \times 100} = 0.0033$. Computing the weighted accuracy in the above example, we will obtain $1500 \times \frac{1}{3 \times 1500} + 1400 \times \frac{1}{3 \times 1400} + 0 \times \frac{1}{3 \times 100} = 0.6666$ instead of 0.9666. The weighted accuracy gives us a better estimate of performance in this particular case.

There is still another limitation with the accuracy metric even in perfectly balanced datasets. Assume that there are 200 different classes in $\mathscr{X}'$ and there are 100 samples in each class yielding 20,000 samples in $\mathscr{X}'$. Assume all the samples belonging to class 1 to 199 are classified correctly and all of the samples belonging to class 200 are classified incorrectly. In this case, the accuracy score will be equal to $\frac{19900}{20000} = 0.995$ showing nearly perfect classification. Even with the above weighting approach the accuracy will be still equal to 0.995.

In general, the accuracy score shows a rough evaluation of the model and it might not be a reliable metric for making final decisions about a model. However, the above examples are hypothetical and it may never happen in practice that all the samples from one class are classified correctly and all the samples from other classes are classified correctly. The above hypothetical example is just to show the limitation of this metric. In practice, the accuracy score is commonly used for assessing models. But, a great care must be taken into account when you are evaluating your model using classification accuracy.

3.5.3.3 Confusion Matrix

Confusion matrix is a powerful metric for accurately evaluating classification models. For a classification problem with C classes, confusion matrix $\mathbf{M}$ is a $C \times C$ matrix where element M_{ij} in this matrix shows the number of samples in $\mathscr{X}'$ whose actual class label are i but they are classified as class j using our ConvNet. Concretely, M_{ii} shows the number of samples which are correctly classified. We first study the confusion matrix on binary classification problems. Then, we will extend it to multiclass classification problems. There are only two classes in binary classification problems. Consequently, the confusion matrix will a 2×2 matrix. Figure 3.15 shows the confusion matrix for a binary classification problem.

Element M_{11} in this matrix shows the number of samples whose actual labels are 1 and they are classified as 1. Technically, this element of matrix is called *true-positive* (TP) samples. Element M_{12} shows the number of samples whose actual label is 1 but they are classified as -1. This element is called *false-negative* (FN) samples. Element M_{21} denotes the number of samples whose actual label is -1 but they are classified as 1. Hence, this element is called *false-positive* (FP) samples. Finally, element M_{22}, that is called *true-negative* (TN) samples, illustrates the number of samples which

Fig. 3.15 For a binary classification problem, confusion matrix is a 2×2 matrix

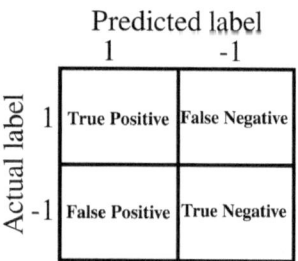

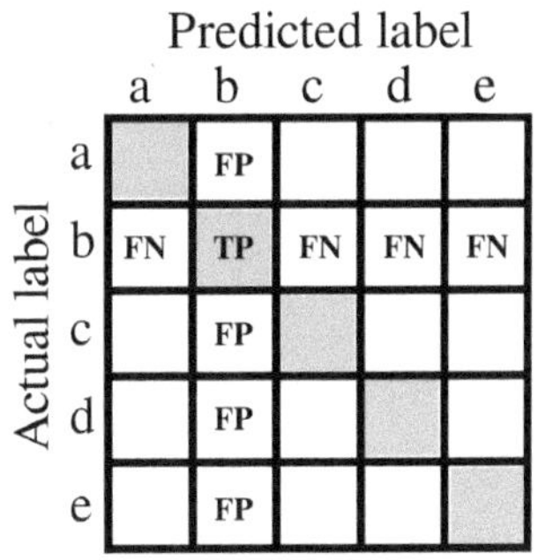

Fig. 3.16 Confusion matrix in multiclass classification problems

are actually -1 and they are classified as -1. Based on this formulation, the accuracy is given by:

$$accuracy = \frac{TP + TN}{TP + TN + FP + FN} \tag{3.18}$$

Concretely, a ConvNet is a perfect classifier if $FP = FN = 0$. The confusion matrix can be easily extended to multiclass classification problem. For example, Fig. 3.16 shows a confusion matrix for five-class classification problems. A ConvNet for this matrix is a perfect classifier if all non-diagonal elements of this matrix are zero. The terms TP, FP, and FN can be extended to this confusion matrix as well.

For any class i in this matrix,

$$FN_i = \sum_{j \neq i} M_{ij} \tag{3.19}$$

returns the number of false-negative samples for the ith class and

$$FP_i = \sum_{j \neq i} M_{ji} \tag{3.20}$$

returns the number of false-positive samples for the ith class. In addition, the accuracy is given by

$$\frac{\sum_i M_{ii}}{\sum_i \sum_j M_{ji}}. \tag{3.21}$$

Studying the confusion matrix tells us how good is our model in practice. Using this matrix, we can see which classes causes trouble in classification. For example if M_{33} is equal to 100 and M_{35} is equal to 80 and all other elements in the same row are zero, this shows that the classifier makes mistake by classifying samples belonging to class 3 as class 5 and it does not make any mistake with other classes in the same row. A similar analysis can be done on columns of a confusion matrix.

In general, confusion matrix is a very powerful tool for assessing a classifier. But, it might be tedious or even impractical to analyze a confusion matrix on a 250-class classification problem. Making sense of a large confusion matrix is a hard task and sometimes nearly impossible. For this reason, we usually extract some quantitative measures from confusion matrix which are more reliable and informative compared with the accuracy score.

3.5.3.4 Precision and Recall

Precision and recall are two important quantitative measures for assessing a classifier. Precision computes the fraction of predicted positives and recall computed the fraction of actual positives. To be more specific, precision is given by

$$precision = \frac{TP}{TP + FP} \qquad (3.22)$$

and recall is computed by

$$recall = \frac{TP}{TP + FN}. \qquad (3.23)$$

Obviously, FP and FN must be zero in a perfect classifier leading to precision and recall scores equal to 1. If precision and recall are both equal to 1, we can say that the classifier is perfect. If these quantities are close to zero, we can imply that the classifier is very inaccurate. Computing precision and recall on binary matrix is trivial. In the case of multiclass classification problem, precision of the ith class is given by

$$precision_i = \frac{M_{ii}}{M_{ii} + \sum_{j \neq i} M_{ji}} = \frac{M_{ii}}{\sum_j M_{ji}} \qquad (3.24)$$

and the recall of the ith class is given by:

$$recall_i = \frac{M_{ii}}{M_{ii} + \sum_{j \neq i} M_{ij}} = \frac{M_{ii}}{\sum_j M_{ij}}. \qquad (3.25)$$

Considering that there are C classes, the overall precision and recall of a confusion matrix can be computed as follows:

$$precision = \sum_{i=1}^{C} w_i \times precision_i \qquad (3.26)$$

$$recall = \sum_{i=1}^{C} w_i \times recall_i. \qquad (3.27)$$

If we set w_i to 1, the above equations will simply compute average of precisions and recalls in a confusion matrix. However, if w_i is equal to the number of samples in the

i^{th} class divided by total number of samples, the above equation will compute the weighted average of precisions and recall taking into account the imbalanced dataset. Moreover, you may also compute the variance of precisions and recalls beside the weighted mean in order to see how much these values are fluctuating in the function matrix.

3.5.3.5 F1 Score

While precision and recall are very informative and useful for assessing a ConvNet, in practice, we are usually interested in designing and evaluating ConvNets based on a single quantity. One effective way to achieve this goal is to combine the values of precision and recall. This can be simply done by computing the average of precision and recall. However, computing the average of these two quantities might not produce accurate results. Instead, we can compute the harmonic mean of precision and recall as follows:

$$F1 = \frac{2}{\frac{1}{precision} + \frac{1}{recall}} = \frac{2TP}{2TP + FP + FN} \qquad (3.28)$$

This harmonic mean is called *F1-score* which is a number in $[0, 1]$ with F1-score equal to 1 showing a perfect classifier. In the case of multiclass classification problems, F1-score can be simply computed by taking the weighted average of class specific F1-scores (similar method that we used for precision and recall in the previous section).

F1-score is a reliable and informative quantity to evaluate a classifier. In practice, we usually evaluate the implemented ideas (ConvNets) using the F1-score on development set and refine the idea until we get a satisfactory F1-score. Then, a complete analysis can be done on the test set using confusion matrix and its related metrics.

3.6 Training a ConvNet

Training a ConvNet can be done in several ways. In this section, we will explain best practices for training a ConvNet. Assume the training set $\mathscr{X}_{train}$. We will use this set for training the ConvNet. However, Coates and Ng (2012) showed that preprocessing data is helpful for training a good model. In the case of ConvNets applied on images, we usually compute the mean image using the samples in $\mathscr{X}_{train}$ and subtract it from each sample in the whole dataset. Formally, the mean image is obtained by computing

$$\bar{\mathbf{x}} = \frac{1}{N} \sum_{\mathbf{x}_i \in \mathscr{X}_{train}} \mathbf{x}_i. \qquad (3.29)$$

Then, each sample in the training set as well as the development set and the test set is replaced by

$$
\begin{aligned}
\mathbf{x}_i = \mathbf{x}_i - \bar{\mathbf{x}} \quad & \forall x_i \in \mathcal{X}_{trian} \\
\mathbf{x}_i = \mathbf{x}_i - \bar{\mathbf{x}} \quad & \forall x_i \in \mathcal{X}_{dev} \\
\mathbf{x}_i = \mathbf{x}_i - \bar{\mathbf{x}} \quad & \forall x_i \in \mathcal{X}_{test}
\end{aligned}
\tag{3.30}
$$

Note that the mean image is only computed on the training set but it is used to preprocess the development and test sets as well. Subtracting mean is very common and helpful in practice. It translates the whole dataset such that the expected value (mean) of the dataset is located very close to origin in the image space. In the case of neural networks designed by hyperbolic tangent activation functions, subtracting mean from data is crucial since it guarantees that the activation of first layer will be close to zero and gradient of network will be close to one. Hence, the network will be able to learn from data.

To further preprocess the dataset, we can compute the variance of every element of $\mathbf{x}_i$. This can be easily obtained by computing

$$
var(\mathcal{X}_{train}) = \frac{1}{N} \sum_{i=1}^{N} (\mathbf{x}_i - \bar{\mathbf{x}})^2
\tag{3.31}
$$

where N is the total number of samples in the training set. The square and division operations in the above equation are applied in the elemenwise fashion. Assuming that $\mathbf{x}_i \in \mathbb{R}^{H \times W \times 3}$, $var(\mathcal{X}_{train})$ will have the same size as $\mathbf{x}_i$. Then, (3.30) can be written as

$$
\begin{aligned}
\mathbf{x}_i = \frac{\mathbf{x}_i - \bar{\mathbf{x}} \quad \forall x_i \in \mathcal{X}_{trian}}{var(\mathcal{X}_{train})} \\
\mathbf{x}_i = \frac{\mathbf{x}_i - \bar{\mathbf{x}} \quad \forall x_i \in \mathcal{X}_{dev}}{var(\mathcal{X}_{train})} \\
\mathbf{x}_i = \frac{\mathbf{x}_i - \bar{\mathbf{x}} \quad \forall x_i \in \mathcal{X}_{test}}{var(\mathcal{X}_{train})}
\end{aligned}
\tag{3.32}
$$

Beside translating the dataset into origin, the above transformation also changes the variance of each element in the input so it will be equal to 1 for each element. This preprocessing technique is commonly known as *mean-variance normalization*. As before, computing the variance is only done using the data in the training set and it is used for transforming data in the development and test sets as well.

3.6.1 Loss Function

Two commonly used loss functions for training ConvNets are the multiclass version of the logistic loss function and the multiclass hinge loss function. It is also possible to define a loss function which is equal to weighted sum of several loss functions.

However, it is not a common approach and we usually train a ConvNet using only one loss function. These two loss functions are throughly explained in Chap. 2.

3.6.2 Initialization

Training a ConvNet successfully using a gradient-based method without a good initialization is nearly impossible. In general, there are two sets of parameter in a ConvNet including weights and biases. We usually set all the biases to zero. In this section, we will describe a few techniques for initializing weights that have produced promising results in practice.

3.6.2.1 All Zero

The trivial method for initializing weights is to set all of them to zero. Concretely, this will not work since all neurons will produce the same signal during backpropagation and weights will be updated using exactly the same rule. This means that the ConvNet will not be trained properly.

3.6.2.2 Random Initialization

The better idea is to initialize the weights randomly. The random values might be drawn from a Gaussian distribution or a uniform distribution. The idea is to generate small random numbers. To this end, the mean of Gaussian distribution is usually fixed at 0 and its variance is fixed at a value such as 0.001. Alternatively, it is also possible to generate random numbers by a uniform distribution where the minimum and maximum value of the distribution are fixed at numbers close to zero such as ± 0.001. Using this technique, each neuron will produce different output in the forward pass. As the result, the update rule of each neuron will be different from other neurons and the ConvNet will be trained properly.

3.6.2.3 Xavier Initialization

As it is illustrated in Sutskever et al. (2013) and Mishkin and Matas (2015), initialization has a great influence in training a ConvNet. Glorot and Bengio (2010) proposed an initialization technique which has been one of the successful methods of initialization so far. This initialization is widely known as *Xavier initialization*.[9]

As we saw in Chap. 2, the output of a neuron in a neural network is given by

$$z = w_1 x_1 + \cdots + w_d x_d \tag{3.33}$$

---

[9]Xavier is the first name of the first author.

where $x_i \in \mathbb{R}$ and $w_i \in \mathbb{R}$ are i^{th} input and its corresponding weight. If we compute the variance of z, we will obtain

$$Var(z) = Var(w_1 x_1 + \cdots + w_d x_d). \tag{3.34}$$

Taking into account the properties of variance, the above equation can be decomposed to

$$Var(z) = \sum_{i=1}^{d} Var(w_i x_i) + \sum_{i \neq j} Cov(w_i x_i, w_j x_j). \tag{3.35}$$

In the above equation, $Cov(.)$ denotes the covariance of inputs. Using the properties of variance, the first term in this equation can be decomposed to

$$var(w_i x_i) = E[w_i]^2 Var(x_i) + E[x_i]^2 Var(y_i) + Var(w_i) Var(x_i) \tag{3.36}$$

where $E[.]$ denotes the expected value of random variable. Assuming that the mean-variance normalization have been applied on the dataset, the second term will be equal to zero in the above equation since $E[x_i] = 0$. Consequently, it will be reduced to

$$var(w_i x_i) = E[w_i]^2 Var(x_i) + Var(w_i) Var(x_i). \tag{3.37}$$

Suppose we want the expected value of weight to be equal to zero. In that case, the above equation will be reduced to

$$var(w_i x_i) = Var(w_i) Var(x_i) \tag{3.38}$$

By plugging the above equation in (3.35), we will obtain

$$Var(z) = \sum_{i=1}^{d} Var(w_i) Var(x_i) + \sum_{i \neq j} Cov(w_i x_i, w_j x_j). \tag{3.39}$$

Assuming that w_i and x_i are independent and identically distributed, the second term in the above equation will be equal to zero. Also, we can assume that $Var(w_i) = Var(w_j), \forall i, j$. Taking into account these two conditions, the above equation will be simplified to

$$Var(z) = d \times Var(w_i) Var(x_i). \tag{3.40}$$

Since the inputs have been normalized using the mean-variance normalization, $Var(x_i)$ will be equal to 1. Then

$$Var(w_i) = \frac{1}{d} \tag{3.41}$$

where d is the number of inputs to the current layer. The above equation tells us that the weights of current layer can be initialized using the Gaussian distribution with mean equal to zero and variance equal to $\frac{1}{d}$. This technique is the default initialization technique in the Caffe library. Glorot and Bengio (2010) carried out a similar analysis on the backpropagation step and concluded that the current layer can be initialized by setting the variance of Gaussian distribution to

$$Var(w_i) = \frac{1}{n_{out}} \tag{3.42}$$

n_{out} is the number of outputs of the layer. Later, He et al. (2015) showed that for a ConvNet with ReLU layers, the variance can be set to

$$Var(w_i) = \frac{2}{n_{in} + n_{out}} \tag{3.43}$$

where $n_{in} = d$ is the number of inputs to the layer. Despite many simplifying assumptions, all three techniques for determining the value of variance works very well with ReLU activations.

3.6.3   Regularization

So far in this book, we explained how to design, train, and evaluate a ConvNet. In this section, we bring up another topic which has to be considered in training a ConvNet. Assume the binary dataset illustrated in Fig. 3.17. The blue solid circles and the red dashed circles show training data of two classes. Also, the dash-dotted red circle is the test data.

This figure shows how the space might be divided into two regions if we fit a linear classifier on the training data. It turns out that the small solid blue circle has been ignored during because any line that classifies this circle correctly will have

Fig. 3.17 A linear model is highly biased toward data meaning that it is not able to model nonlinearities in the data

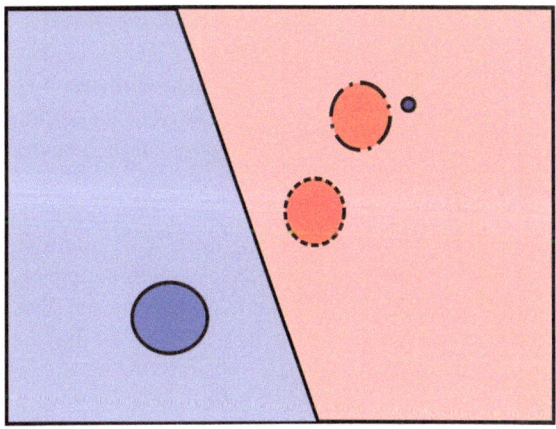

Fig. 3.18 A nonlinear model
is less biased but it may
model any small nonlinearity
in data

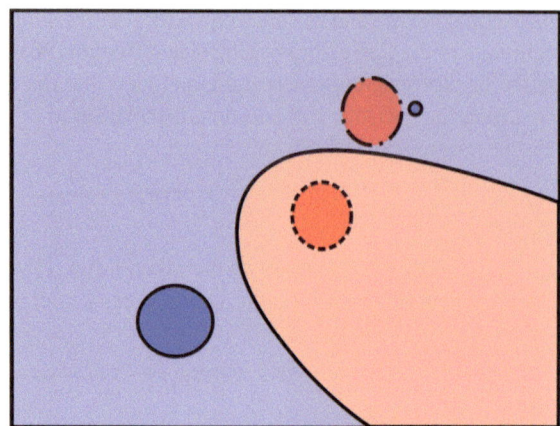

Fig. 3.19 A nonlinear model
may still overfit on a training
set with many samples

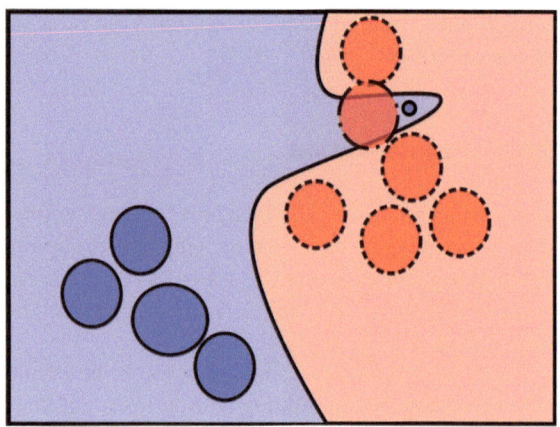

a higher loss compared with the line in this figure. If the linear model is evaluated
using the test data it will perfectly classify all its samples.

However, suppose that we have created a feedforward neural network with one
hidden layer in order to make this dataset linearly separable. Then, a linear classifier
is fitted on the transformed data. As we saw earlier, this is equivalent to a nonlinear decision boundary in the original space. The decision boundary, may look like
Fig. 3.18.

Here, we see that model is able to perfectly distinguish the training samples.
However, if it is assessed using the test set, none of the samples in this set will be
classified correctly. Technically, we say the model is *overfitted* on the training data
and it has not been *generalized* on the test set. The obvious cure for this problem
seems to be gathering more data. Figure 3.19 illustrates a scenario where the size of
training set is large.

Clearly, the system works better but it still classifies most of test samples incorrectly. One reason is that the feedforward neural network may have many neurons

in the hidden layer and, hence, it is a highly nonlinear function. For this reason, it is able to model even small nonlinearities in the dataset. In contrast, a linear classifier trained on the original data is not able to model nonlinearities in the data.

In general, if a model highly nonlinear and it is able to learn any small nonlinearities in the data, we say that the model has high *variance*. In contrast, if a model is not able to learn nonlinearities in a data we say it has a high *bias*. A model with high variance is prone to overfit on data which can adversely reduce the accuracy on test set. In contrary, a highly biased model is not able to deal with nonlinear datasets. Therefore, it is not able to accurately learn from data.

The important point in designing and training is to find a trade-off between model bias and model variance. But, what does cause a neural network to have high variance/bias? This mainly depends on two factors. These two factors are the number of neurons/layer in a neural network and the magnitude of weights. Concretely, a neural network with many neurons/layers is capable of modeling very complex functions. As the number of neurons increases its ability to model highly nonlinear function increases as well. In opposite, by reducing the number of neurons, the ability of a neural network for modeling highly nonlinear functions decreases.

A highly nonlinear function has different characteristics. One of them is that a highly nonlinear function is differentiable several times. In the case of neural networks with sigmoid activation, the neural networks are infinitely differentiable. A neural network with ReLU activations could be also differentiable several times. Assume a neural network with sigmoid activations. If we compute the derivative of output with respect to its input, it will depend on the values of weights. Since the neural network is differentiable several times (infinitely in this case), the derivative is also a nonlinear function.

It turns out that the derivative of the function for a given input will be higher if the weights are also higher. As the magnitude of weights increases, the neural network become more capable to model sudden variations. For example, Fig. 3.20 shows the two decision boundaries generated by a feedforward neural network with four hidden layers.

The neural network is initialized with random numbers between -1 and 1. The decision boundary associated with these values is shown in the left. The decision boundary in the right plot is obtained using the same neural network. We have only multiplied the weights of the third layer with 10 in the right plot. As we can see, the decision boundary in the left plot is smooth but the decision boundary in the right plot is spiky with sharp changes. These sharp changes sometimes causes a neural network to overfit on the training set.

For this reason, we have to keep the magnitude of weights close to zero in order to control the variance of our model. This is technically called *regularization* and it is an important step in training a neural network. There are different ways to regularize a neural network. In this section, we will only explain the methods that are already implemented in the Caffe library.

3.6.3.1 L_2 Regularization

Let us denote the weights of all layers in a neural network using $\mathbf{W}$. A simple but effective way for regularizing a neural network is to compute L_2 norm of the weights and add it to the loss function. This regularization technique is called L_2 regularization. Formally, instead of minimizing $\mathscr{L}(\mathbf{x})$, we define the loss function as

$$\mathscr{L}_{l2}(\mathbf{x}) = \mathscr{L}(\mathbf{x}) + \lambda \|W\|^2 \qquad (3.44)$$

where $\|W\|^2$ is the L_2 norm of the weights and λ is a user-defined value showing that how much the regularization term can penalize the loss function. The regularization term will be minimized when all the weights are zero. Consequently, the second term encourages the weights to have small values. If λ is high, the weights will be very close to zero which means we reduce the variance of our model and increase its bias. In contrast, if λ is small, we let the weights to take higher values. Therefore, the variance of model increases.

A nice property of L_2 regularization is that it does not produce spiky weights where a few of weights might be much higher than other weights. Instead, it distributes the weights evenly so the weight vector is smooth.

3.6.3.2 L_1 Regularization

Instead of L_2 norm, L_1 regularization penalizes the loss function using L_1 norm of weights vectors. Formally, the penalized loss function is given by

$$\mathscr{L}_{l1}(\mathbf{x}) = \mathscr{L}(\mathbf{x}) + \lambda |W| \qquad (3.45)$$

where $|W|$ is the L_1 norm of the weights and λ is a user-defined value and has the same effect as in L_2 regularization. In contrast to L_2 regularization, L_1 regularization

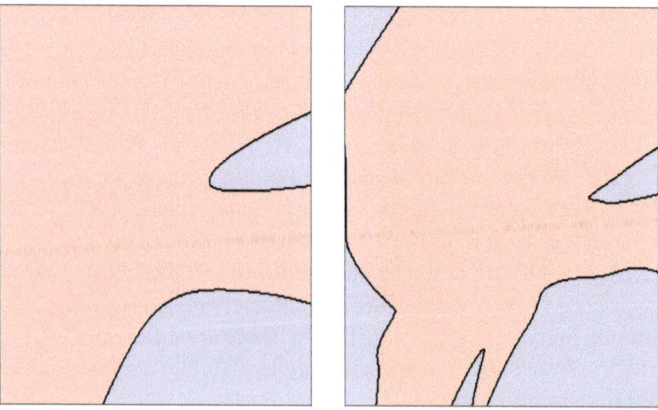

Fig. 3.20 A neural network with greater weights is capable of modeling sudden changes in the output. The right decision boundary is obtained by multiplying the third layer of the neural network in left with 10

can produce sparse weight vectors in which some of the weights are very close to or exactly zero. However, this property is not guaranteed if we optimize the L_1 regularized loss function using the gradient descend algorithm.

From another perspective, L_1 regularization select features that are useful for the classification task in hand. This is done by making weights of irrelevant features. However, if there is no need to do a feature selection, L_2 regularization is preferred over L_1 regularization. It is also possible to combine L_2 and L_1 regularizations and obtain

$$\mathscr{L}_{l1l2}(\mathbf{x}) = \mathscr{L}(\mathbf{x}) + \lambda_1 |W| + \lambda_2 \|W\|^2. \tag{3.46}$$

The above regularization is called *elastic net*. But, training a ConvNet using the above combined regularization is not common and in practice we mainly use L_2 regularization.

3.6.3.3   Max-Norm Regularization

The previous two regularization methods are applied by adding a penalizing term to the loss function. *Max-norm regularization* does not penalize the loss function. Instead, it always keep $\|W\|$ within a ball of radius c. Formally, after computing the gradient and applying the update rule on the weights, we compute $\|W\|$ (L_2 norm of weights) and if they exceed the user-defined threshold c, the weights are projected to the surface of the ball with radius c using

$$W = \frac{W}{\|W\|} \times c \tag{3.47}$$

One interesting property of the max-norm regularization is that it prevents the neural network to *explode*. In other words, we previously see that gradient may vanish in deep networks during backpropagation in which case the deep network does learn properly. This phenomena is called *gradient vanishing problem*. In contrast, gradients might be greater than one in a deep neural network. In that case, gradient become higher as backpropagation moves to the first layers. In this case, the weights suddenly explodes and become very large. This phenomena is called *exploding gradient problem*. In addition, if learning rate in the gradient descend algorithm is set to a high value, the network may explode. However, applying max-norm regularization on the weight prevents the network to explode since it always keep the norm of weights below the threshold c.

3.6.3.4   Dropout

Dropout (Hinton 2014) is another technique for regularizing a neural network and preventing it from overfitting. For each neuron in the network, it generates a number between 0 and 1 using the uniform distribution. If the probability of a neuron is less than p, the neuron will be dropped out from the network along with all its connections. Then, the forward and backward passes will be computed on the new network. This

Fig. 3.21 If dropout is activated on a layer, each neuron in the layer will be attached to a blocker. The blocker blocks information flow in the forward pass as well as the backward pass (i.e., backpropagation) with probability p

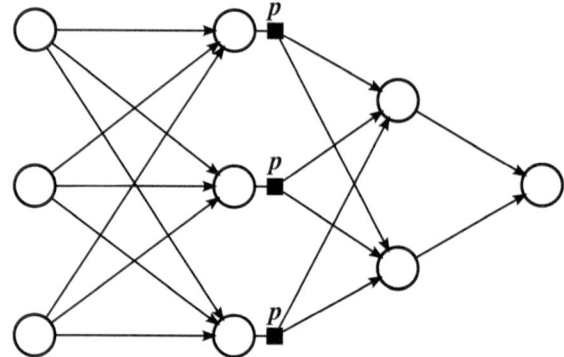

process of dropping some neurons from the original network and computing the forward and backward pass on the new network is repeated for every sample in the training set.

In other words, for each sample in the training set, a subset of neuron from the original network is selected to form a new network and the forward pass is computed using this smaller network. Likewise, the backward pass is computed on the small network and the weights of the smaller network are updated. Then, the weights are copied to the original network. The above procedure seems complicated. But, it can be efficiently implemented. This is illustrated in Fig. 3.21.

First, we can define a new layer called dropout. This layer can be connected to any other layers such as convolution or fully connected layers. The number of elements in this layer is equal to the number of outputs in the previous layer. There is only one parameter in this layer which is called *dropout ratio* which is denoted by p and it is defined by user during designing the network. This layers has been shown using black squares in this Figure. For each element in this layer a random number between 0 and 1 is generated using the uniform distribution. If the generated number for the ith is greater than p, it will pass the output of the neuron from previous layer to the next layer. Otherwise, it will block the output of the previous neuron and send 0 to the next layer. Since the blocker is activated for the ith neuron during the forward pass, it will also block the signals coming to this element during backpropagation and will pass (backward pass) 0 to the neuron in the previous layer. This way, the gradient of the neuron in the previous layer will be equal to zero. Consequently, the ith will have no effect on the forward or backward pass which is similar to dropping out this neuron from the network.

In the test time, if we execute the forward pass several times we are likely to get different outputs from the network. This is due to the fact that the dropout layer blocks the signal going out from some of the neuron in the network. To get a stable output, we can execute the forward pass many times on the same test sample and compute the average of the outputs. For instance, we can run the forward pass 1000 times. This way, we will get 1000 outputs for the same test sample. Then, we can simply compute the average of 1000 outputs and obtain the final output.

However, this method is not practical since obtaining a result for one samples requires running the network many times. The efficient way is to run the forward pass only one time but scale the output of dropout gates in the test time. To be more specific, the dropout gates (black squares in the figure) act as scalers rather than blockers. They simply get the output of the neuron and pass it to the next layer after rescaling by factor β. Determining value of β is simple.

Assume a single neuron attached to a dropout blocker. Assume that the output of the neuron for the given input $\mathbf{x}_i$ is z. Since there is no randomness in a neuron, it will always return z for the input $\mathbf{x}_i$. However, when it passes through a dropout blocker, it will be blocked with probability p. In other words, if we perform the forward pass N times, we expect that $(1 - p) \times N$ times z is passed by the blocker and $p \times N$ times it is blocked (0 passed through the blocker). The average value of the dropout gate will be equal to $\frac{(1-p) \times N \times z + p \times N \times 0}{N} = (1 - p) \times z$.

Consequently, instead of running the network many times in the test time, we can simply set $\beta = 1 - p$ and rescale the output neuron connected to the dropout layer by this factor. In this case, dropout gates will act as scalers instead of blockers.[10]

Dropout is an effective way of regularizing neural networks including ConvNets. Commonly, dropout layers are placed after fully connected layers in a ConvNet. However, it is not a golden rule. One can attach a dropout layer to the input in order to generate noisy inputs! Also, the dropout ratio p is usually set to 0.5 but there is no theoretical proof to tell what should be the value of dropout ratio. We can start from $p = 0.5$ and adjust it using the development set.

3.6.3.5 Mixed Regularization

We can incorporate several methods for regularizing a ConvNet. For example, using both L_2 regularization and dropout are common. But, you can combine all the regularizations methods we explained in this section and train your network.

3.6.4 Learning Rate Annealing

Stochastic gradient descent have a user-defined parameter called *learning rate* which we denote it by α. The training usually starts with an initial value for the learning rate such as $\alpha = 0.001$. The learning rate can be kept constant all the time during the training. Ideally, if the initial value of the learning rate is chosen properly, we expect that the loss function is decreased at each iteration.

In other words, the algorithm gets closer to a local minimum at each iteration. Depending on the shape of loss function in high-dimensional space, the optimization

---

[10]For interested readers: More efficient way of implementing dropout is to rescale the signals by factor $\frac{1}{1-p}$ if they pass through dropout gates during the training. Then, in the test time, we can simply remove the dropout layer from the network and compute the forward pass as we do in a network without dropout layers. This technique is incorporated by the Caffe library.

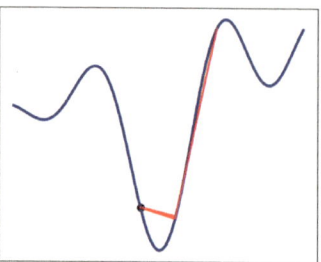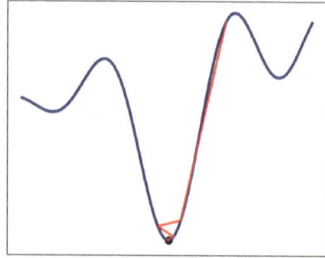

Fig. 3.22 If the learning rate is kept fixed it may jump over local minimum (*left*). But, annealing the learning rate helps the optimization algorithm to converge to a local minimum

algorithm may fluctuate near local minimum and it may not converge to the local minimum. One possible cause of fluctuations could be the learning rate. The reason is that a gradient based method moves toward local minimum based on the gradient of loss function. When the learning rate is kept constant and the current location is close to a local minimum, the algorithm may jump over the local minimum after multiplying the gradient with the learning rate and updating the location based on this value.

This problem is illustrated in Fig. 3.22. In the left, the learning rate is kept constant. We see that the algorithm jumps over the local minimum and it may or may not converge to the local minimum in finite iterations. In contrary, in the right plot, the learning rate is reduced linearly at each iteration. We see that the algorithm is able to converge to the local minimum in finite iterations.

In general, it is a good practice to reduce the learning rate over time. This can be done in different ways. Denoting the initial learning rate by $\alpha_{initial}$, the learning rate at iteration t can be obtained by:

$$\alpha_t = \alpha_{initial} \times \gamma^t \tag{3.48}$$

Fig. 3.23 Exponential learning rate annealing

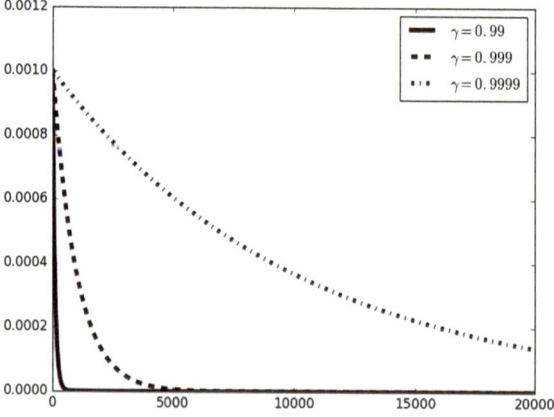

Fig. 3.24 Inverse learning rate annealing

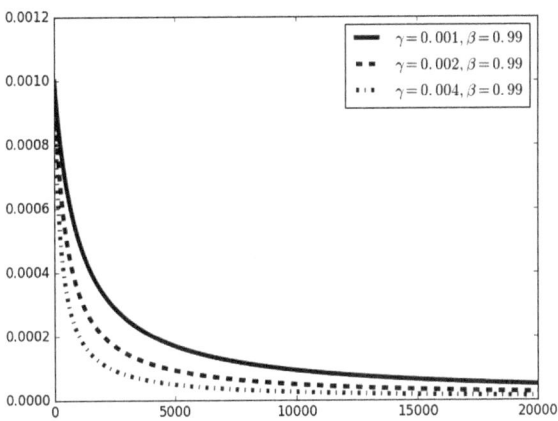

where $\gamma \in [0, 1]$ is a user-defined value. Figure 3.23 shows the plot of this function for different values of γ with $\alpha_{initial} = 0.001$. If the value of γ is close to zero, the learning rate will approach zero quickly. The value of gamma is chosen based on maximum number of iterations of the optimization algorithm. For example, if the maximum number of iterations is equal 20,000, γ may take a value smaller than but close to 0.9999. In general, we have to adjust γ such that the learning rate becomes smaller in the last iterations. If the maximum number of iterations is equal to 20,000 and we set γ to 0.99, it is likely that the ConvNet will not learn because the learning rate has become almost zero after 1000 iterations. This learning annealing method is known as *exponential annealing*.

Learning rate can be also reduced using

$$\alpha_t = \alpha_{initial} \times (1 + \gamma \times t)^{-\beta} \tag{3.49}$$

where γ and β are user-defined parameters. Figure 3.24 illustrates the plot of this function for different values of γ and $\beta = 0.99$. This annealing method is known as *inverse annealing*. Similar to the exponential annealing, the parameters of the inverse annealing method should be chosen such that the learning rate becomes smaller when it reaches to the maximum number of iterations.

The last annealing method which is commonly used in training neural networks is called *step annealing* and it is given by

$$\alpha_t = \alpha_{initial} \times \gamma^{t \, d} \tag{3.50}$$

In the above equation, denotes the integer division operator and $\gamma \in [0, 1]$ and $d \in \mathscr{Z}^+$ are user-defined parameters. The intuition behind this method is that instead of constantly reducing the learning rate, we can multiply the learning rate with γ every d iterations. Figure 3.25 shows the plot of this function for different values of γ and $d = 5000$.

Fig. 3.25 Step learning rate annealing

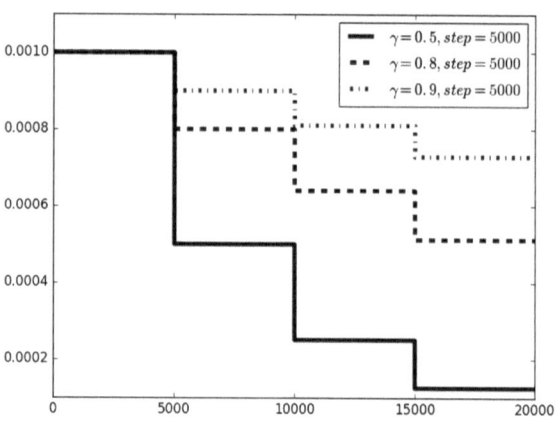

In contrast to other two methods, adjusting parameters of the step annealing is straightforward. The step parameters d is usually equal to the number of training samples or fraction/multiple of this number. For example, if there are 10,000 samples in the training set, we may consider setting d to 5,000 meaning that the learning rate will be reduced every 5,000 samples. The amount of reduction can be chosen based on the maximum number of iterations and step size d. Also, in the case of mini-batch gradient descend with batch size 50, setting d to 100 will exactly reduce the learning rate every 5,000 samples.

3.7 Analyzing Quantitative Results

Throughout this chapter, we discussed about designing, implementing, and evaluating ConvNets. So far, we see that the dataset is divided into three disjoint sets namely training, development, and test. Then, the idea is implemented, trained, and evaluated. We also explained that assessing the idea is done using a single quantitative number computed by a metric such as accuracy or F1-score. Based on the results on these three sets of data, we decide whether the model must be refined or it is satisfactory.

Table 3.1 Four different scenarios that may happen in practice

	Scenario 1 (%)	Scenario 2 (%)	Scenario 3 (%)	Scenario 4 (%)
Goal	99	99	99	99
Train	80	98	98	98
Development		80	97	97
Test			80	97

Typically, we may encounter four scenarios illustrated in Table 3.1. Assume that our evaluation metric is accuracy. We are given a dataset and we have already split the dataset into training, development, and test sets. The goal is to design a ConvNet with 99% accuracy. Assume we have designed a ConvNet and trained it. Then, the accuracy of the ConvNet on the training set is 80% (Scenario 1). Without even assessing the accuracy on the development and test sets, we conclude that this idea is not good for our purpose. The possible actions in this scenario are

- Train the model longer
- Make the model bigger (*e.g.,* increasing number of filters in each layer, increasing number of neuron in fully connected layers)
- Design a new architecture
- Make the regularization coefficient λ smaller (closer to zero)
- Increase the threshold of norm in the max-norm constraint
- Reduce the dropout ratio
- Check the learning rate and learning rate annealing
- Plot the value of loss function in all the iterations to see if the loss is decreasing or it is fluctuating or it is constant.

In other words, if we are sure that the ConvNet is trained for enough number of iterations and the learning rate is correct we can conclude that the current ConvNet is not flexible enough to capture the nonlinearity of data. This means that the current model has a high bias. So, we have to increase the flexibility of our model. Paying attention to the above solutions, we realize that most of them try to increase the flexibility of the model.

We may apply the above solutions and increase the accuracy on the training set to 98%. However, when the model is evaluated on the development set, the accuracy is 80% (Scenario 2). This is mainly a high variance problem meaning that our model might be very flexible and it captures every detail in the training set. In other words, it overfits on the training set. Possible actions in this scenario are

- Make the regularization coefficient λ bigger
- Reduce the threshold in max-norm constraint
- Increase the dropout ratio
- Collect more data
- Synthesize new data on the training set (we will discuss this method in the next chapters)
- Change the model architecture.

If we decide to change the model architecture, we have to keep this in mind that the new architecture must be less flexible (*e.g.,* shallower, fewer neurons/filters) since our current model is very flexible and it overfits on the training set. After applying these changes, we may find a model with 98 and 97% accuracies on the training set and development set, respectively. But, after evaluating the model on the test set we realize that its accuracy is 80% (Scenario 3).

At this point, one may consider changing the model architecture or tweaking the model parameters in order to increase the accuracy on the test set as well. But, this approach is wrong and the model trained this way may not work in real world. The reason is that, we have tried to adjust our models on both development set and test set. However, the main problem in Scenario 3 is that our model is overfit on the development set. If we try to adjust it on the test set, we cannot be sure that the high accuracy on the test set is because the model is generalized well or it is because the model is overfit on the test set. So, the best solution in this case is to collect more development data. By collecting data we mean new and fresh data.

The Scenario 4 is what we usually expect to achieve in practice. In this scenario, we have adjusted our model on the development set but it also produces good results on the test set. In this case, we can be confident that our model is ready to be used in real world.

There are other serious issues about data such as what happens if the distribution of test set is different from training and development set. Solutions for addressing this problem are not within the scope of this book. Interested reader can refer to textbooks about data science for more details.

3.8 Other Types of Layers

ConvNets that we will design for detecting and classifying traffic signs are composed of convolution, pooling, activation, and fully connected layers. However, there are other types of layers that have been proposed recently and there are several works utilizing these kind of layers. In this section, we will explain some of these layers.

3.8.1 Local Response Normalization

Local response normalization (LRN) (Krizhevsky et al. 2012) is a layer which is usually placed immediately after the activation of a convolution layer. In the reminder of this section, when we say a convolution layer we refer to the activation of the convolution layer. Considering that feature maps of a convolution layer has N channels of size $H \times W$, the LRN layer will produce a new N channel feature maps of size $H \times W$ (exactly the same size as feature maps of the convolution layer) where the element $h^i_{m,n}$ at location (m, n) in the ith channel is given by

$$b^i_{m,n} = \frac{a^i_{m,n}}{\left(k + \alpha \sum_{j=max(0,i-n/2)}^{min(N-1,i+n/2)} \left(a^j_{m,n}\right)^2\right)^\beta} \tag{3.51}$$

In the above equation, $a^i_{m,n}$ denotes the value of the feature map of the convolution layer at spatial location (m, n) from i^{th} channel. Also, k, n, α and β are user defined parameters. Their default value are $k = 2$, $n = 5$, $\alpha = 10^{-4}$ and $\beta = 0.75$. The LRN layer normalizes the activations in the same spatial location using neighbor channels. This layer does not have a trainable parameter.

3.8.2 Spatial Pyramid Pooling

Spatial pyramid pooling (He et al. 2014) is proposed to generate fixed-length feature vectors for input images with arbitrary size. A spatial pyramid pooling layer is placed just before the first fully connected layer. Instead of pooling a feature map with a fixed size, it divides the feature maps into fixed number of regions and pool all elements inside each region. Also, as it is illustrated in the figure, it does this in several scales. In the first scale, it pools over whole feature map. In the second scale, it divides each feature map into four regions. In the third scale, it divides the feature map into 16 regions. Then, it concatenates all these vectors and connects it to the fully connected layer.

3.8.3 Mixed Pooling

Basically, we put one pooling layer after a convolution layer. Lee et al. (2016) proposed an approach which is called *mixed* pooling. The idea behind mixed pooling is to combine max-pooling and average pooling. It turns out that mixed pooling combines the output of a max pooling and average pooling as follows:

$$pool_{mix} = \alpha pool_{max} + (1 - \alpha)pool_{avg} \tag{3.52}$$

In the above equation, $\alpha \in [0, 1]$ is a trainable parameter which can be trained using the standard backpropagation algorithm.

3.8.4 Batch Normalization

Distribution of each layer in a ConvNet changes during training and it varies from one layer to another. This reduces the convergence speed of the optimization algorithm. Batch normalization (Ioffe and Szegedy 2015) is a technique to overcome this problem. Denoting the input of a batch normalization layer with x and its output using z, batch normalization applies the following transformation on x:

$$z = \frac{x - \mu}{\sqrt{\sigma^2 + \epsilon}}\gamma + \beta. \tag{3.53}$$

Basically, it applies the mean-variance normalization on the input x using μ and σ and linearly scales and shifts it using γ and β. The normalization parameters μ and σ are computed for the current layer over the training set using a method called exponential moving average. In other words, they are not trainable parameters. In contrast, γ and β are trainable parameters.

In the test time, the μ and σ that are computed over the training set are used for doing the forward pass and they remain unchanged. The batch normalization layer is usually placed between the fully connected/convolution layer and its activation function.

3.9 Summary

Understanding the underlying process in a convolutional neural networks is crucial for developing reliable architectures. In this chapter, we explained how convolution operations are derived from fully connected layers. For this purpose, weight sharing mechanism of convolutional neural networks was discussed. Next basic building block in convolutional neural network is pooling layer. We saw that pooling layers are intelligent ways to reduce dimensionality of feature maps. To this end, a max pooling, average pooling, or a mixed pooling is applied on feature maps with a stride bigger than one.

In order to explain how to design a neural network, two classical network architectures were illustrated and explained. Then, we formulated the problem of designing network in three stages namely idea, implementation, and evaluation. All these stages were discussed in detail. Specifically, we reviewed some of the libraries that are commonly used for training deep networks. In addition, common metrics (i.e., classification accuracy, confusion matrix, precision, recall, and F1 score) for evaluating classification models were mentioned together with their advantages and disadvantages.

Two important steps in training a neural network successfully are initializing its weights and regularizing the network. Three commonly used methods for initializing weights were introduced. Among them, Xavier initialization and its successors were discussed thoroughly. Moreover, regularization techniques such as L_1, L_2, max-norm, and dropout were discussed. Finally, we finished this chapter by explaining more advanced layers that are used in designing neural networks.

3.10 Exercises

3.1 How can we compute the gradient of convolution layer when the convolution stride is greater than 1?

3.2 Compute the gradient of max pooling with overlapping regions.

3.3 How much memory is required by LeNet-5 for feed forward an image and keep the information of all layers?

3.4 Show that the number of parameters of AlexNet is equal to 60,965,224.

3.5 Assume that there are 500 different classes in $\mathscr{X}'$ and there are 100 samples in each class yielding 50,000 samples in $\mathscr{X}'$. In which situations the accuracy score is a reliable metric for assessing the model? In which situations the accuracy score might be very close to 1 but it model might not be practically accurate?

3.6 Consider the trivial example where precision is equal to 0 and recall is equal to 1. Show that why computing harmonic mean is preferable over simple averaging.

3.7 Plot the logistic loss function and L_2 regularized logistic loss function with different values for λ and compare the results. Repeat the procedure using L_1 regularization and elastic nets.

References

Aghdam HH, Heravi EJ, Puig D (2016) Computer vision ECCV 2016 workshops, vol 9913, pp 178–191. doi:10.1007/978-3-319-46604-0

Coates A, Ng AY (2012) Learning feature representations with K-means. Lecture notes in computer science (lecture notes in artificial intelligence and lecture notes in bioinformatics), vol 7700. LECTU, pp 561–580. doi:10.1007/978-3-642-35289-8-30

Deng J, Dong W, Socher R, Li LJ, Li K, Fei-Fei L (2009) ImageNet: a large-scale hierarchical image database. In: IEEE conference on computer vision and pattern recognition, pp 2–9. doi:10.1109/CVPR.2009.5206848

Dong C, Loy CC, He K (2014) Image super-resolution using deep convolutional networks, vol 8828(c), pp 1–14. doi:10.1109/TPAMI.2015.2439281, arXiv:1501.00092

Glorot X, Bengio Y (2010) Understanding the difficulty of training deep feedforward neural networks. In: Proceedings of the 13th international conference on artificial intelligence and statistics (AISTATS), vol 9, pp 249–256. doi:10.1.1.207.2059. http://machinelearning.wustl.edu/mlpapers/paper_files/AISTATS2010_GlorotB10.pdf

He K, Zhang X, Ren S, Sun J (2014) Spatial pyramid pooling in deep convolutional networks for visual recognition, cs.CV, pp 346–361. doi:10.1109/TPAMI.2015.2389824, arXiv:abs/1406.4729

He K, Zhang X, Ren S, Sun J (2015) Delving deep into rectifiers: surpassing human-level performance on ImageNet classification. arXiv:1502.01852

Hinton G (2014) Dropout: a simple way to prevent neural networks from overfitting. J Mach Learn Res (JMLR) 15:1929–1958

Ioffe S, Szegedy C (2015) Batch normalization: accelerating deep network training by reducing internal covariate shift. In: Proceedings of the 32nd international conference on machine learning (ICML), Lille, pp 448–456. doi:10.1007/s13398-014-0173-7.2, http://www.JMLR.org

Krizhevsky A, Sutskever I, Hinton G (2012) Imagenet classification with deep convolutional neural networks. In: Advances in neural information processing systems. Curran Associates, Inc., pp 1097–1105

LeCun Y, Bottou L, Bengio Y, Haffner P (1998) Gradient-based learning applied to document recognition. Proc IEEE 86(11):2278–2323. doi:10.1109/5.726791, arXiv:1102.0183

Lee CY, Gallagher PW, Tu Z (2016) Generalizing pooling functions in convolutional neural networks: mixed, gated, and tree. Aistats 51. arXiv:1509.08985

Mishkin D, Matas J (2015) All you need is a good init. In: ICLR, pp 1–8. doi:10.1016/0898-1221(96)87329-9, arXiv:1511.06422

Scherer D, Müller A, Behnke S (2010) Evaluation of pooling operations in convolutional architectures for object recognition. In: International conference on artificial neural networks, vol 6354. LNCS, pp 92–101

Schmidhuber J (2015) Deep Learning in neural networks: an overview. Neural Networks 61:85–117. doi:10.1016/j.neunet.2014.09.003, arXiv:1404.7828

Springenberg JT, Dosovitskiy A, Brox T, Riedmiller M (2015) Striving for simplicity: the all convolutional net. In: ICLR-2015 workshop track, pp 1–14. arXiv:1412.6806

Sutskever I, Martens J, Dahl G, Hinton G (2013) On the importance of initialization and momentum in deep learning. JMLR W&CP 28(2010):1139–1147. doi:10.1109/ICASSP.2013.6639346

Zeiler MD, Fergus R (2013) Stochastic pooling for regularization of deep convolutional neural networks. In: ICLR, pp 1–9. arXiv:1301.3557

Caffe Library

4

4.1 Introduction

Implementing ConvNets from scratch is a tedious task. Especially, implementing the backpropagation algorithm correctly requires to calculate the gradient of each layer correctly. Even after implementing the backward pass, it has to be validated by computing the gradient numerically and comparing it with the result of backpropagation. This is called *gradient check*. Moreover, efficient implementation of each layer on GPU is another hard work. For these reasons, it might be more practical to use a library for this purpose.

As we discussed in the previous chapter, there are many libraries and frameworks that can be used for training ConvNets. Among them, there is one library which is suitable for development as well as applied research. This library is called *Caffe*.[1] Figure 4.1 illustrates the structure of Caffe.

The Caffe library is developed in C++ and it utilizes CUDA library for performing computations on GPU.[2] There is a library which is developed by NVIDIA and it is called *cuDNN*. It has implemented common layers found in ConvNets as well as their gradients. Using cuDNN, it is possible to design and train ConvNets which are only executed on GPUs. Caffe makes use of cuDNN for implementing some of layers on GPU. It has also implemented some other layers directly using CUDA. Finally, besides providing interfaces for Python and MATLAB programming languages, it also provides a command tool that can be used for training and testing ConvNets.

One beauty of Caffe is that designing and training a network can be done by employing text files which are later parsed using Protocol Buffers library. But, you are not limited to design and train using only text files. It is possible to also design and

---

[1] http://caffe.berkeleyvision.org.
[2] There are some branches of Caffe that use OpenCL for communicating with GPU.

© Springer International Publishing AG 2017   131
H. Habibi Aghdam and E. Jahani Heravi, *Guide to Convolutional Neural Networks*, DOI 10.1007/978-3-319-57550-6_4

Fig. 4.1 The Caffe library
uses different third-party
libraries and it provides
interfaces for C++, Python,
and MATLAB programming
languages

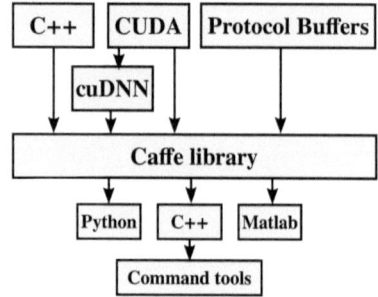

train ConvNets by writing a computer program in C++, Python or MATLAB. However, a detailed analysis of ConvNets has to be done by writing compute programs or special softwares.

In this chapter, we will first explain how to use text files and the command tools for designing and training ConvNets. Then, we will explain how to do it in Python. Finally, methods for analyzing ConvNets using Python will be also discussed.

4.2 Installing Caffe

Installation of Caffe requires installing CUDA and some third-party libraries on your system. The list of required libraries can be found in *caffe.berkeleyvision.org*. If you are using Ubuntu, Synaptic Package Manager can be utilized for installing these libraries. Next, CUDA drivers must be installed on the system. Try to download the latest CUDA driver compatible with Caffe from NIVIDIA website. Installing CUDA drivers can be as simple as just running the installation file. In worst case scenario, it may take some time to figure out what are the error messages and to finally install it successfully.

After that, cuDNN library must be downloaded and copied into the CUDA folder which is by default located in */usr/local/cuda..* You must copy the *cudnn*.h* into the include folder and *libcudnn*.so** into *lib/lib64* folder. Finally, you must follow the instructions provided in the Caffe's website for installing this library.

4.3 Designing Using Text Files

A ConvNet and its training procedure can be defined using two text files. The first text file defines architecture of the neural network including ConvNets and the second file defines the optimization algorithm as well as its parameters. These text files are usually stored with *.prototxt* extension. This extension shows that the text inside these files follows the syntax defined by the Protocol Buffers (protobuf) protocol.[3]

A protobuf is composed of *messages* where each message can be interpreted as a struct in a programming language such as C++. For example, the following protobuf contains two messages namely *Person* and *Group*.

```
message Person {                                                1
    required string name = 1;                                   2
    optional int32 age = 2;                                     3
    repeated string email = 3;                                  4
}                                                               5
                                                                6
message Group {                                                 7
    required string name = 1;                                   8
    repeated Person member = 3;                                 9
}                                                               10
```

Listing 4.1 A protobuf with two messages.

The field rule *required* shows that specifying this field in the text file is mandatory. In contrast, the rule *optional* shows that specifying this field in the text file is optional. As it turns out, the rule *repeated* states that this filed can be repeated zero or more times in the text file. Finally, numbers after the equal signs are unique tag numbers which are assigned to each field in a message. The number has to be unique inside the message.

From programming perspective, these two messages depict two data structures namely *Person* and *Group*. The Person struct is defined using three fields including one required, one optional and one repeated (array) field. The Group struct also is defined using one required filed and one repeated filed, where each element in this field is an instance of Person.

You can write the above definition in a text editor and save it with *.proto* extension (e.g. sample.proto). Then, you can open the terminal in Ubuntu and execute the following command:

```
protoc −I=SRC_DIR −−python_out=DST_DIR SRC_DIR/sample.proto          1
```

If the command is executed successfully, you should find a file named *sample_pb2.py* in directory *DST_DIR*. Instantiating Group can be done in a programming language. To this end, you should import *sample_pb2.py* to python environment and run the following code:

```
g = sample_pb2.Group()                                          1
g.name ='group 1'                                               2
                                                                3
m = g.member.add()                                              4
m.name = 'Ryan'                                                 5
m.age=20                                                        6
m.email.append('mail1@sample.com')                             7
m.email.append('mail1@sample.com')                             8
                                                                9
m = g.member.add()                                              10
m.name = 'Harold'                                               11
m.age=23                                                        12
```

---

[3]Implementations of the methods in this chapter are available at *github.com/pcnn/*.

Using the above code, we create a group called "group 1" with two members. The age of the first member is 20, his name is "Ryan" and he has two email addresses. Moreover, the name of second member is "Harold". He is 23 years old and he does not have any email.

The appealing property of protobuf is that you can instantiate the Group structure using a plain text file. The following plain text is exactly equivalent to the above Python code:

```
name: "group 1"                                                              1
member {                                                                     2
    name: "member1"                                                          3
    age: 20                                                                  4
    email: "mail1@sample.com"                                                5
    email: "mail1@sample.com"                                                6
}                                                                            7
member {                                                                     8
    name: "member2"                                                          9
    age: 23                                                                 10
}                                                                           11
```

This method has some advantages over instantiating using programming. First, it is independent of programming language. Second, its readability is higher. Third, it can be easily edited. Fourth, it is more compact. However, there might be some cases that instantiating is much faster when we write a computer program rather than a plain text file.

There is a file called *caffe.proto* inside the source code of the Caffe library which defines several protobuf messages.[4] We will use this file for designing a neural network. In fact, *caffe.proto* is the reference file that you must always refer to it when you have a doubt in your text file. Also, it is constantly updated by developers of the library. Hence, it is a good idea to always keep studying the changes in the newer version so you will have a deeper knowledge about what can be implemented using the Caffe library. There is a message in *caffe.proto* called "NetParameter" and it is currently defined as follows[5]:

```
message NetParameter {                                                       1
    optional string name = 1;                                                2
    optional bool force_backward = 5 [default = false];                      3
    optional NetState state = 6;                                             4
    optional bool debug_info = 7 [default = false];                          5
    repeated LayerParameter layer = 100;                                     6
}                                                                            7
```

We have excluded deprecated fields marked in the current version from the above message. The architecture of a neural network is defined using this message. It contains a few fields with basic data types (e.g., string, int32, bool). It has also one field of type NetState and an array (repeated) of LayerParameters. Arguably, one can learn Caffe just by throughly studying NetParameter. The reason is illustrated in Fig. 4.2.

---

[4]All the explanations for the Caffe library in this chapter are valid for the commit number 5a201dd.
[5]This definition may change in next versions.

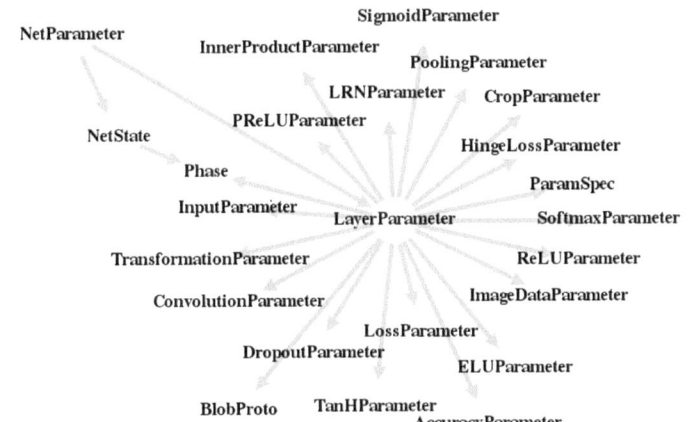

Fig. 4.2 The NetParameter is indirectly connected to many other messages in the Caffe library

It is clear from the figure that NetParameter is indirectly connected to different kinds of layers through LayerParameter. It turns outs that NetParameter is a container to hold layers. Also, there are several other kind of layers in the Caffe library that we have not included in the figure. The message LayerParamter has many fields. Among them, following are the fields that we may need for the purpose of this book:

```
message LayerParameter {                                              1
    optional string name = 1;                                        2
    optional string type = 2;                                        3
    repeated string bottom = 3;                                      4
    repeated string top = 4;                                         5
                                                                     6
    optional ImageDataParameter image_data_param = 115;             7
    optional TransformationParameter transform_param = 100;         8
                                                                     9
    optional AccuracyParameter accuracy_param = 102;                10
    optional ConvolutionParameter convolution_param = 106;          11
    optional CropParameter crop_param = 144;                        12
    optional DropoutParameter dropout_param = 108;                  13
    optional ELUParameter elu_param = 140;                          14
    optional InnerProductParameter inner_product_param = 117;       15
    optional LRNParameter lrn_param = 118;                          16
    optional PoolingParameter pooling_param = 121;                  17
    optional PReLUParameter prelu_param = 131;                      18
    optional ReLUParameter relu_param = 123;                        19
    optional ReshapeParameter reshape_param = 133;                  20
    optional SigmoidParameter sigmoid_param = 124;                  21
    optional SoftmaxParameter softmax_param = 125;                  22
    optional TanHParameter tanh_param = 127;                        23
                                                                     24
    optional HingeLossParameter hinge_loss_param = 114;             25
                                                                     26
    repeated ParamSpec param = 6;                                   27
    optional LossParameter loss_param = 101;                        28
                                                                     29
    optional Phase phase = 10;                                      30
}                                                                    31
```

Fig. 4.3 A computational
graph (neural network) with
three layers

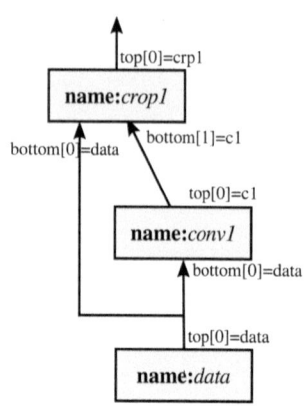

Each layer has a name. Although entering a name for a layer is optional but it is
highly recommended to give each layer a *unique* name. This increases readability of
your model. It has also another function. Assume you want to have two convolution
layers with exactly the same parameters. In other words, these two convolution layers
share the same set of weights. This can be easily specified in Caffe by giving these
two layers an identical name.

The string filed "type" specifies the type of the layer. For example, by assigning
"Convolution" to this field, we tell Caffe that the current layer is a convolution layer.
Note that the type of layer is case-sensitive. This means that, assigning "convolution" (small letter c instead of capital letter C) to type will raise an error telling that
"convolution" is not a valid type for a layer.

There are two arrays of strings in LayerParameter called "top" and "bottom". If
we assume that a layer (an instance of LayerParameter) is represented by a node
in computational graphs, the bottom variable shows the tag of incoming nodes to
the current node and the top variable shows the tag of outgoing edges. Figure 4.3
illustrates a computational graph with three nodes.

This computational graph is composed of three layers namely *data*, *conv*$_1$ and
crop$_1$. For now, assume that the node *data* reads images along with their labels
from a disk and stores them in memory. Apparently, the node *data* does not get its
information from another node. For this reason, it does not have any bottom (the
length of bottom is zero). The node *data* passes this information to other nodes in
the graph. In Caffe, the information produced by a node is recognized by unique
tags. The variable *top* stores the name of these tags. A tag and name of a node could
be identical. As we can see in node *data*, it produces only one output. Hence, the
length of array *top* will be equal to 1. The first (and only) element in this array shows
the tag of the first output of the node. In the case of *data*, the tag has been also called
data. Now, any other node can have access to information produced by the node
data using its tag.

The second node is a convolution node named *conv*$_1$. This node receives information from node *data*. The convolution node in this example has only one incoming

node. Therefore, length of bottom array for $conv_1$ will be 1. The first (and only) element in this array refers to the tag, where the information from this tag will come to $conv_1$. In this example, the information comes from $data$. After convolving $bottom[0]$ with filters in $conv_1$ (the value of filter are stored in node itself), it produces only one output. So, length of array top for $conv_1$ will be equal to 1. The tag of output for $conv_1$ has been called $c1$. In this case, the name of node and top of node are not identical.

Finally, the node $crop_1$ receives two inputs. One from $conv_1$ and one from $data$. For this reason, the bottom array in this node has two elements. The first element is connected to $data$ and the second element is connected to $c1$. Then, $crop_1$, crops the first element of bottom (bottom[0]) to make its size identical to the second element of bottom (bottom[1]). This node also generates a single output. The tag of this output is $crp1$.

In general, passing information between computational nodes is done using array of bottoms (incoming) and array of tops (outgoing). Each node stores information about its bottoms and tops as well as its parameters and hyperparameters. There are many other fields in LayerParameter all ending with phrase "Parameter". Based the $type$ of a node, we may need to instantiate some of these fields.

4.3.1 Providing Data

The first thing to put in a neural network is at least one layer that provides data for the network. There are a few ways in Caffe to do this. The simplest approach is to provide data using a layer with $type="ImageData"$. This type of layer requires instantiating the field $image\_data\_param$ from LayerParameter. ImageDataParameter is also a message with the following definition:

```
message ImageDataParameter {                                        1
    optional string source = 1;                                     2
                                                                    3
    optional uint32 batch_size = 4 [default = 1];                   4
    optional bool shuffle = 8 [default = false];                    5
                                                                    6
    optional uint32 new_height = 9 [default = 0];                   7
    optional uint32 new_width = 10 [default = 0];                   8
                                                                    9
    optional bool is_color = 11 [default = true];                  10
    optional string root_folder = 12 [default = ""];               11
}                                                                   12
```

Again, deprecated fields have been removed from this list. This message is composed of fields with basic data types. An ImageData layer needs a text file with the following structure:

```
ABSOLUTE_PATH_OF_IMAGE1  LABEL1                                      1
ABSOLUTE_PATH_OF_IMAGE2  LABEL2                                      2
. . .                                                               3
ABSOLUTE_PATH_OF_IMAGEN  LABELN                                      4
```

Listing 4.2 Structure of train.txt

An ImageData layer assumes that images are stored on the disk using a regular image format such as jpg, bmp, ppm, png, etc. Images could be stored on different locations and different disks on your system. In the above structure, there is one line for each image in the training set. Each line is composed of two parts separated by a *space* character (ASCII code 32). The left part shows the absolute path of the image and the right part shows the class label of that image.

The current implementation of Caffe identifies class label from image using the space character in the line. Consequently, if the path of the image contains space characters, Caffe will not able to decode this line and it may raise an exception. For this reason, avoid space characters in the name of folders and files when you are creating a text file for an ImageData layer.

Moreover, class labels have to be integer numbers and they have to always start from zero. That said, if there are 20 classes in your dataset, the class labels have to be integer numbers between 0 and 19 (19 included). Otherwise, Caffe may raise an exception during training. For example, the following sample shows a small part of a text file that is prepared for an ImageData layer.

```
/home/pc/Desktop/GTSRB/Training_CNN/00019/00000_00006.ppm 19          1
/home/pc/Desktop/GTSRB/Training_CNN/00029/00003_00021.ppm 29          2
/home/pc/Desktop/GTSRB/Training_CNN/00010/00054_00008.ppm 10          3
/home/pc/Desktop/GTSRB/Training_CNN/00023/00010_00027.ppm 23          4
/home/pc/Desktop/GTSRB/Training_CNN/00033/00022_00008.ppm 33          5
/home/pc/Desktop/GTSRB/Training_CNN/00021/00000_00005.ppm 21          6
/home/pc/Desktop/GTSRB/Training_CNN/00005/00020_00022.ppm 5           7
/home/pc/Desktop/GTSRB/Training_CNN/00025/00026_00018.ppm 25          8
...                                                                    9
```

Suppose that our dataset contains 3,000,000 images and they are all located in a common folder. In the above sample, all files are stored at */home/pc/Desktop/GT-SRB/Training_CNN*. However, this common address is repeated in the text file 3 million times since we have provided absolute path of images. Taking into account the fact that Caffe loads all the paths and their labels into memory once, this means $3,000,000 \times 35$ characters are repeated in the memory which is equal to about 100 MB memory. If the common path is longer or the number of samples is higher, more memory will be needed to store the information.

To use the memory more efficiently, ImageDataParameter has provided a filed called *root_folder*. This field points to the path of the common folder in the text file. In the above example, this will be equal to */home/pc/Desktop/GTSRB/Training_CNN*. In that case, we can remove the common path from the text file as follows:

```
/00019/00000_00006.ppm 19          1
/00029/00003_00021.ppm 29          2
/00010/00054_00008.ppm 10          3
/00023/00010_00027.ppm 23          4
/00033/00022_00008.ppm 33          5
/00021/00000_00005.ppm 21          6
/00005/00020_00022.ppm 5           7
/00025/00026_00018.ppm 25          8
...                                9
```

Caffe will always add the *root_folder* to the beginning of path in each line. This way, redundant information are not stored in the memory.

The variable *batch_size* denotes the size of mini-batch to be forwarded and back-propagated in the network. Common values for this parameter vary between 20 and 256. This also depends on the available memory on your GPU. The Boolean variable *shuffle* shows whether or not Caffe must shuffle the list of files in each epoch or not. Shuffling could be useful for having diverse mini-batches at each epoch. Considering the fact that one epoch refers to processing whole dataset, the list of files is shuffled when the last mini-batch of dataset is processed. In general, setting shuffle to true could be a good practice. Especially, setting this value to true is essential when the text file containing the training samples is ordered based on the class label. In this case, shuffling is an essential step in order to have diverse mini-batches. Finally, as it turns out from their name, if *new_height* and *new_width* have a value greater than zero, the loaded image will be resized to the new size based on the value of these parameters. Finally, the variable *is_color* tells Caffe to load images in color format or grayscale format.

Now, we can define a network containing only an ImageData layer using the protobuf grammar. This is illustrated below.

```
name:"net1"                                                       1
layer{                                                            2
    name:"data"                                                   3
    type:"ImageData"                                              4
    top:"data"                                                    5
    top:"label"                                                   6
    image_data_param{                                             7
        source:"/home/pc/Desktop/train.txt"                       8
        batch_size:30                                             9
        root_folder:"/home/pc/Desktop/"                          10
        is_color:true                                            11
        shuffle:true                                             12
        new_width:32                                             13
        new_height:32                                            14
    }                                                            15
}                                                                16
```

In Caffe, a tensor is a $mini-batch \times Channel \times Height \times Width$ array. Note that an ImageData layer produces two tops. In other words, the length of *top* array for this layer is 2. The first element of the top array stores loaded images. Therefore, the first top of the above layer will be a $30 \times 3 \times 32 \times 32$ tensor. The second element of the top array stores labels of each image in the first top and it will be an array with $mini-batch$ integer elements. Here, it will be a 30-element array of integers.

4.3.2 Convolution Layers

Now, we want to add a convolution layer to the network and connect it to the Image-Data layer. To this end, we must create a layer with *type="Convolution"* and then configure the layer by instantiating *convolution_param*. The type of this variable is ConvolutionParameter which is defined as follows:

```
message ConvolutionParameter {                                          1
    optional uint32 num_output = 1;                                     2
    optional bool bias_term = 2 [default = true];                      3
                                                                        4
    repeated uint32 pad = 3;                                            5
    repeated uint32 kernel_size = 4;                                    6
    repeated uint32 stride = 6;                                         7
                                                                        8
    optional FillerParameter weight_filler = 7;                        9
    optional FillerParameter bias_filler = 8;                          10
}                                                                      11
```

The variable *num_output* determines the number of convolution filters. Recall from the previous chapter that the activation of neuron basically is given by $\}(\mathbf{w}\mathbf{x} + bias)$. The variable *bias_term* states that whether or not the bias term must be considered in the neuron computation. The variable *pad* denotes the zero-padding size and it is 0 by default. Zero padding is used to handle the borders during convolution. Zero-Padding a $H \times W$ image with *pad=2* can be thought as creating a zero matrix of size $(H + 2pad) \times (W + 2pad)$ and copying the image into this matrix such that is placed exactly in the middle of the zero matrix. Then, if the size of convolution filters is $(2pad + 1) \times (2pad + 1)$, the result of convolution with zero-padded image will be $H \times W$ images which is exactly equal to the size of input image. Padding is usually done for keeping the size of input and output of convolution operations constant. But, it is commonly set to zero.

As it turns out, the variable *kernel_size* determines the spatial size (width and height) of convolution filters. It should be noted that a convolution layer must have the same number of bottoms and tops. It convolves each bottom separately with the filter and passes it to the corresponding top. The third dimension of filters is automatically computed by Caffe based on the number of channels coming from the bottom node. Finally, the variable *stride* illustrates the stride of convolution operation and it is set to 1 by default. Now, we can update the protobuf text and add a convolution layer to the network.

```
name:"net1"                                                             1
layer{                                                                  2
    name:"data"                                                         3
    type:"ImageData"                                                    4
    top:"data"                                                          5
    top:"label"                                                         6
    image_data_param{                                                   7
        source:"/home/hamed/Desktop/train.txt"                          8
        batch_size:30                                                   9
        root_folder:"/home/hamed/Desktop/"                             10
        is_color:true                                                  11
        shuffle:true                                                   12
        new_width:32                                                   13
        new_height:32                                                  14
    }                                                                  15
}                                                                      16
layer{                                                                 17
    name:"conv1"                                                       18
    type:"Convolution"                                                 19
    bottom:"data"                                                      20
    top:"conv1"                                                        21
    convolution_param{                                                 22
        num_output: 6                                                  23
        kernel_size:5                                                  24
    }                                                                  25
}                                                                      26
```

The convolution layer has six filters of size 5×5 and it is connected to a data layer that produces mini-batches of images. Figure 4.4 illustrates the diagram of the neural network created by the above protobuf text.

4.3.3 Initializing Parameters

Any layer with trainable parameters including convolution layers has to be initialized before training. Concretely, convolution filters (weights) and biases of convolution layer have to be initialized. As we explained in the previous chapter, this can be done by setting each weight/bias to a random number. However, generating random number can be done using different distributions and different methods. The *weight_filler* and *bias_filler* parameters in LayerParameter specify the type of initialization method. They are both instances of FillerParameter which are defined as follows:

```
message FillerParameter {                                                       1
    optional string type = 1 [default = 'constant'];                            2
    optional float value = 2 [default = 0];                                     3
    optional float min = 3 [default = 0];                                       4
    optional float max = 4 [default = 1];                                       5
    optional float mean = 5 [default = 0];                                      6
    optional float std = 6 [default = 1];                                       7
                                                                                8
    enum VarianceNorm {                                                         9
        FAN_IN = 0;                                                            10
        FAN_OUT = 1;                                                           11
        AVERAGE = 2;                                                           12
    }                                                                          13
    optional VarianceNorm variance_norm = 8 [default = FAN_IN];                14
}                                                                              15
```

The string variable *type* defines the method that will be used for generating number. Different values can be assigned to this variable. Among them, "constant", "gaussian", "uniform", "xavier" and "mrsa" are commonly used in classification networks. Concretely, a constant filler sets the parameters to a constant value specified by the floating point variable *value*.

Also, a "gaussian" filler assigns random numbers generated by a Gaussian distribution specified by *mean* and *std* variables. Likewise, "uniform" filler assigns random

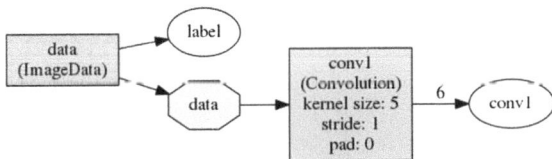

Fig. 4.4 Architecture of the network designed by the protobuf text. *Dark rectangles* show nodes. *Octagon* illustrates the name of the top element. The number of *outgoing arrows* in a node is equal to the length of top array of the node. Similarly, the number of *incoming arrows* to a node shows the length of bottom array of the node. The *ellipses* show the tops that are not connected to another node

number generated by the uniform distribution within a range determined by *min* and *max* variables.

The "xavier" filler generates uniformly distributed random numbers within $[-\sqrt{\frac{3}{n}}, \sqrt{\frac{3}{n}}]$, where depending on the value of *variance_norm* variable n could be the number of inputs (FAN_IN), the number of output (FAN_OUT) or average of them. The "msra" filler is like "xavier" filler. The difference is that it generates Gaussian distributed random number with standard deviation equal to $\sqrt{\frac{2}{n}}$.

As we mentioned in the previous chapter, filters are usually initialized using "xavier" or "mrsa" methods and biases are initialized using constant value zero. Now, we can also define weight and bias initializer for the convolution layer. The updated protobuf text will be:

```
name:"net1"                                                   1
layer{                                                        2
    name:"data"                                               3
    type:"ImageData"                                          4
    top:"data"                                                5
    top:"label"                                               6
    image_data_param{                                         7
        source:"/home/hamed/Desktop/train.txt"                8
        batch_size:30                                         9
        root_folder:"/home/hamed/Desktop/"                    10
        is_color:true                                         11
        shuffle:true                                          12
        new_width:32                                          13
        new_height:32                                         14
    }                                                         15
}                                                             16
layer{                                                        17
    name:"conv1"                                              18
    type:"Convolution"                                        19
    bottom:"data"                                             20
    top:"conv1"                                               21
    convolution_param{                                        22
        num_output: 6                                         23
        kernel_size:5                                         24
        weight_filler{                                        25
            type:"xavier"                                     26
        }                                                     27
        bias_filler{                                          28
            type:"constant"                                   29
            value:0                                           30
        }                                                     31
    }                                                         32
}                                                             33
```

4.3.4 Activation Layer

Each output of convolution layer is given by $\mathbf{wx} + b$. Next, these values must be passed through a nonlinear activation function. In the Caffe library, ReLU, Leaky ReLU, PReLU, ELU, sigmoid, and hyperbolic tangent activations are implemented. Setting *type="ReLU"* will create a Leaky ReLU activation. If we set the leak value to zero, this is equivalent to the ReLU activation. The other activations are created by setting *type="PReLU"*, *type="ELU"*, *type="Sigmoid"* and *type="TanH"*. Then, depending on the type of activation function, we can also adjust their hyperparameters. The messages for these activations are defined as follows:

```
message ELUParameter {                                              1
    optional float alpha = 1 [default = 1];                        2
}                                                                   3
message ReLUParameter {                                             4
    optional float negative_slope = 1 [default = 0];               5
}                                                                   6
message PReLUParameter {                                            7
    optional FillerParameter filler = 1;                           8
    optional bool channel_shared = 2 [default = false];            9
}                                                                  10
```

Clearly, the sigmoid and hyperbolic tangent activation do not have parameters to set. However, as it is mentioned in (2.93) and (2.96) the family of the ReLU activation in Caffe has hyperparameters that should be configured. In the case of Leaky ReLU and ELU activations, we have to determine the value of α in (2.93) and (2.96). In Caffe, α for Leaky ReLU is illustrated by *negative_slope* variable. In the case of PReLU activation, we have to tell Caffe how to initialize the α parameter using the *filler* variable. Also, the Boolean variable *channel_shared* determines whether Caffe should share the same α for all activations (*channel_shared=true*) in the same layer or it must find separate α for each channel in the layer. We can add this activation to the protobuf as follows:

```
name:"net1"                                                         1
layer{                                                              2
    name:"data"                                                     3
    type:"ImageData"                                                4
    top:"data"                                                      5
    top:"label"                                                     6
    image_data_param{                                               7
        source:"/home/hamed/Desktop/train.txt"                      8
        batch_size:30                                               9
        root_folder:"/home/hamed/Desktop/"                         10
        is_color:true                                              11
        shuffle:true                                               12
        new_width:32                                               13
        new_height:32                                              14
    }                                                              15
}                                                                  16
layer{                                                             17
    name:"conv1"                                                   18
    type:"Convolution"                                             19
    bottom:"data"                                                  20
    top:"conv1"                                                    21
    convolution_param{                                             22
        num_output: 6                                              23
        kernel_size:5                                              24
        weight_filler{                                             25
            type:"xavier"                                          26
        }                                                          27
        bias_filler{                                               28
            type:"constant"                                        29
            value:0                                                30
        }                                                          31
    }                                                              32
}                                                                  33
layer{                                                             34
    type."ReLU"                                                    35
    bottom:"conv1"                                                 36
    top:"relu_c1"                                                  37
}                                                                  38
```

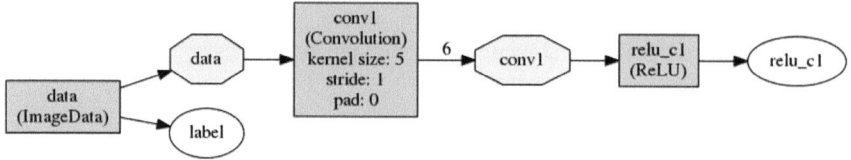

Fig. 4.5 Diagram of the network after adding a ReLU activation

After adding this layer to the network, the architecture will look like Fig. 4.5.

4.3.5 Pooling Layer

A pooling layer is created by setting *type="Pooling"*. Similar to a convolution layer, a pooling layer must have the same number of bottoms and tops. It applies pooling on each bottom separately and passes it to the corresponding top. Parameters of the pooling operation are also determined using an instance of PoolingParameter.

```
message PoolingParameter {                                              1
    enum PoolMethod {                                                   2
    MAX = 0;                                                            3
    AVE = 1;                                                            4
    STOCHASTIC = 2;                                                     5
    }                                                                   6
    optional PoolMethod pool = 1 [default = MAX];                       7
    optional uint32 pad = 4 [default = 0];                              8
                                                                        9
    optional uint32 kernel_size = 2;                                    10
    optional uint32 stride = 3 [default = 1];                           11
    optional bool global_pooling = 12 [default = false];               12
}                                                                       13
```

Similar to Convolutionparameter, the variables *pad*, *kernel_size* and *stride* determine the amount of zero padding, size of pooling window, and stride of pooling, respectively. The variable *pool* determines the type of pooling. Currently, Caffe supports max pooling, average pooling, and stochastic pooling. However, we often choose max pooling and it is the default option in Caffe. The variable *global_pooling* pools over the entire spatial region of bottom array. It is equivalent to setting *kernel_size* to the spatial size of the bottom blob. We add a max-pooling layer to our network. The resulting protobuf will be:

```
name:"net1"                                                             1
layer{                                                                  2
    name:"data"                                                         3
    type:"ImageData"                                                    4
    top:"data"                                                          5
    top:"label"                                                         6
    image_data_param{                                                   7
        source:"/home/hamed/Desktop/train.txt"                          8
        batch_size:30                                                   9
        root_folder:"/home/hamed/Desktop/"                              10
        is_color:true                                                   11
        shuffle:true                                                    12
        new_width:32                                                    13
        new_height:32                                                   14
```

```
        }                                                                    15
}                                                                            16
layer{                                                                       17
    name:"conv1"                                                             18
    type:"Convolution"                                                       19
    bottom:"data"                                                            20
    top:"conv1"                                                              21
    convolution_param{                                                       22
        num_output: 6                                                        23
        kernel_size:5                                                        24
        weight_filler{                                                       25
            type:"xavier"                                                    26
        }                                                                    27
        bias_filler{                                                         28
                type:"constant"                                              29
        value:0                                                              30
        }                                                                    31
    }                                                                        32
}                                                                            33
layer{                                                                       34
    name:"relu_c1"                                                           35
    type:"ReLU"                                                              36
    bottom:"conv1"                                                           37
    top:"relu_c1"                                                            38
    relu_param{                                                              39
        negative_slope:0.01                                                  40
    }                                                                        41
}                                                                            42
layer{                                                                       43
    name:"pool1"                                                             44
    type:"Pooling"                                                           45
    bottom:"relu_c1"                                                         46
    top:"pool1"                                                              47
    pooling_param{                                                          48
        kernel_size:2                                                        49
        stride:2                                                             50
    }                                                                        51
}                                                                            52
```

The pooling will be done over 2×2 regions with stride 2. This will halve the spatial size of the input. Figure 4.6 shows the diagram of the network.

We added another convolution layer with 16 filters of size 5×5, a ReLU activation and a max-pooling with 2×2 region and stride 2 to the network. Figure 4.7 illustrates the diagram of the network.

4.3.6   Fully Connected Layer

A fully connected layer is defined by setting *type="InnerProduct"* in the definition of layer. The number of bottoms and tops must be equal in this type of layer. It computes

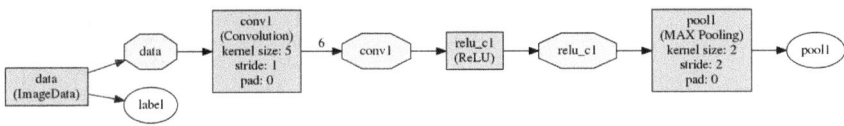

Fig. 4.6 Architecture of network after adding a pooling layer

Fig. 4.7 Architecture of network after adding a pooling layer

the top for each bottom separately using the same set of parameters. Hyperparameters
of a fully connected layer are specified using an instance of InnerProductParameter
which is defined as follows.

```
message InnerProductParameter {                                          1
    optional uint32 num_output = 1;                                      2
    optional bool bias_term = 2 [default = true];                        3
    optional FillerParameter weight_filler = 3;                          4
    optional FillerParameter bias_filler = 4;                            5
}                                                                        6
```

The variable *num_output* determines the number of neurons in the layer. The
variable *bias_term* tells Caffe whether or not to consider the bias term in neuron
computations. Also, *weight_filler* and *bias_filler* are used to specify how to initialize
the parameters of the fully connected layer.

4.3.7 Dropout Layer

A dropout layer can be placed anywhere in a network. But, it is more common to put it
immediately after an activation layer. However, it is mainly placed after activation of
fully connected layers. The reason is that fully connected layers increase nonlinearity
of a model and they apply final transformations on the extracted features by previous
layers. Our model may over fit because of the final transformations. For this reason,
we try to regularize the model using dropout layers in fully connected layers. A
dropout layer is defined by setting *type="Dropout"*. Then, hyperparameter of a
dropout layer is determined using an instance of DropoutParameter which is defined
as follows:

```
message DropoutParameter {                                               1
    optional float dropout_ratio = 1 [default = 0.5];                    2
}                                                                        3
```

As we can see, a dropout layer only has one hyperparameter which is the ratio of
dropout. Since this ratio shows the probability of dropout, it has to be set to a floating
point number between 0 and 1. The default value in Caffe is 0.5. We added two fully
connected layers to our network and placed a dropout layer after each of these layers.
The diagram of network after applying these changes is illustrated in Fig. 4.8.

4.3.8 Classification and Loss Layers

The last layer in a classification network is a fully connected layer, where the number
of neurons in this layer is equal to number of classes in the dataset. Training a

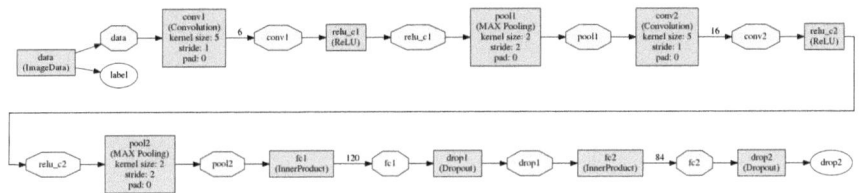

Fig. 4.8 Diagram of network after adding two fully connected layers and two dropout layers

neural network is done by minimizing a loss function. In this book, we explained hinge loss and logistic loss functions for multiclass classification problems. These two loss functions accept at least two bottoms. The first bottom is the output of the classification layer and the second bottom is actual labels produced by the ImageData layer. The loss layer computes the loss based on these two bottoms and returns a scaler in its top.

The hinge loss function is created by setting *type="HingeLoss"* and multiclass logistic loss is created by setting *type="SoftmaxWithLoss"*. Then, we mainly need to enter the bottoms and top of the loss layer. We added a classification layer and a multiclass logistic loss to the protobuf. The final protobuf will be:

```
layer{                                                                      1
    name:"data"                                                             2
    type:"ImageData"                                                        3
    top:"data"                                                              4
    top:"label"                                                             5
    image_data_param{                                                       6
        source:"/home/hamed/Desktop/GTSRB/Training_CNN/train.txt"           7
        batch_size:30                                                       8
        root_folder:"/home/hamed/Desktop/GTSRB/Training_CNN/"               9
        is_color:true                                                      10
        shuffle:true                                                       11
        new_width:32                                                       12
        new_height:32                                                      13
    }                                                                      14
}                                                                          15
layer{                                                                     16
    name:"conv1"                                                           17
    type:"Convolution"                                                     18
    bottom:"data"                                                          19
    top:"conv1"                                                            20
    convolution_param{                                                     21
        num_output: 6                                                      22
        kernel_size:5                                                      23
        weight_filler{ type:"xavier" }                                     24
        bias_filler{ type:"constant" value:0 }                            25
    }                                                                      26
}                                                                          27
layer{                                                                     28
    name:"relu_c1"                                                         29
    type:"ReLU"                                                            30
    bottom:"conv1"                                                         31
    top:"relu_c1"                                                          32
    relu_param{ negative_slope:0.01 }                                      33
}                                                                          34
layer{                                                                     35
    name:"pool1"                                                           36
    type:"Pooling"                                                         37
```

```
    bottom:"relu_c1"                                              38
    top:"pool1"                                                   39
    pooling_param{ kernel_size:2 stride:2 }                       40
}                                                                 41
layer{                                                            42
    name:"conv2"                                                  43
    type:"Convolution"                                            44
    bottom:"pool1"                                                45
    top:"conv2"                                                   46
    convolution_param{                                            47
        num_output: 16                                            48
        kernel_size:5                                             49
        weight_filler{ type:"xavier" }                            50
        bias_filler{ type:"constant" value:0 }                    51
    }                                                             52
}                                                                 53
layer{                                                            54
    name:"relu_c2"                                                55
    type:"ReLU"                                                   56
    bottom:"conv2"                                                57
    top:"relu_c2"                                                 58
    relu_param{ negative_slope:0.01 }                             59
}                                                                 60
layer{                                                            61
    name:"pool2"                                                  62
    type:"Pooling"                                                63
    bottom:"relu_c2"                                              64
    top:"pool2"                                                   65
    pooling_param{ kernel_size:2 stride:2 }                       66
}                                                                 67
layer{                                                            68
    name:"fc1"                                                    69
    type:"InnerProduct"                                           70
    bottom:"pool2"                                                71
    top:"fc1"                                                     72
    inner_product_param{                                          73
        num_output:120                                            74
        weight_filler{ type:"xavier" }                            75
        bias_filler{ type:"constant" value:0 }                    76
    }                                                             77
}                                                                 78
layer{                                                            79
    name:"relu_fc1"                                               80
    type:"ReLU"                                                   81
    bottom:"fc1"                                                  82
    top:"relu_fc1"                                                83
    relu_param{ negative_slope:0.01 }                             84
}                                                                 85
layer{                                                            86
    name:"drop1"                                                  87
    type:"Dropout"                                                88
    bottom:"relu_fc1"                                             89
    top:"drop1"                                                   90
    dropout_param{ dropout_ratio:0.4 }                            91
}                                                                 92
layer{                                                            93
    name:"fc2"                                                    94
    type:"InnerProduct"                                           95
    bottom:"drop1"                                                96
    top:"fc2"                                                     97
    inner_product_param{                                          98
        num_output:84                                             99
        weight_filler{ type:"xavier" }                            100
        bias_filler{ type:"constant" value:0 }                    101
    }                                                             102
}                                                                 103
layer{                                                            104
```

```
    name: "relu_fc2"                                                       105
    type: "ReLU"                                                           106
    bottom: "fc2"                                                          107
    top: "relu_fc2"                                                        108
    relu_param{ negative_slope:0.01 }                                      109
}                                                                          110
layer{                                                                     111
    name: "drop2"                                                          112
    type: "Dropout"                                                        113
    bottom: "relu_fc2"                                                     114
    top: "drop2"                                                           115
    dropout_param{ dropout_ratio:0.4 }                                     116
}                                                                          117
layer{                                                                     118
    name: "fc3_classification"                                             119
    type: "InnerProduct"                                                   120
    bottom: "drop2"                                                        121
    top: "classifier"                                                      122
    inner_product_param{                                                   123
        num_output:43                                                      124
        weight_filler{type:"xavier"}                                       125
        bias_filler{ type:"constant" value:0 }                            126
    }                                                                      127
}                                                                          128
layer{                                                                     129
    name: "loss"                                                           130
    type: "SoftmaxWithLoss"                                                131
    bottom: "classifier"                                                   132
    bottom: "label"                                                        133
    top: "loss"                                                            134
}                                                                          135
```

Considering that there are 43 classes in the GTSRB dataset, the number of neurons in the classification layer must be also equal to 43. The diagram of final network is illustrated in Fig. 4.9.

The above protobuf text is stored in a text file on disk. In this example, we store the above text file in "*/home/pc/cnn.prototxt*". The above definition reads the training samples and feeds them to the network. However, in practice, the network must be evaluated using a *validation set* during training in order to assess how good the network is.

To achieve this goal, the network can be evaluated every K iterations of the training algorithm. As we will see shortly, this can be easily done by setting a parameter. Assume, K iterations have been finished and Caffe wants to evaluate the network. So far, we have only fetched data from a training set. Obviously, we have to tell Caffe where to look for validation samples. To this end, we add another ImageData layer right after the first ImageData layer and specify the location of the validation samples instead of the training samples. In other words, the first layer in the above network definition will be replaced by:

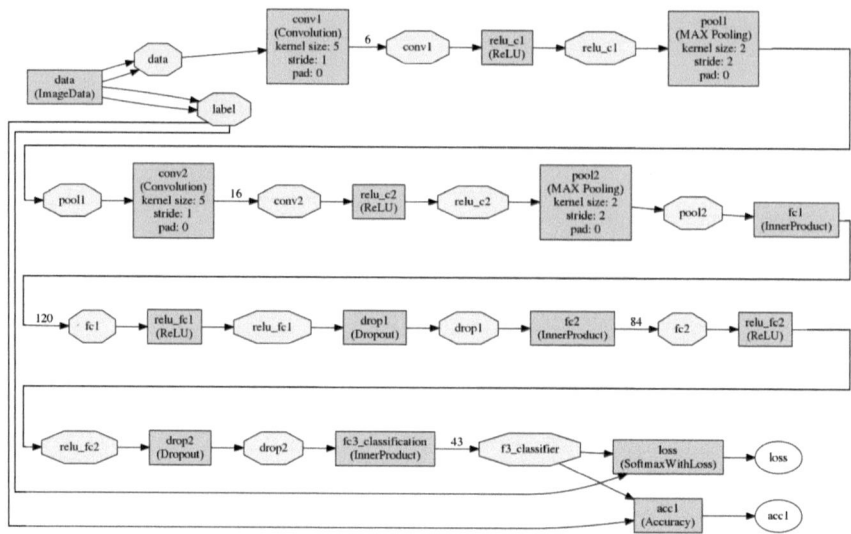

Fig. 4.9 Final architecture of the network. The architecture is similar to the architecture of LeNet-5 in nature. The differences are in activations functions, dropout layer, and connection in middle layers

```
layer{                                                                                  1
    name:"data"                                                                         2
    type:"ImageData"                                                                    3
    top:"data"                                                                          4
    top:"label"                                                                         5
    image_data_param{                                                                   6
        source:"/home/hamed/Desktop/GTSRB/Training_CNN/train.txt"                       7
        batch_size:30                                                                   8
        root_folder:"/home/hamed/Desktop/GTSRB/Training_CNN/"                           9
        is_color:true                                                                   10
        shuffle:true                                                                    11
        new_width:32                                                                    12
        new_height:32                                                                   13
    }                                                                                   14
}                                                                                       15
layer{                                                                                  16
    name:"data"                                                                         17
    type:"ImageData"                                                                    18
    top:"data"                                                                          19
    top:"label"                                                                         20
    image_data_param{                                                                   21
        source:"/home/hamed/Desktop/GTSRB/Training_CNN/validation.txt"                  22
        batch_size:10                                                                   23
        root_folder:"/home/hamed/Desktop/GTSRB/Validation_CNN/"                         24
        is_color:true                                                                   25
        shuffle:false                                                                   26
        new_width:32                                                                    27
        new_height:32                                                                   28
    }                                                                                   29
}                                                                                       30
```

First, the tops of these two layers have to be identical. This is due to the fact the first convolution layer is connected to a top called *data*. If we set top in the second ImageData layer to another name, the convolution layer will not receive any data during validation. Second, the variable *source* in the second layer points to the validation set. Third, the batch sizes of these two layers can be different. Usually, if memory on the GPU device is limited, we usually set the batch size of training set to the appropriate value and then set the batch size of the validation set according to the memory limitations. For instance, we have to set the batch size of validation samples to 10. Fourth, shuffle must be set to false in order to prevent unequal validations sets. In fact, the parameters that we will explain in the next section are adjusted such that the validation set is only scanned once in every test.

However, a user may forget to adjust this parameter properly and some of samples in validation set are fetched more than one time to the network. In that case, if *shuffle* is set to true it is very likely that some samples in two validation steps are not identical. This makes the validation result inaccurate. We alway want to test/validate the different models or same models in different time on exactly identical datasets.

During testing, the data has to **only** come from the first ImageData layer. During validation, the data has to **only** come from the second ImageData layer. One missing piece in the above definition is that how should Caffe understand when to switch from one ImageData layer to another. There is a variable in definition of LayerParameter called *include* which is an instance of NetStateRule.

```
message NetStateRule {                                                             1
    optional Phase phase = 1;                                                      2
}                                                                                  3
```

When this variable is specified, Caffe will include the layer based on the state of training. This can be explained better in an example. Let us update the above two ImageData layers as follows:

```
layer{                                                                             1
name:"data"                                                                        2
    type:"ImageData"                                                               3
    top:"data"                                                                     4
    top:"label"                                                                    5
    include{                                                                       6
        phase:TRAIN                                                                7
    }                                                                              8
    image_data_param{                                                              9
        source:"/home/hamed/Desktop/GTSRB/Training_CNN/train.txt"                 10
        batch_size:30                                                             11
        root_folder:"/home/hamed/Desktop/GTSRB/Training_CNN/"                     12
        is_color:true                                                             13
        shuffle:true                                                              14
        new_width:32                                                             15
        new_height:32                                                            16
    }                                                                             17
    phase:                                                                        18
}                                                                                  19
layer{                                                                            20
    name:"data"                                                                  21
    type:"ImageData"                                                             22
    top:"data"                                                                   23
    top:"label"                                                                  24
    include{                                                                     25
        phase:TRAIN                                                              26
```

```
        }                                                                        27
    image_data_param{                                                            28
        source:"/home/hamed/Desktop/GTSRB/Training_CNN/validation.txt"            29
        batch_size:10                                                             30
        root_folder:"/home/hamed/Desktop/GTSRB/Validation_CNN/"                   31
        is_color:true                                                             32
        shuffle:false                                                             33
        new_width:32                                                              34
        new_height:32                                                             35
    }                                                                            36
}                                                                                37
```

During training a network, Caffe alternatively changes its state between TRAIN and
TEST based on a parameter called *test_interval* (this parameters will be explain in
the next section). In the TRAIN phase, the second ImageData layer will be discarded
by Caffe. In contrast, the first layer will be discarded and the second layer will be
included in the TEST phase. If the variable *include* is not instantiated in a layer, the
layer will be included in both phases. We apply the above changes on the text file
and save it

Finally, we add a layer to our network in order to compute the accuracy of the
network on test samples. This is simply done by adding the following definition right
after the loss layer.

```
layer{                                                                            1
    name:"acc1"                                                                   2
    type:"Accuracy"                                                               3
    bottom:"classifier"                                                           4
    bottom:"label"                                                                5
    top:"acc1"                                                                    6
    include{ phase:TEST }                                                         7
}                                                                                 8
```

4.4 Training a Network

In order to train a neural network in Caffe, we have to design another text file and
instantiate a SolverParameter inside this file. All required rules for training a neural
network will be specified using SolverParameter.

```
message SolverParameter {                                                         1
    optional string net = 24;                                                     2
    optional float base_lr = 5;                                                   3
                                                                                  4
    repeated int32 test_iter = 3;                                                 5
    optional int32 test_interval = 4 [default = 0];                               6
    optional int32 display = 6;                                                   7
                                                                                  8
    optional int32 max_iter = 7;                                                  9
    optional int32 iter_size = 36 [default = 1];                                 10
                                                                                 11
    optional string lr_policy = 8;                                              12
    optional float gamma = 9;                                                   13
    optional float power = 10;                                                  14
    optional int32 stepsize = 13;                                               15
                                                                                 16
    optional float momentum = 11;                                               17
```

```
optional float weight_decay = 12;                                          18
optional string regularization_type = 29 [default = "L2"];                 19
optional float clip_gradients = 35 [default = −1];                         20
                                                                           21
optional int32 snapshot = 14 [default = 0];                                22
optional string snapshot_prefix = 15;                                      23
                                                                           24
enum SolverMode {                                                          25
    CPU = 0;                                                               26
    GPU = 1;                                                               27
}                                                                          28
optional SolverMode solver_mode = 17 [default = GPU];                      29
optional int32 device_id = 18 [default = 0];                               30
                                                                           31
optional string type = 40 [default = "SGD"];                              32
}                                                                          33
```

The string variable *net* points to the *.prototxt* file that includes the definition of the network. In our example, this variable is set to *net="/home/pc/cnn.prototxt"*. The variable *base_lr* denotes the base learning rate. The effective learning rate at each iteration is defined based on the value of *lr_policy*, *gamma*, *power*, and *step-size*. Recall from Sect. 3.6.4 that the learning rate is usually decreased over time. We explained different methods for decreasing the learning rate. In Caffe, setting *lr_policy="exp"* will decrease the learning rate using exponential rule. Likewise, setting this parameter to "step" and "inv" will decrease the learning rate using step method and the inverse method.

The parameter *test_iter* tells Caffe how many mini-batches it should use during test phase. The total number of samples that is used in the test phase will be equal to *test_iter × batch size of test ImageData layer*. The variable *test_iter* is usually set such that the test phase covers all samples of validation set without using a sample twice. Caffe will change its phase from TRAIN to TEST every *test_interval* iterations (mini-batches). Then, it will run the TEST phase for *test_iter* mini-batches and changes its phase to TRAIN again.

While Caffe is training the network, it produces human-readable output. The variable *display* will show this information in the console and write them into a log file for every *display* iterations. Also, the variable *max_iter* shows the maximum number of iterations that must be performed by the optimization algorithm. The log file is accessible in director */tmp* in Ubuntu.

Sometimes, because images are large or memory on GPU device is limited, it is not possible to set mini-batch size of training samples to an appropriate value. On the other hand, if the size of mini-batch is very small, gradient descend is likely to have a very zigzag trajectory and in some cases it may even jump over a (local) minimum. This makes the optimization algorithm more sensitive to the learning rate. Caffe alleviates this problem by first accumulating gradients of *iter_size* mini-batches and updating parameters based on accumulated gradients. This makes it possible to train large networks when memory on the GPU device is not sufficient.

As it turns out, the variable *momentum* determines the value of momentum in the momentum gradient descend. It is usually set to 0.9. The variable *weight_decay* shows the value of λ in the L_1 and L_2 regularizations. The type of regularization is also defined using the string variable *regularization_type*. This variable can be only set to "L1" or "L2". The variable *clip_gradients* defines the threshold in the max-norm regularization method (Sect. 3.6.3.3).

Caffe stores weights and state of optimization algorithm inside a folder at *snapshot_prefix* for every *snapshot* iteration. Using these files, you can load the parameters of the network after training or resume training from a specific iteration.

The optimization algorithm can be executed on CPU of a GPU. This is specified using the variable *solver_mode*. In the case that you have more than one graphic cards, the variable *device_id* tells Caffe which graphic must be used for computations.

Finally, the string variable *type* determines the type of optimization algorithm. In the rest of this book, we will always use "SGD" which refers to mini-batch gradient descend. Other optimization algorithms such as Adam, AdaGrad, Nesterov, RMSProp, and AdaDelta are also implemented in the Caffe library. For our example, we write the following protobuf text in a file called *solver.prototxt*.

```
net:'/tmp/cnn.prototxt'                                          1
type:"SGD"                                                       2
                                                                 3
base_lr : 0.01                                                   4
                                                                 5
test_iter : 50;                                                  6
test_interval :500;                                              7
display : 50                                                     8
                                                                 9
max_iter : 30000                                                10
                                                                11
lr_policy: "step"                                               12
stepsize :3000                                                  13
gamma : 0.98                                                    14
                                                                15
momentum :0.9                                                   16
weight_decay :0.00001                                           17
                                                                18
snapshot: 1000                                                  19
snapshot_prefix: 'cnn'                                          20
```

After creating the text files for the network architecture and for the optimization algorithm, we can use command tools of the Caffe library to train and evaluate the network. Specifically, running the following command in Terminal of Ubuntu will train the network:

```
./caffe-master/build/tools/caffe train —solver "/PATH_TO_SOLVER/solver.prototxt"   1
```

4.5 Designing in Python

Assume we have 100 GPUs in which we can train a big neural network on each of them, separately. With these resources available, our aim is to generate 1000 different architectures and train/validate each of them on one of these GPUs. Obviously, it is not tractable for a human to create 1000 different architectures in text files. The situation gets even more impractical if our aim is to generate 1000 significantly different architectures.

The more efficient solution is to generate these files using a computer program. The program may use heuristics to create different architectures or it may generate them randomly. Regardless, the program must generate text files including the definition of network.

The Caffe library provides a Python interface that makes it possible to use Caffe functions in a Python program. The Python interface is located at *caffe-master/python*. If this path is not specified in the *PYTHONPATH* environment variable, importing the Python module of Caffe will cause an error. To solve this problem, you can either set the environment variable or write the following code before importing the module:

```
import sys                                                    1
sys.path.insert(0, "/home/pc/caffe-master/python")           2
import caffe                                                  3
```

In the above script, we have considered that the Caffe library is located at "*/home/pc/caffe-master/*". If you open *__init__.py* from *caffe-master/python/caffe/* you will find the name of functions, classes, objects, and attributes that you can use in your Python script. Alternatively, you can run the following code to obtain the same information:

```
import sys                                                    1
sys.path.insert(0, "/home/pc/caffe-master/python")           2
import caffe                                                  3
                                                             4
print dir(caffe)                                             5
```

In order to design a network, we should work with two attributes called *layers* and *params* and a class called *NetSpec*. The following Python script creates a ConvNet identical to the network we created in the previous section.

```
import sys                                                                       1
sys.path.insert(0, "/home/hamed/caffe-master/python")                           2
import caffe                                                                      3
                                                                                 4
L = caffe.layers                                                                 5
P = caffe.params                                                                 6
                                                                                 7
def conv_relu(bottom, ks, nout, stride=1, pad=0):                               8
    c = L.Convolution(bottom, kernel_size=ks, num_output=nout,                  9
                      stride=stride, pad=pad,                                   10
                      weight_filler={'type':'xavier'},                         11
                      bias_filler={'type':'constant',                          12
                                   'value':0})                                 13
    r = L.ReLU(c)                                                              14
    return c, r                                                               15
                                                                                16
def fc_relu_drop(bottom, nout):                                               17
    fc = L.InnerProduct(bottom, num_output=nout,                              18
                        weight_filler={'type':'xavier'},                      19
                        bias_filler={'type':'constant',                       20
                                     'value':0})                              21
    r = L.ReLU(fc)                                                            22
    d = L.Dropout(r, dropout_ratio=0.4)                                       23
    return fc, r, d                                                           24
                                                                                25
net = caffe.net_spec.NetSpec()                                                26
                                                                                27
net.data, net.label = L.ImageData(source='/home/hamed/Desktop/train.txt',    28
                                  batch_size=30, is_color=True,               29
```

```
                        shuffle=True, new_width=32,                            30
                        new_height=32, ntop=2)                                 31
                                                                               32
net.conv1, net.relu1 = conv_relu(net.data, 5, 16)                              33
net.pool1 = L.Pooling(net.relu1, kernel_size=2,                                34
                        stride=2, pool=P.Pooling.MAX)                          35
                                                                               36
net.conv2, net.relu2 = conv_relu(net.pool1, 5, 16)                             37
net.pool2 = L.Pooling(net.relu2, kernel_size=2,                                38
                        stride=2, pool=P.Pooling.MAX)                          39
                                                                               40
net.fc1, net.fc_relu1, net.drop1 = fc_relu_drop(net.pool2, 120)                41
net.fc2, net.fc_relu2, net.drop2 = fc_relu_drop(net.drop1, 84)                 42
net.f3_classifier = L.InnerProduct(net.drop2, num_output=43,                   43
                        weight_filler={'type':'xavier'},                       44
                        bias_filler={'type':'constant',                        45
                                     'value':0})                               46
net.loss = L.SoftmaxWithLoss(net.classifier, net.label)                        47
                                                                               48
with open('cnn.prototxt', 'w') as fs:                                          49
    fs.write(str(net.to_proto()))                                             50
    fs.flush()                                                                 51
```

In general, creating a layer can be done using the following template:

```
net.top1, net.top2,...., net.topN = L.LAYERTYPE(bottom1, bottom2,...., bottomM, 1
    kwarg1=value, kwarge=value, kwarg=dict(kwarg=value,...),..., ntop=N)
```

The number of tops in a layer is determined using the argument *ntop*. Using this method, the function will generate *ntop* top(s) in the output. Hence, there have to be N variables in the left side assignment operator. The name of tops in the text file will be "top1", "top2" and so on. That said, if the first top of the function is assigned to *net.label*, it is analogous to putting *top="label"* in the text file.

Also, note that the assignments have to be done on *net.**. If you study the source code of NetSpec, you will find that the *__setattr__* of this class is designed in a special way such that executing:

```
net.DUMMY_NAME = value                                                         1
```

will actually create an entry in a dictionary with key *DUMMY_NAME*.

The next point is that calling *L.LAYERTYPE* will actually create a layer in the text file where type of the layer will be equal to *type="LAYERTYPE"*. Therefore, if we want to create a convolution layer, we have to call L.Convolution. Likewise, creating pooling, loss and ReLU layers is done by calling L.Pooling, L.SoftmaxWithLoss, and L.ReLU, respectively.

Any argument that is passed to L.LAYERTYPE function will be considered as the bottom of the layer. Also, any keyword argument will be treated as the parameters of the layer. In the case that there is a parameter in a layer such as *weight_filler* with a data type other than basic types, the inner parameters of this parameter can be defined using a dictionary in Python.

After that the architecture of network is defined, it can be simply converted to a string by calling *str(net.to_proto())*. Then, this text can be written into a text file and stored on disk.

4.6 Drawing Architecture of Network

The Python interface provides a function for generating a graph for a given network definition text file. This can be done by calling the following function:

```
import sys                                                           1
sys.path.insert(0, "/home/hamed/caffe-master/python")               2
import caffe                                                          3
import caffe.draw                                                     4
from caffe.proto import caffe_pb2                                     5
from google.protobuf import text_format                              6
                                                                     7
def drawcaffe(def_file, save_to, direction ='TB'):                   8
    net = caffe_pb2.NetParameter()                                   9
    text_format.Merge(open(def_file).read(), net)                   10
                                                                    11
    caffe.draw.draw_net_to_file(net, save_to, direction)           12
```

This function uses the *GraphViz* Python module to generate the diagram. The parameter *direction* shows the direction of the graph and it might be called by passing 'TB' (top-bottom), 'BT' (bottom-top), 'LR' (left-right), 'RL' (right-left). The diagrams indicated in this chapter are created by calling this function.

4.7 Training Using Python

After creating the *solver.prototxt* file, we can use it for training the network by writing a Python script rather than command tools. The Python script for training a network might look like:

```
caffe.set_mode_gpu()                                                 1
solver = caffe.get_solver('/tmp/solver.prototxt')                   2
solver.step(25)                                                      3
```

The first line in this code tells Caffe to use GPU instead of CPU. If this command is not executed, Caffe will use CPU by default. The second line in this code loads the solver definition. Because the path of network is also mentioned inside the solver definition, the network is also automatically loaded. Then, calling the *step(25)* function, runs the optimization algorithm for 25 iterations and stops. Assume that *test_interval=100* and we call *solver.step(150)*. If the network is trained using command tools, Caffe will switch from TRAIN to TEST when immediate after 100th iteration. This will also happen when *solver.step(150)* is called. Hence, if you want that the test phase is not automatically invoked by Caffe, the variable *test_interval* must be set to a large number (larger than the variable *max_iter*).

4.8   Evaluating Using Python

Any neural network must be evaluated in three stages. The first evaluation is done during training using training set. The second evaluation is done during training using validation set and the third evaluation is done using test set after that designing and training the network is completely done.

Recall from Sect. 3.5.3 that a network is usually evaluated using a classification metric. All the classification metrics that we explained in that section are based on actual labels and predicted labels of samples. Actual labels of samples are already available in the dataset. However, predicted labels are obtained using the network. That means in order to evaluate a network using one of the classification metrics, it is necessary to predict labels of samples. These samples may come from the training set, the validation set or the test set.

In the case of neural network, we have to feed the samples to the network and forward them through the network. The output layer shows the score of samples for each class. For example, the output layer of the network in Sect. 4.5 is called f3_classifier. We can access the value of the network computed for a sample using the following command:

```
solver = caffe.get_solver('/tmp/solver.prototxt')               1
net = solver.net                                                2
print net.blobs['classifier'].data                              3
```

In the above script, the first line loads a solver along with the network. The filled *solver.net* returns the network that is used for training. In Caffe, a tensor that retains data is encapsulated in objects of type Blob. The field *net.blobs* is a dictionary where keys of this dictionary are the *tops* of network that we have specified in the network definition and value of each entry in this dictionary is an instance of Blob. For example, the top of the classification layer in Sect. 4.5 is called "classifier". The command *net.blobs['classifier']* returns the blob associated with this layer.

The tensor of a blob is accessible through the field *data*. Hence, *net.blobs['KEY'].data* returns the numerical data in a 4D matrix (tensor). This matrix is in fact a Numpy array. The shape of tensors in Caffe is $N \times C \times H \times W$, where N denotes the number of samples in mini-batch and C illustrates the number of channels. As it turns out, H and W also denote the height and width, respectively.

The batch size of the layer "data" in Sect. 4.5 is equal to 30. Also, this layer loads color images (3 channels) of size 32×32. Therefore, the command *net.blobs['data'].data* returns a 4D matrix of shape $40 \times 3 \times 32 \times 32$. Taking into account the fact that layer "classifier" in this network contains 43 neurons, the command *net.blobs['classifier'].data* will return a matrix of size $40 \times 43 \times 1 \times 1$, where each row in this matrix shows class specific score of the first samples in the mini-batch. Each sample belongs to the class with the highest score.

Assume we want to classify a single image which is stored at */home/sample.ppm*. This means that, the size of mini-batch is equal to 1. To this end, we have to load the image in RGB format and resize it to 32×32 pixels. Then, transpose the axis such that the shape of image becomes $3 \times 32 \times 32$. Finally, this matrix has to be converted to a $1 \times 3 \times 32 \times 32$ matrix in order to make it compatible with tensors in Caffe. This can be easily done using the following commands:

```
import numpy as np                                                          1
im = caffe.io.load_image('/home/sample.ppm', color=True)                    2
im = caffe.io.resize(im, (32, 32))                                          3
im = np.transpose(im, [2,0,1])                                              4
im = im[np.newaxis, ...]                                                    5
```

Next, this image has to be fed into the network and the output layers must be computed one by one. Technically, this is called forwarding the samples throughout the network. Assuming that *net* is an instance of Caffe.Net, forwarding the above sample can be easily done by calling:

```
net.blobs['data'].data[...] = im[...]                                       1
net.forward()                                                               2
```

It should be noted that [...] in the above code the image into the memory of field *data*. Removing this from the above line will raise an error since it will mean that we are assigning a new memory to the field *data* rather than updating its memory. At this point, *net.blobs[top].data* returns the output of a top in network. In order to classify the above image in our network, we only need to run the following line:

```
label = np.argmax(net.blobs['classifier'].data, axis=1)                     1
```

This will return the index of the class with maximum score. The general procedure for training a ConvNet is illustrated below.

```
Givens:                                                                     1
        X_train: A dataset containing N images of size WxHx3                2
        Y_train: A vector of length N containing labels of each samples in X_train  3
                                                                            4
        X_valid: A dataset containing K images of size WxHx3                5
        Y_valid: A vector of length K containing labels of each samples in X_valid  6
                                                                            7
FOR t=1 TO MAX                                                              8
    TRAIN THE CONVNET FOR m ITERATIONS USING X_train and Y_train            9
                                                                            10
    EVALUATE THE CONVNET USING X_valid and Y_valid                         11
END FOR                                                                     12
```

The training procedure involves constantly updating parameters using the training set and evaluating the network using the validation set. More specifically, the network is trained for m iterations using the training samples. Then, validations samples are fetched into the network and a classification metric such as accuracy is computed for the samples in the validation set. The above procedure is repeated MAX times and the training is finished. One may wonder why the network must be evaluated during training. As we will see in the next chapter, validation is a crucial step in training a classification model such as neural networks. The following code shows how to implement the above procedure in Python:

```
solver = caffe.get_solver('solver.prototxt')                               1
                                                                            2
with open('validation.txt','r') as file_id:                                3
    valid_set = csv.reader(file_id, delimiter=' ')                         4
    valid_set = [(row[0], int(row[1])) for row in valid_set]               5
                                                                            6
```

```
net_valid = solver.test_nets[0]                                                  7
data_val = np.zeros(net_valid.blobs['data'].data.shape, dtype='float32')         8
label_actual = np.zeros(net_valid.blobs['label'].data.shape, dtype='int8')       9
for i in xrange(500):                                                           10
    solver.step(1000)                                                           11
                                                                                12
    print 'Validating ...'                                                      13
    acc_valid = []                                                              14
    net_valid.share_with(solver.net)                                            15
                                                                                16
    batch_size = net_valid.blobs['data'].data.shape[0]                          17
    cur_ind = 0                                                                 18
                                                                                19
    for _ in xrange(800):                                                       20
        for j in xrange(batch_size):                                            21
            rec = valid_set[cur_ind]                                            22
            im = cv2.imread(rec[0], cv2.cv.CV_LOAD_IMAGE_COLOR).astype('float32') 23
            im = im / 255.                                                      24
            im = cv2.resize(im, (32, 32))                                       25
            im = np.transpose(im, [2,0,1])                                      26
                                                                                27
            data_val[j, ...] = im                                               28
            label_actual[j, ...] = rec[1]                                       29
            cur_ind = cur_ind + 1 if ((cur_ind+1) < len(valid_set)) else 0      30
                                                                                31
        net_valid.blobs['data'].data[...] = data_val                           32
        net_valid.blobs['label'].data[...] = label_actual                      33
        net_valid.forward()                                                    34
                                                                                35
        class_score = net_valid.blobs['classifier'].data.copy()               36
        label_pred = np.argmax(class_score, axis=1)                           37
        acc = sum(label_actual.ravel() == label_pred) / float(label_pred.size) 38
        acc_valid.append(acc)                                                  39
    mean_acc = np.asarray(acc_valid).mean()                                    40
    print 'Validation accuracy: {}'.format(mean_acc)                           41
```

First line loads the solver together with the train and test networks associated with this solver. Line 3 to Line 5 read the validation dataset into a list. Line 8 and Line 9 create containers for validation samples and their labels. The training loop starts at Line 10 and it will be repeated 500 times. The first statement in this loop (Line 11) is to train the network using training samples for 1000 iterations.

After that, validating the network starts at Line 13. The idea is to load 800 minibatches of validation samples, where each mini-batch contains *batch_size* samples. The loop from Line 21 to Line 30, loads color images and resize them using OpenCV functions. It also rescales the pixel intensities to [0, 1]. Rescaling is necessary since the training samples are also rescaled by setting *scale:0.0039215* in the definition of the ImageData layer.[6]

The loaded images are transposed and copied to *data_val* tensor. Label of each sample is also copied into *label_actual* tensor. After filling the mini-batch, it is copied into the first layer of the network in Line 32 and Line 33. Then, it is forwarded throughout the network at Line 34.

---

[6]It is possible to load and scale images using functions in caffe.io module. However, it should be noted that the *imread* function from OpenCV loads color images in BGR order rather than RGB. This is similar to the way the ImageData layer loads images using OpenCV. In the case of using *caffe.io.load_image* function, we must swap R and B channel before feeding them to the network.

Line 36 and Line 37 finds the class of each samples and the accuracy of classification is computed on the mini-batch and it is stored in a list. Finally, the mean accuracy of 800 mini-batches is computed and stored in *mean_acc*. The above code can be used as a basic template for training and validating neural network in Python using Caffe library. It is also possible to keep history of training and validation accuracies in the above code.

However, there are a few points to bear in mind. First, the same transformations must be applied on the validation/test samples as we have used for training samples. Second, the validation samples must be identical every time the network is evaluated. Otherwise, it might not be trivial to assess the network properly. Third, as we discussed earlier, F1-score can be computed over all validation samples rather than accuracy.

4.9 Save and Restore Networks

During training, we might want to save and restore the parameters of the network. In particular, we will need the value of trained parameters in order to load them into the network and use the network in real-world applications. This can be done by writing a customized function to read the value of *net.params* dictionary and save them in a file. Later, we can load the same values to *net.params* dictionary.

Another way is to use the built-in functions in Caffe library. Specifically, the *net.save(string filename)* and the *net.copy_from(string filename)* functions saves the parameters into a binary file and loads them into the network, respectively.

In some cases, we may also want to save information related to the optimizer such as current iteration, current learning rate, current momentum, etc., besides the parameters of network. Later, this information can be loaded into the optimizer as well as the network in order to resume the training from the last stopped point. Caffe provides *solver.snapshot()* and *solver.restore(string filename)* functions for these purposes.

Assume the field *snapshot_prefix* is set to *"/tmp/cnn"* in the solver definition file. Calling *solver.snapshot()* will create two files as follows:

```
/tmp/cnn_iter_X.caffemodel                                                       1
/tmp/cnn_iter_X.solverstate                                                      2
```

where X is automatically replaced by Caffe with the current iteration of the optimization algorithm. In order to restore the state of the optimization algorithm from a disk, we only need to call *solver.restore(filename)* with a path to a valid *.solverstate* file.

4.10   Python Layer in Caffe

One limitation of the Caffe library is that we are obliged to only utilize the implemented layers of this library. For example, the softplus activation function is not implemented in the current version of the Caffe library. In some cases, we may want to add a layer with a new function that is not implemented in the Caffe library. The obvious solution is to implement this layer directly in C++ by inheriting our classes from classes of the Caffe library. This could be a tedious task especially when the goal is to quickly implement and test our idea.

A more likely scenario in which having a special layer could be advantageous when we work with different datasets. For instance, there are thousands of samples in the GTSRB dataset for the task of traffic sign classification. The bounding box information of each image is provided using a text file. Apparently, these images have to be cropped to exactly fit the bounding box before feeding to a classification network.

This can be done in three ways. The first way is to process whole dataset and crop each image based on their bounding box information and store them on the disk. Then, the processed dataset can be used for training/validation/testing the network. The second solution is to process images on the fly and fill each mini-batch after processing the images. Then these mini-batches can be used for training/validation/testing. However, it should be noted that using this method we will not be longer able to call the *solver.step(int)* function with an argument greater than one or set *iter_size* to a value greater than one. The reason is that, each mini-batch must be filled manually using our code. The third method is to develop a new layer which automatically reads images from the dataset, processes, and passes them to the output (top) of the layer. Using this method, the *solver.step(int)* function can be called with any arbitrary positive number.

The Caffe library provides a special type of layer called PythonLayer. Using this layer, we are able to develop new layers in Python which can be accessed by Caffe. A Python layer is configured using an instance of PythonParameter which is defined as follows:

```
message PythonParameter {                                                    1
    optional string module = 1;                                              2
    optional string layer = 2;                                               3
    optional string param_str = 3 [default = ''];                            4
}                                                                            5
```

Based on this definition, a Python layer might look like:

```
layer {                                                                      1
    name: "data"                                                             2
    type: "Python"                                                           3
    top: "data"                                                              4
    python_param {                                                           5
        module: "python_layer"                                               6
        layer: "mypythonlayer"                                               7
        param_str: "{\'param1\':1, \'param2\':2.5}"                          8
    }                                                                        9
}                                                                           10
```

The variable *type* of a Python layer must be set to *Python*. Upon reaching to this layer, Caffe will look for *python_layer.py* file next to the *.prototxt* file. Then, it will look for a class called *mypythonlayer* inside this file. Finally, it will pass *"param1':1, param2':2.5"* into this class. Caffe will interact with *mypythonlayer* using four methods inside this class. Below is the template that must be followed in designing a new layer in Python.

```
class mypythonlayer(caffe.Layer):                                    1
    def setup(self, bottom, top):                                    2
        pass                                                         3
                                                                     4
    def reshape(self, bottom, top):                                  5
        pass                                                         6
                                                                     7
    def forward(self, bottom, top):                                  8
        pass                                                         9
                                                                     10
    def backward(self, top, propagate_down, bottom):                 11
        pass                                                         12
```

First, the class must be inherited from *Caffe.Layer*. The *setup* method will be called only once when Caffe creates train and test networks. The *backward* method is only called during the backpropagation step. Computing the output of each layer given an input is done by calling *net.forward()* method. Whenever this method is called, the *reshape* and *forward* methods of the layer will be called automatically. The *reshape* method is always called before *forward* method.

It is noteworthy to draw you attention to the prototype of the *backward* method. In contrast to the other three methods, where the first argument is *bottom* and the last argument is *top*, in the *backward* method the places of these two arguments are switched. So, a great care must be taken into account in defining the prototype of the method. Otherwise, you may end up with a layer, where the gradients are not computed correctly. For instance, let us implement the PReLU activation using the Python layer. In this implementation, we consider a distinct PReLU activation for each feature map.

```
class prelu(caffe.Layer):                                            1
    def setup(self, bottom, top):                                    2
        params = eval(self.param_str)                                3
        shape = [1]*len(bottom[0].data.shape)                        4
        shape[1] = bottom[0].data.shape[1]                           5
        self.axis = range(len(shape))                                6
        del self.axis[1]                                             7
        self.axis = tuple(self.axis)                                 8
                                                                     9
        self.blobs.add_blob(*shape)                                  10
        self.blobs[0].data[...] = params['alpha']                    11
                                                                     12
    def reshape(self, bottom, top):                                  13
        top[0].reshape(*bottom[0].data.shape)                        14
                                                                     15
    def forward(self, bottom, top):                                  16
        top[0].data[...] = np.where(bottom[0].data > 0,              17
                              bottom[0].data,                        18
                              self.blobs[0].data*bottom[0].data)     19
                                                                     20
    def backward(self, top, propagate_down, bottom):                21
        self.blobs[0].diff[...] = np.sum(np.where(bottom[0].data > 0,  22
                              np.zeros(bottom[0].data.shape),        23
```

```
                                    bottom[0].data) * top[0].diff,                    24
                                 axis=self.axis, keepdims=True)                       25
          bottom[0].diff[...] = np.where(bottom[0].data > 0,                          26
                                 np.ones(bottom[0].data.shape),                       27
                                 self.blobs[0].data) * top[0].diff                    28
```

The *setup* method converts the *param_str* value specified in the network definition into a dictionary. Then, the shape of parameter vector is determined. Specifically, if the shape of bottom layer is $N \times C \times H \times W$, the shape of parameter vector must be $1 \times C \times 1 \times 1$. The dimensions of array with length 1 will be broadcasted during operations by Numpy. Since there are C feature maps in the bottom layer, there must also be C PReLU activations with different values of α.

In the case of fully connected layers, the bottom layer might be a two-dimensional array instead of four-dimensional array. The *shape* variable in this method ensures that the parameter vector will have a shape consistent with the bottom layer.

The variable *axis* indicates the axis to which the summation of gradient must be performed. Again, this axis also must be consistent with the shape of bottom layer.

Line 10 creates a parameter array in which the shape of this array is determined using the variable *shape*. Note the unpacking operator in this line. Line 11 initializes α of all PReLU activations with a constant number. The setup method is called once and it initializes all parameters of the layer.

The reshape method, determines the shape of top layer in Line 14. The channel-wise PReLU activations are applied on the bottom layer and assigned to the top layer. Note how we have utilized broadcasting of Numpy arrays in order to multiply parameters with the bottom layer. Finally, the backward method computes the gradient with respect to parameters and gradient with respect to the bottom layer.

4.11 Summary

There are various powerful libraries such as Theano, Lasagne, Keras, mxnet, Torch, and TensorFlow that can be used for designing and training neural networks including convolutional neural networks. Among them, Caffe is a library that can be used for both doing research and developing real-world applications. In this chapter, we explained how to design and train neural networks using the Caffe library. Moreover, the Python interface of Caffe was discussed using real examples. Then, we mentioned how to develop new layers in Python and use them in neural networks.

4.12 Exercises

4.1 Suppose the following text files:

```
/sample1.jpg 0                                                                        1
/sample2.jpg 0                                                                        2
/sample3.jpg 0                                                                        3
```

/sample4.jpg 0	4
/sample5.jpg 1	5
/sample6.jpg 1	6
/sample7.jpg 1	7
/sample8.jpg 1	8

/sample7.jpg 1	1
/sample1.jpg 0	2
/sample3.jpg 0	3
/sample6.jpg 1	4
/sample4.jpg 0	5
/sample5.jpg 1	6
/sample2.jpg 0	7
/sample8.jpg 1	8

From optimization algorithm perspective, which one of the above files is appropriate for passing to an ImageData layer? Also, which of these files hast to be shuffled before starting the optimization? Why?

4.2 Shifted ReLU activation is given by Clevert et al. (2015):

$$f(x) = \begin{cases} x - 1 & x > 0 \\ -1 & otherwise \end{cases} \tag{4.1}$$

This activation function is not basically implemented in Caffe. However, you can implement it using current layers in this library. Use a ReLU layer together with Bias layer to implement this activation function in Caffe. A bias layer basically adds a constant to bottom blobs. You can find more information in *caffe.proto* about this layer.

4.3 Why and when shuffle of an ImageData layer in TEST phase must be set to false.

4.4 When setting shuffle to true or false in TEST phase does not matter?

4.5 What happens if we add *include* to the first convolution layer in the network we mentioned in this chapter and set *phase=TEST* for this layer?

4.6 Add codes to the Python script in order to keep the history of training and validation accuracies and plot them using Python.

4.7 How we can check the gradient of the implemented PReLU layer using numerical methods?

4.8 Implement the softplus activation function using a Python layer.

Reference

Clevert DA, Unterthiner T, Hochreiter S (2015) Fast and accurate deep network learning by exponential linear units (ELUs) 1997:1–13. arXiv:1511.07289.pdf

Classification of Traffic Signs

5

5.1 Introduction

Car industry has significantly progressed in the last two decades. Today's car not only are faster, more efficient, and more beautiful they are also safer and smarter. Improvements in safety of cars are mainly due to advances in hardware and software. From software perspective, cars are becoming smarter by utilizing artificial intelligence. The component of a car which is basically responsible for making intelligent decisions is called Advanced Driver Assistant System (ADAS).

In fact, ADASs are indispensable part of smart cars and driverless cars. Tasks such as adaptive cruise control, forward collision warning, and adaptive light control are performed by an ADAS.[1] These tasks usually obtain information from sensors other than a camera.

There are also tasks which may directly work with images. Drivers fatigue (drowsiness) detection, pedestrian detection, blind spot monitor, drivable lane detection are some of the tasks that chiefly depends on images obtained by cameras. There is one task in ADASs which is the main focus of the next two chapters in this book. This task is recognizing vertical traffic signs.

A driver-less car might not be considered smart if it is not able to automatically recognize traffic signs. In fact, traffic signs help a driver (human or autonomous) to conform with road rules and drive the car, safely. In the near future when driver-less cars will be common, a road might be shared by human drivers as well as driver-less cars. Consequently, it is rational to expect that driver-less cars at least perform as good as a human driver.

Humans are good at understanding scene and recognizing traffic signs using their vision. There are two major goals in designing traffic signs. First, they must be easily distinguishable from rest of objects in the scene and, second, their meaning must be

---

[1]An ADAS may perform many other intelligent tasks. The list here is just a few examples.

© Springer International Publishing AG 2017

H. Habibi Aghdam and E. Jahani Heravi, *Guide to Convolutional Neural Networks*, DOI 10.1007/978-3-319-57550-6_5

easily perceivable and independent from spoken language. To this end, traffic signs are designed with a simple geometrical shape such as triangle, circle, rectangle, or polygon. To be easily detectable from the rest of objects, traffic signs are painted using basic colors such as red, blue, yellow, black, and white. Finally, the meaning of traffic signs are mainly carried out by pictographs in the center of traffic signs. It should be noted that some signs heavily depend on text-based information. However, we can still think of the texts in traffic signs as pictographs.

Even though classification of traffic signs is an easy task for a human, there are some challenges in developing an algorithm for this purpose. This challenges are illustrated in Fig. 5.1. First, the image of traffic signs might be captured from different perspectives. Second, whether condition may dramatically affect the appearance of traffic signs. An example is illustrated in the figure where the "no stopping" sign is covered by snow. Third, traffic signs are being impaired during time. Because of that color of traffic sign is affected and some artifacts may appear on the sign which might have a negative impact on the classification score. Fourth, traffic signs might be partially occluded by another signs or objects. Fifth, the pictograph area might be manipulated by human which in some cases might change the shape of the pictograph. The last issue shown in this figure is the pictograph differences of

Fig. 5.1 Some of the challenges in classification of traffic signs. The signs have been collected in Germany and Belgium

the same traffic sign from one country to another. More specifically, we observe that the "danger: bicycle crossing" sign posses a few important differences between two countries.

Beside the aforementioned challenges, motion blur caused by sudden camera movements, shadow on traffic signs, illumination changes, weather condition, and daylight variations are other challenges in classifying traffic signs.

As we mentioned earlier, traffic sign classification is one of the tasks of an ADAS. Consequently, their classification must be done in real time and it must consume as few CPU cycles as possible in order to release the CPU immediately. Last but not the least, the classification model must be easily scalable so the model can be adjusted to new classes in the future with a few efforts. In sum, any model for classifying traffic signs must be *accurate*, *fast*, *scalable*, and *fault-tolerant*.

Traffic sign classification is a specific case of object classification where objects are more rigid and two dimensional. Recently, ConvNets surpassed human on classification of 1000 natural objects (He et al. 2015). Moreover, there are other ConvNets such as Simonyan and Zisserman (2015) and Szegedy et al. (2014a) with close performances compared to He et al. (2015). However, the architecture of the ConvNets is significantly different from each other. This suggests that the same problem might be solved using ConvNets with different architectures and various complexities. Part of the complexity of ConvNets is determined using activation functions. They play an important role in neural networks since they apply nonlinearities on the output of the neurons which enable ConvNets to apply series of nonlinear functions on the input and transform the input into a space where classes are linearly separable. As we discuss in the next sections, selecting a highly computational activation function can increase the number of required arithmetic operations of the network which in sequel increases the response-time of a ConvNet.

In this chapter, we will first study the methods for recognizing traffic signs and then we will explain different network architectures for the task of traffic signs classification. Moreover, we will show how to implement and train these networks on a challenging dataset.

5.2 Related Work

In general, efforts for classifying traffic signs can be divided into *traditional classification* and *convolutional neural network*. In the former approach, researchers have tried to design hand-crafted features and train a classifier on top of these features. In contrast, convolutional neural networks learn the representation and classification automatically from data. In this section, we first review the traditional classification approaches and then we explain the previously proposed ConvNets for classifying traffic signs.

5.2.1   Template Matching

Early works considered a traffic sign as a rigid object and classified the query image by comparing it with all templates stored in the database (Piccioli et al. 1996). Later, Gao et al. (2006) matched shape features instead of pixel intensity values. In this work, matching features is done using Euclidean distance function. The problem with this matching function is that they consider every pixel/feature equally important. To cope with this problem, Ruta et al. (2010) learned a similarity measure for matching the query sign with templates.

5.2.2   Hand-Crafted Features

More accurate and robust results were obtained by learning a classification model over a feature vector. Paclík et al. (2000) produce a binary image depending on the color of the traffic sign. Then, moment invariant features are extracted from the binary image and fetched into a one-versus-all Laplacian kernel classifier. One problem with this method is that the query image must be binarized before fetching into the classifier. Maldonado-Bascon et al. (2007) addressed this problem by transforming the image into the HSI color space and calculating histogram of Hue and Saturation components. Then, the histogram is classified using a multiclass SVM. In another method, Maldonado Bascón et al. (2010) classified traffic signs using only the pictograph of each sign. Although the pictograph is a binary image, however, accurate segmentation of a pictogram is not a trivial task since automatic thresholding methods such as Otsu might fail taking into account the illumination variation and unexpected noise in real-world applications. For this reason, Maldonado Bascón et al. (2010) trained SVM where the input is a 31×31 block of pixels in a gray-scale version of pictograph. In a more complicated approach, Baró et al. (2009) proposed an Error Correcting Output Code framework for classification of 31 traffic signs and compared their method with various approaches.

Before 2011, there was not a public and challenging dataset of traffic signs. Radu Timofte (2011), Larsson and Felsberg (2011) and Stallkamp et al. (2012) introduced three challenging datasets including annotations. These databases are called Belgium Traffic Sing Classification (BTSC), Swedish Traffic Sign, and German Traffic Sign Recognition Benchmark (GTSRB), respectively. In particular, the GTSRB was used in a competition and, as we will discuss shortly, the winner method classified 99.46% of test images correctly (Stallkamp et al. 2012). Zaklouta et al. (2011) and Zaklouta and Stanciulescu (2012, 2014) extracted Histogram of Oriented Gradient (HOG) descriptors with three different configurations for representing the image and trained a Random Forest and a SVM for classifying traffic sings in the GTSRB dataset. Similarly, Greenhalgh and Mirmehdi (2012), Moiseev et al. (2013), Huang et al. (2013), Mathias et al. (2013) and Sun et al. (2014) used the HOG descriptor. The main difference between these works lies in the utilized classification model (e.g., SVM, Cascade SVM, Extreme Learning Machine, Nearest Neighbour, and LDA). These works except (Huang et al. 2013) use the traditional classification approach.

In contrast, Huang et al. (2013) utilize a two level classification. In the first level, the image is classified into one of super-classes. Each super-class contains several traffic signs with similar shape/color. Then, the perspective of the input image is adjusted based on its super-class and another classification model is applied on the adjusted image. The main problem of this method is sensitivity of the final classification to the adjustment procedure. Timofte et al. (2011) proposed a framework for recognition and the traffic signs in the BTSC dataset and achieved 97.04% accuracy on this dataset.

5.2.3   Sparse Coding

Hsu and Huang (2001) coded each traffic sign using the Matching Pursuit algorithm. During testing, the input image is projected to different set of filter bases to find the best match. Lu et al. (2012) proposed a graph embedding approach for classifying traffic signs. They preserved the sparse representation in the original space by using $L_{1,2}$ norm. Liu et al. (2014) constructed the dictionary by applying k-means clustering on the training data. Then, each data is coded using a novel coding input similar to Local Linear Coding approach (Wang et al. 2010). Recently, a method based on visual attributes and Bayesian network was proposed in Aghdam et al. (2015). In this method, we describe each traffic sign in terms of visual attributes. In order to detect visual attributes, we divide the input image into several regions and code each region using the Elastic Net Sparse Coding method. Finally, attributes are detected using a Random Forest classifier. The detected attributes are further refined using a Bayesian network.

Fleyeh and Davami (2011) projected the image into the principal component space and find the class of the image by computing the Euclidean distance of the projected image with the images in the database. Yuan et al. (2014) proposed a novel feature extraction method to effectively combine color, global spatial structure, global direction structure, and local shape information. Readers can refer to Møgelmose et al. (2012) to study traditional approaches of traffic sign classification.

5.2.4   Discussion

Template matching approaches are not robust against perspective variations, aging, noise, and occlusion. Hand-crafted features has a limited representation power and they might not scale well if the number of classes increases. In addition, they are not robust against irregular artifacts caused by motion blurring and weather condition. This can be observed by the results reported in the GTSRB competition (Stallkamp et al. 2012) where the best performed solution based on hand-crafted feature was only able to correctly classify 97.88% of test cases.[2] Later, Mathias et al. (2013) improved

---

[2]http://benchmark.ini.rub.de/.

the accuracy based on hand-crafted features up to 98.53% on the GTSRB dataset. Notwithstanding, there are a few problems with this method. Their raw feature vector is a 9000 dimensional vector constructed by applying five different methods. This high dimensional vector is later projected to a lower dimensional space. For this reason, their method is time consuming when they are executed on a multi-core CPU. Note that Table V in Mathias et al. (2013) have only reported the time on classifiers and it has disregarded the time required for computing feature vectors and projecting them into a lower dimension space. Considering that the results in Table V have been computed on the test set of the GTSRB dataset (12630 samples), only classification of a feature vector takes 48 ms.

5.2.5 ConvNets

ConvNets were utilized by Sermanet and Lecun (2011) and Ciresan et al. (2012a) in the field of traffic sign classification during the GTSRB competition where the ConvNet of (Ciresan et al. 2012a) surpassed human performance and won the competition by correctly classifying 99.46% of test images. Moreover, the ConvNet of (Sermanet and Lecun 2011) ended up in the second place with a considerable difference compared with the third place which was awarded for a method based on the traditional classification approach. The classification accuracies of the runner-up and the third place were 98.97 and 97.88%, respectively.

Ciresan et al. (2012a) constructs an ensemble of 25 ConvNets each consists of 1,543,443 parameters. Sermanet and Lecun (2011) create a single network defined by 1,437,791 parameters. Furthermore, while the winner ConvNet uses the *hyperbolic* activation function, the runner-up ConvNet utilizes the *rectified sigmoid* as the activation function. Both methods suffer from the high number of arithmetic operations. To be more specific, they use highly computational activation functions. To alleviate these problems, Jin et al. (2014) proposed a new architecture including 1,162,284 parameters and utilizing the *rectified linear unit* (ReLU) activations (Krizhevsky et al. 2012). In addition, there is a Local Response Normalization layer after each activation layer. They built an ensemble of 20 ConvNets and classified 99.65% of test images correctly. Although the number of parameters is reduced using this architecture compared with the two networks, the ensemble is constructed using 20 ConvNets which is not still computationally efficient in real-world applications. It is worth mentioning that a ReLU layer and a Local Response Normalization layer together needs approximately the same number of arithmetic operations as a single hyperbolic layer. As the result, the run-time efficiency of the network proposed in Jin et al. (2014) might be close to Ciresan et al. (2012a).

Recently, Zeng et al. (2015) trained a ConvNet to extract features of the image and replaced the classification layer of their ConvNet with an Extreme Learning Machine (ELM) and achieved 99.40% accuracy on the GTSRB dataset. There are two issues with their approach. First, the output of last convolution layer is a 200 dimensional vector which is connected to 12,000 neurons in the ELM layer. This layer is solely defined by $200 \times 12,000 + 12,000 \times 43 = 2,916,000$ parameters which makes it impractical. Besides, it is not clear why their ConvNet reduces the dimension of the feature vector from $250 \times 16 = 4000$ in Layer 7 to 200 in Layer 8 and then map their lower dimensional vector to 12,000 dimensions in the ELM layer (Zeng et al. 2015, Table 1). One reason might be to cope with calculation of the matrix inverse during training of the ELM layer. Finally, since the input connections of the ELM layer is determined randomly, it is probable that their ConvNets do not generalize well on other datasets.

5.3   Preparing Dataset

In the rest of this book, we will design different ConvNets for classification of traffic signs in the German Traffic Sign Recognition Benchmark (GTSRB) dataset (Stallkamp et al. 2012). The dataset contains 43 classes of traffic signs. Image of traffic signs are in RGB format and their are stored in Portable Pixel Map (PPM) format. Furthermore, each image contains only one traffic sign and they vary from 15×15 to 250×250 pixels. The training set consists of 39,209 images and the test set contains 12,630 images. Figure 5.2 shows one sample for each class from this dataset.

It turns out that images of this dataset are collected in real-world conditions. They possess some challenges such as blurry images, partially occluded signs, low

Fig. 5.2 Sample images from the GTSRB dataset

resolution, and poor illumination. The first thing to do in any dataset including the GTSRB dataset is to split the dataset into *training* set, *validation* set, and *test* set. Fortunately, the GTSRB dataset comes with a separate test and training set. However, it does not contain a validation set.

5.3.1 Splitting Data

Given any dataset, our first task is to divide it into one *training* set, **one or more** *validation* sets, and one *test* set. In the case, that the test set and training set are drawn from the same distribution we do not usually need more than one validation set. Simply speaking, set of images are drawn from the same distribution if they are collected under the same condition. The term condition here may refer to model of camera, pose of camera with respect to the reference coordinate system, geographical information of collection images, illumination, and etc. For example, if we collect training images during daylight and the test images at night, these two sets are now drawn from the same distribution. As another example, if the training images are collected in Spain and the test images are collected in Germany, it is likely that images are not drawn from the same distribution. If training and test are not drawn from the same distribution, we usually need more than one validation set to assess our models.

However, for the sake of simplicity, we consider that whole dataset is drawn from the same distribution. Our task is to divide this dataset into the three sets that we mentioned above. Before doing that, we have to decide the ratio of each set with respect to whole dataset. For example, one may split the dataset such that 80% of samples are assigned to training set, 10% to validation set, and 10% to test set. Other common choices are $60 - 20 - 20\%$ and $70 - 15 - 15\%$ for training, validation and test sets, respectively.

The main idea behind splitting data into different sets is to evaluate whether or not the trained model is generalized on unseen samples. We have discussed in detail about this in Sect. 3.7. If the number of samples in the dataset is very high and they are diverse, splitting data with ratio of $80 - 10 - 10\%$ is a good choice. One can take 100 photos from the same traffic sign with slight changes in camera pose. Then, if this process is repeated for 10 signs the collected dataset will contain 1000 samples. Even though the number of samples is high the samples might not be diverse. When the number of samples is very high and they are diverse, these samples adequately cover the input space so the chance of generalization increases. For this reason, we might not need a lot of validation or test samples to assess how well the model is generalized.

Notwithstanding, when the number of samples is low, $60 - 20 - 20\%$ split ratio might be a better choice since with smaller number of training samples the model might overfit on training data which can dramatically reduce its generalization. However, when the number of validation and test samples is high, it is possible to assess the model more accurately.

After deciding about the split ratio, we have to assign each sample in the dataset into one and only one of these sets. Note that a sample cannot be assigned to more than one set. Next, we explain three ways for splitting dataset $\mathscr{X}$ into disjoint sets $\mathscr{X}_{train}$, $\mathscr{X}_{validation}$, and $\mathscr{X}_{test}$.

5.3.1.1 Random Sampling

In random sampling, samples are selected using uniform distribution. Specifically, all samples have the same probability to be assigned to one of the sets without replacement. This method is not deterministic meaning that if we run the algorithm 10 times, we will end up with 10 different training, validation, and test sets. The easiest way to make this algorithm deterministic is to always seed the random number generator with a constant value.

Implementing this method is trivial and its complexity is a linear function of number of samples in the original set. However, if $|\mathscr{X}_{train}| \ll |\mathscr{X}|$ it is likely that the training set does not cover the input space properly so the model may learn the training data accurately but it does not generalize well on the test samples. Technically, this may lead to a model with high variance. Notwithstanding, random sampling is very popular approach and it works well in practice especially when $\mathscr{X}$ is large.

5.3.1.2 Cluster Based Sampling

In cluster based sampling, the input space is first partitioned into K clusters. The partitioning can be done using common clustering methods such as *k-means*, *c-mean*, and *hierarchical clustering*. Then, for each cluster, some of the samples are assigned to $\mathscr{X}_{train}$, some of them are assigned to $\mathscr{X}_{validation}$, and the rest are assigned to $\mathscr{X}_{test}$. This method ensures that each of these three sets covers the whole space represented by $\mathscr{X}$. Assigning samples from each cluster to any of these sets can be done using the uniform sampling approach. Again, the sampling has to be without replacement.

The advantage of this method is that each set adequately covers the input space. Nonetheless, this method might not be computationally tractable to be applied on large and high dimensional datasets. This is due to the fact that that clustering algorithms are iterative methods and applying them on large datasets may need a considerable time in order to minimize their cost function.

5.3.1.3 DUPLEX Sampling

DUPLEX is a deterministic method that selects samples based on their mutual Euclidean distance in the input space. The DUPLEX sampling algorithm is as follows:

```
Input:                                                                              1
    Set of samples 𝒳                                                               2
Outputs:                                                                            3
    Training set 𝒳_train                                                           4
    validation set 𝒳_validation                                                    5
    Test set 𝒳_test                                                                6
                                                                                    7
𝒳_train = ∅                                                                        8
𝒳_validation = ∅                                                                   9
𝒳_test = ∅                                                                         10
FOR 𝒳_t ∈ {𝒳_train, 𝒳_validation, X_test} REPEAT                                   11
    x_1, x_2 = max_{x_i, x_j ∈ 𝒳} ||x_i − x_j||                                     12
        𝒳_t = 𝒳_t ∪ {x_1, x_2}                                                     13
        𝒳 = 𝒳 − {x_1, x_2}                                                         14
END FOR                                                                             15
While 𝒳 ≠ ∅ REPEAT:                                                                16
    FOR 𝒳_t ∈ {𝒳_train, 𝒳_validation, X_test} REPEAT                               17
        IF |𝒳_t| == n_t THEN                                                        18
            continue                                                                19
        END IF                                                                      20
        x = max_{x_i ∈ 𝒳} min_{x_j ∈ 𝒳_t} ||x_i − x_j||                            21
        𝒳_t = 𝒳_t ∪ {x}                                                           22
        𝒳 = 𝒳 − {x}                                                               23
    END FOR                                                                         24
END WHILE                                                                           25
```

Listing 5.1 DUPLEX algorithm

In the above algorithm, n_t denotes the maximum number of samples in the set. It is computed based on split ratio of samples. First, it finds the two samples with maximum Euclidean distance and assigns them to the training set. Then, these samples are removed from the original set. This process is repeated for the validation and test sets as well (Lines 11–15).

The second loop (Lines 16–22) is repeated until the original set is empty. At each iteration, it finds the sample from $\mathcal{X}$ with maximum distance from the closest sample in $\mathcal{X}_t$. This sample is added to $\mathcal{X}_t$ and removed from $\mathcal{X}$. This procedure is repeated for $\mathcal{X}_{train}$, $\mathcal{X}_{validation}$, and $\mathcal{X}_{test}$, respectively as each iteration.

The DUPLEX algorithm guarantees that each of the sets will cover the input space. However, as it turns out, this algorithm is not computationally efficient and it is not feasible to apply it on large and high dimensional datasets.

5.3.1.4 Cross-Validation

In some cases, we may have a special set for testing our models. Alternatively, the test set $\mathcal{X}_{test}$ might be extracted from the original set $\mathcal{X}$ using one of the methods in the previous section. Let $\mathcal{X}'$ denotes the set after subtracting $\mathcal{X}_{test}$ from $\mathcal{X}$ without ($\mathcal{X}' = \mathcal{X} - \mathcal{X}_{test}$).

The aim of cross-validation is to split $\mathcal{X}'$ into training and validation sets. In the previous section, we mentioned how to divide $\mathcal{X}'$ into only one training and one validation set. This is a method that is called *hold-out* cross-validation where there are only one training, one validation, and one test set.

Cross-validation techniques are applied on $\mathcal{X}'$ rather than $\mathcal{X}$ (The test set is never modified). If number of data in $\mathcal{X}'$ is high, the hold-out cross-validation might be the first choice. However, one can use the random sampling technique and create more than one training/validation sets. Then, training and validating the model will be done using each pair of training/validation sets separately and the average of evaluation will be reported. It should be noted that training the model starts from scratch with each training/validation pair. This method is called *repeated random sub-sampling* cross-validation. This method can be useful in practical applications since it provides better estimate of generalization of the model. We encourage the reader to study other cross-validation techniques such as *K-fold* cross-validation.

5.3.2   Augmenting Dataset

Assume a training image $\mathbf{x}_i \in \mathbb{R}^{H \times W \times 3}$. To human eye, the class of any sample $\mathbf{x}_j \in \mathbb{R}^{H \times W \times 3}$ where $\|x_i - x_j\| < \varepsilon$ is exactly the same as $\mathbf{x}_i$. However, in the case of ConvNets these slightly modified samples might be problematic (Szegedy et al. 2014b). Techniques for augmenting a dataset try to generate several $\mathbf{x}_j$ for each samples in the training set $\mathcal{X}_{train}$. This way the model will be able to be adjusted not only on a sample but also on its neighbors.

In the case of datasets composed of only images, this is analogous to slightly modifying $\mathbf{x}_i$ and generating $\mathbf{x}_j$ in its close neighborhood. Augmenting dataset is important since it usually improves the accuracy of the ConvNets and makes them more robust to small changes in the input. We explain the reason on Fig. 5.3.

The middle image is the flipped version of the left image and the right image is another sample from training set. Denoting the left, middle, and right images with $\mathbf{x}_l$, $\mathbf{x}_m$ and $\mathbf{x}_r$, respectively; their pairwise Euclidean distances are equal to:

$$\|\mathbf{x}_l - \mathbf{x}_m\| = 25{,}012.5$$
$$\|\mathbf{x}_l - \mathbf{x}_r\| = 27{,}639.4$$
$$\|\mathbf{x}_r - \mathbf{x}_m\| = 26{,}316.0. \tag{5.1}$$

Fig. 5.3 The image in the *middle* is the flipped version of the image in the *left*. The image in the *right* another sample from dataset. Euclidean distance from the *left* image to the *middle* image is equal to 25,012.461 and the Euclidean distance from the *left* image to the *right* image is equal to 27,639.447

In other words, in terms of Euclidean distance in image space (a $\mathbb{R}^{H \times W \times 3}$ space), these images are located in approximately close distances from each other. We might have expected that $\|\mathbf{x}_l - \mathbf{x}_m\|$ to be much smaller than $\|\mathbf{x}_l - \mathbf{x}_r\|$. However, computing the pairwise Euclidean distance between these three samples revels not it is not always the case.[3] Augmenting the training set with flipped images will help the training set to cover the input space better and this way, it improves the accuracy.

It should be noted that a great care must be taken into account when the dataset is augmented by flipping images. The reason is that flipping an image may completely change the class of object. For example, flipping image of "danger: curve to the right" sign will alter its meaning to "danger: curve to left" sign. There are many other techniques for augmenting training set with slightly modified samples. Next, we will explain some of these techniques.

5.3.2.1 Smoothing

Samples can be smoothed using blurring filters such as average filters or Gaussian filters. Smoothing images mimics the out-of-focus effect in cameras. Augmenting dataset using this technique makes the model more tolerant to blurring effect of cameras. Smoothing an image using a Gaussian filter can be simply done using the OpenCV library:

```
import cv2                                                                          1
import numpy as np                                                                  2
                                                                                    3
def smooth_gaussian(im, ks):                                                        4
    sigma_x = (ks[1] // 2.) / 3.                                                     5
    sigma_y = (ks[0] // 2.) / 3.                                                     6
    return cv2.GaussianBlur(im, ksize=ks, sigmaX=sigma_x, sigmaY=sigma_y)           7
```

Concretely, an image can be smoothed using different kernels sizes. Bigger kernels makes blurring effect stronger. It is worth mentioning that *cv2.GaussianBlur* returns an image with the same size as its input by default. It internally manages the borders of images. Also, depending on how much you may want to simulate the out-of-focus effect you may apply the above function on the same sample with different kernel sizes.

5.3.2.2 Motion Blur

A camera mounted on a car is in fact a moving camera. Because of that stationary objects in road appears as moving objects in sequence of images. Depending on the shutter speed, ISO speed an speed of car, images taken from these objects might be degraded by camera motion effect. Accurate simulation of this effect might not be trivial. However, there is a simple approach for simulating motion blur effect using linear filters. Assume we want to simulate a linear motion where the camera

---

[3]Note that Euclidean distance in high-dimensional spaces might be close even for far samples.

is moved along a line with orientation θ. To this end, a filter must be created where all the elements of this filter is zero except the elements lying on the line with orientation θ. These elements will be assigned 1 and finally the elements of matrix will be normalized to ensure that the result of convolution will be always within the valid range of pixel intensities. This function can be implemented in Python as follows:

```
def motion_blur(im, theta, ks):                                              1
    kernel = np.zeros((ks, ks), dtype='float32')                             2
                                                                             3
    half_len = ks // 2                                                       4
    x = np.linspace(-half_len, half_len, 2*half_len+1,dtype='int32')         5
    slope = np.arctan(theta*np.pi/180.)                                      6
    y = -np.round(slope*x).astype('int32')                                   7
    x += half_len                                                            8
    y += half_len                                                            9
                                                                            10
    kernel[y, x] = 1.0                                                      11
    kernel = np.divide(kernel, kernel.sum())                                12
    im_res = cv2.filter2D(im, cv2.CV_8UC3, kernel)                          13
                                                                            14
    return im_res                                                           15
```

Note that control statements such as "if the size of filter is odd" or "if it is bigger than a specific size" are removed from the above code. The *cv2.filter2D* function handles the border effect internally by default and it returns an image with the same size as its input. Motion filters might be applied on the same sample with different orientations and sizes in order to simulate wide range of motion blur effects.

5.3.2.3 Median Filtering

Median filters are edge preserving filters which are used for smoothing images. Basically, for each pixel on image, all neighbor pixels in a small window are sorted based on their intensity. Then, value of current pixel is replaced with the median of the sorted intensities. This filtering approach can be implemented as follows:

```
def blur_median(im, ks):                                                     1
    return cv2.medianBlur(im, ks)                                            2
```

The second parameter in this function is a scaler defining the size of square window around each pixel. In contrast to the previous smoothing methods, it is not common to apply median filter with large windows. The reason is that it may not generate real images taken in real scenarios. However, depending on the resolution of input images, you may use median filtering with a 7×7 kernels size. For low-resolution images such as traffic signs a 3×3 kernel usually produce realistic images. Applying a median filter with larger kernel sizes may not produce realistic images.

5.3.2.4 Sharpening

Contrary to smoothing, it is also possible to sharpen an image in order to make their finer details stronger. For example, edges and noisy pixel are two examples of fine details. In order to sharpen an image, a smoothed version of the image is subtracted from the original image. This will give an image where fine detail of image have higher intensities. The sharpened image is obtained by adding the fine image with the original image. This can be implemented as follows:

```
def sharpen(im, ks=(3, 3), alpha=1):                                            1
    sigma_x = (ks[1] // 2.) / 3.                                                2
    sigma_y = (ks[0] // 2.) / 3.                                                3
    im = im.astype('float32') * 0.0039215                                       4
    im_coarse = cv2.GaussianBlur(im, ks, sigmaX=sigma_x, sigmaY=sigma_y)        5
    im_fine = im - im_coarse                                                    6
    im += alpha * im_fine                                                       7
    return np.clip(im * 255, 0, 255).astype('uint8')                           8
```

Here, the fine image is added using a weight called α. Also, the size of smoothing kernel affects the resulting sharpened image. This function can be applied with different values of kernel sizes and α on a sample in order to generate different sharpened images. Figure 5.4 illustrates examples of applying smoothing and sharpening techniques with different configuration of parameters on an image from the GTSRB dataset.

5.3.2.5 Random Crop

Another effective way for augmenting a dataset is to generate random crops for each sample in the dataset. This may generate samples that are far from each other in the input space but they belong to the same class of object. This is a desirable property since it helps to cover some gaps in the input space. This method is already implemented in Caffe using the parameter called *crop_size*.[4] However, in some cases you may develop a special Python layer for your dataset or may want to store random copies of each samples on disk. In these cases, the random cropping method can be implemented as follows:

```
def crop(im, im_shape, rand):                                                   1
    dx = rand.randint(0, im.shape[0]-im_shape[0])                               2
    dy = rand.randint(0, im.shape[1]-im_shape[1])                               3
    im_res = im[dy:im_shape[0]+dy, dx:im_shape[1]+dx, :].copy()                 4
    return im_res                                                               5
```

In the above code, the argument *rand* is an instance of *numpy.random.RandomState* object. You may also directly call *numpy.random.randint* function instead. However, we can use the above argument in order to seed the random number generator with a desired value.

5.3.2.6 Saturation/Value Changes

Another technique for generating similar images in far distances is to manipulate *saturation* and *value* components of image in the HSV color space. This can be

---

[4]You can refer to the *caffe.proto* file in order to see how to use this parameter.

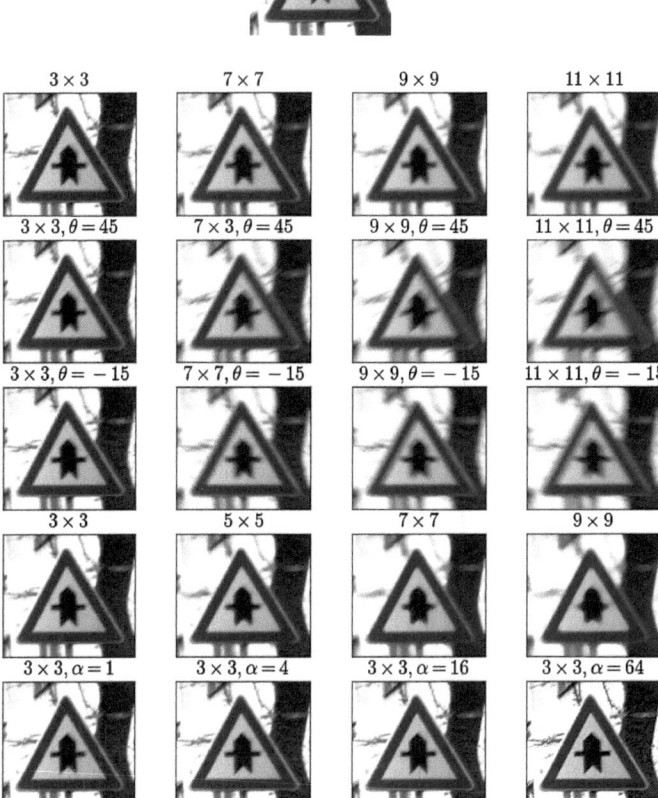

Fig. 5.4 The original image in *top* is modified using Gaussian filtering (*first row*), motion blur (*second* and *third rows*), median filtering (*fourth row*), and sharpening (*fifth row*) with different values of parameters

done by first transforming the image from RGB space to HSV space. Then, the saturation and value components are manipulated. Finally, the manipulated image is transformed back into the RGB space. Manipulating the saturation and value components can be done in different ways. In the following code, we have changed these components using a simple nonlinear approach:

```
def hsv_augment(im,scale, p,component):                                      1
    im_res = im.astype('float32')/255.                                       2
    im_res = cv2.cvtColor(im_res, cv2.COLOR_BGR2HSV)                          3
    im_res[:, :, component] = np.power(im_res[:, :, component]*scale, p)      4
    im_res = cv2.cvtColor(im_res, cv2.COLOR_HSV2BGR)                          5
    return im_res                                                            6
```

Setting the argument p to 1 will change the component linearly based on the value of $scale$. These two arguments usually take a value within [0.5, 1.5]. A sample might be modified using different combinations of these parameters. It is worth mentioning that manipulating the *hue* component might not produce realistic results. The reason is that it may change the color of image and produce unrealistic images.

Similarly, you may manipulate the image in other color spaces such as YUV color space. The algorithm is similar to the above code. The only different is to set the second argument on *cvtColor* function to the desired color space and manipulate the correct component in this space.

5.3.2.7 Resizing

In order to simulate the images taken from a distant object, we can resize a sample with a scaler factor less than one. This way, the size of image will be reduced. Likewise, a sample might be upscaled using interpolation techniques. Moreover, the scale factor along each axis might be different but close to each other. The following code shows how to implement this method using OpenCV in Python:

```
def resize(im, scale_x, scale_y, interpolation=cv2.INTER_NEAREST):     1
    im_res = cv2.resize(im, None, fx=scale_x, fy=scale_y, interpolation=interpolation)     2
    return im_res                                                          3
```

Augmenting datasets with this technique is also a good practice. Especially, if the number of low-resolution images is low, we can simply augment them by resizing high-resolution images with a small scaler factor.

5.3.2.8 Mirroring

Another effective way for augmenting datasets is to mirror images. This technique is already implemented in the Caffe library using a parameter called *mirror*. It can be also easily implemented as follows:

```
def flip(im):                        1
    return im[:, -1::-1, :].copy()   2
```

Mirroring usually generates instances in far distances from the original sample. However, as we mentioned earlier, a great care must be taken into account in using this technique. While flipping the "give way" or "priority road" signs does not change their meaning, flipping a "mandatory turn left" signs will completely change its meaning. Also, flipping "speed limit 100" signs will generate an image without any meaning from traffic sign perspective. However, flipping images of objects such as animals and foods are totally a valid approach.

5.3.2.9 Additive Noise

Adding noisy samples are beneficial for two reasons. First, they generate samples of the same class in relatively far distances from each sample. Second, they teach our model how to make correct predictions in presence of noise. In general, given an image **x**, the degraded image $\mathbf{x}_{noisy}$ can be obtained using the vector v by adding this vector to the original image ($\mathbf{x}_{noisy} = \mathbf{x} + v$). Due to addition operator used for degrading the image, the vector v is called an additive noise. It turns out that the size of v is identical to the size of **x**.

Here, they key of degradation is to generate the vector v. To common ways for generating this vector is to generate random numbers using uniform or Gaussian distributions. This can be implemented using the Numpy library in Python as follows:

```
def gaussian_noise(im, mu=0, sigma=1, rand_object=None):          1
    noise_mask = rand_object.normal(mu, sigma, im.shape)          2
    return cv2.add(im, noise_mask, dtype=cv2.CV_8UC3)             3
                                                                  4
def uniform_noise(im, d_min, d_max, rand_object=None):            5
    noise_mask = rand_object.uniform(d_min, d_max, im.shape)      6
    return cv2.add(im, noise_mask, dtype=cv2.CV_8UC3)             7
```

In the above code, a separate noise vector is generated for each channel. The noisy images generated with this approach might not be very realistic. Nonetheless, they are useful for generating samples in relatively far distances. The noise vector can be shared between the channels. This will produce more realistic images.

```
def gaussian_noise_shared(im, mu=0, sigma=1, rand_object=None):                    1
    noise_mask = rand_object.normal(mu, sigma, (im.shape[0], im.shape[1],1))       2
    noise_mask = np.dstack((noise_mask, noise_mask, noise_mask))                   3
    return cv2.add(im, noise_mask, dtype=cv2.CV_8UC3)                              4
                                                                                  5
def uniform_noise_shared(im, d_min, d_max, rand_object=None):                      6
    noise_mask = rand_object.uniform(d_min, d_max, (im.shape[0], im.shape[1],1))   7
    noise_mask = np.dstack((noise_mask,noise_mask,noise_mask))                     8
    return cv2.add(im, noise_mask, dtype=cv2.CV_8UC3)                              9
```

The addition operator can be implemented using *numpy.add* function. In the case of using this function, the type of inputs must be appropriately selected to avoid the overflow problem. Also, the outputs must be clipped withing a valid range using *numpy.clip* function. The *cv2.add* function from OpenCV takes care of all these conditions internally. Due to random nature of this method, we can generate millions of different noisy samples for each sample in the dataset.

5.3.2.10 Dropout

The last technique that we explain in this section is to generate noisy samples by randomly zeroing some of pixels in the image. This can be done using two different way. The first way is to connect a Dropout layer to an input layer in the network definition. Alternatively, this can be done by generating a random binary mask using the *binomial* distribution and multiplying the mask with input image.

```
def dropout(im, p=0.2, rand_object=None):                         1
    mask = rand_object.binomial(1, 1 - p, (im.shape[0], im.shape[1]))  2
    mask = np.dstack((mask,mask,mask))                            3
    return np.multiply(im.astype('float32'), mask).astype('uint8')  4
```

Using the above implementation all channels of the selected pixels are zeroed making them completely darks pixels. You may want to dropout channels of selected pixels randomly. In other words, instead of sharing the same mask between all channel, a separate mask for each channel can be generated. Figure 5.5 shows a few examples of augmentations with different configurations applied on the sample from the previous figure.

Fig. 5.5 Augmenting the sample in Fig. 5.4 using random cropping (*first row*), hue scaling (*second row*), value scaling (*third row*), Gaussian noise (*fourth row*), Gaussian noise shared between channels (*fifth row*), and dropout (*sixth row*) methods with different configuration of parameters

5.3.2.11 Other Techniques

The above methods are common techniques used for augmenting datasets. There are many other methods that can be used for this purpose. Contrast stretching, histogram equalization, contrast normalization, rotating and shearing are some of these methods that can be used for this purpose. In general, depending on the application, you can design new algorithms for synthesizing new images and augmenting datasets.

5.3.3 Static Versus One-the-Fly Augmenting

There are two ways for augmenting a dataset. In the first technique, all images in the dataset are processed and new images are synthesized using above methods. Then, the synthesized images are stored on disk. Finally, a text file containing the path to original image as well as synthesized images along with their class labels is created to pass to the ImageData layer in the network definition file. This method is called static augmenting. Assume that 30 images are synthesized for each sample in dataset. Storing these images on disk will make the dataset 30 times larger meaning that its required space on disk will be 30 times larger.

Another method is to create a PythonLayer in the network definition file. This layer connects to the database and loads the images into memory. Then, the loaded images are synthesized using the above methods and fed to the network. This method of synthesizing is called *on-the-fly* augmenting. The advantage of this method is that it does not require more space on disk. Also, in the case of adding new methods for synthesizing images, we do not need to process the dataset and store them on disk. Rather, it is simply used in the PythonLayer to synthesize loaded images. The problem with this method is that it increases the size of mini-batch considerably if the synthesized images are directly concatenated to the mini-batch. To alleviate this problem, we can alway keep the size of mini-batch constant by randomly picking the synthesizing methods or randomly selecting N images from pool of original and synthesized images in order to fill the mini-batch of size N.

5.3.4 Imbalanced Dataset

Assume that you are asked to develop a system for recognizing traffic signs. The first task is to collect images of traffic signs. For this purpose, a camera can be attached to a car and images of traffic signs can be stored during driving the car. Then, these images are annotated and used for training classification models. Traffic signs such as "speed limit 90" might be very frequent. In contrast, if images are collected from coastal part of Spain, the traffic sign "danger: snow ahead" might be very scarce. Therefore, while the "speed limit 90" sign will appear frequently in the database, the "danger: snow ahead" sign may only appear few times in the dataset.

Technically, this dataset is *imbalanced* meaning that number of samples in each class varies significantly. A classification model trained on an imbalanced dataset is likely not to generalize well on the test set. The reason is that classes with more samples contribute to the loss more than classes with a few samples. In this case, the model is very likely to learn to correctly classify the classes with more samples so that the loss function is minimized. As the result, the model might not generalize well on the classes with much less samples.

There are different techniques for partially solving this problem. The obvious solution is to collect more data from classes with less samples. However, this might be a very costly and impractical approach in terms of time and resources. There are other approaches which are commonly used for training a model on imbalanced datasets.

5.3.4.1   Upsampling
In this approach, samples of smaller classes are copied in order to match the number of samples of the largest class. For example, if there are 10 samples in class 1 and 85 samples in class 2, samples of class 1 are copied so that there will be 85 samples in class 1 as well. Copying samples can be done by random sampling with replacement. It may be also done using a deterministic algorithm. This method is called *upsampling* since it replicates samples in the minority class.

5.3.4.2   Downsampling
Downsampling is the opposite of upsampling. In this method, instead of copying samples from minority class, some samples from majority class are removed from dataset in order to match number of samples in minority class. Downsampling can be done by randomly removing samples or applying a deterministic algorithm. One disadvantage of this method is that important information might be removed from dataset by removing samples from majority classes.

5.3.4.3   Hybrid Sampling
Hybrid sampling is combination of the two aforementioned methods. Concretely, some of majority classes might be downsampled and some of minority classes might be upsampled such that they all have a common number of samples. This is more practical approach than just using one of the above sampling methods.

5.3.4.4   Weighted Loss Function
Another approach is to add a penalizing mechanism to the loss function such that a sample from minority class will contribute more than the sample from majority class to the loss function. In other words, assume that error of a sample from minority class is equal to e_1 and the error of sample from majority class is equal to e_2. However, because the number of samples in the minority class is less, we want that e_1 has more impact on loss function compared with e_2. This can be simply done by incorporating

a specific weight for each class in dataset and multiplying error of each sample with their corresponding weight. For example, assuming that the weight of minority class is w_1 and the weight of majority class is w_2, the error terms the above samples will be equal to $w_1 \times e_1$ and $w_2 \times e_2$, respectively. Apparently, $w_1 > w_2$ so that the error of one sample from minority class will contribute to the loss more than the error of one sample from majority class. Notwithstanding, because the number of samples in majority class is higher, the overall contribution of samples from minority and majority classes on loss function will be approximately equal.

5.3.4.5 Synthesizing Data

The last method that we describe in this section is synthesizing data. To be more specific, minority classes can be balanced with majority classes by synthesizing data on minority classes. Any of the methods for augmenting dataset might be used for synthesizing data on minority classes.

5.3.5 Preparing the GTSRB Dataset

In this book, the training set is augmented using some of the methods mentioned in previous sections. Then, 10% of the augmented training set is used for validation. Also, the GTSRB dataset comes with a specific test set. The test set is not augmented and it remains unchanged. Next, all samples in the training, validation and test sets are cropped using the bounding box information provided in the dataset. Then, all these samples are resized according to the input size of ConvNets that we will explain in the rest of this book.

Next, mean image is obtained over the training set[5] and it is subtracted from each image in order to shift the training set to origin. Also, the same transformation is applied on the validation and test sets using the mean image learned from the training set. Previous study in Coates and Ng (2012) suggests that subtracting mean image increases the performance of networks. Subtracting mean is commonly done on-the-fly by storing the mean image as a *.bindaryproto* file and setting the *mean_file* parameter in the network definition file.

Assume that the mean image is computed and stored in a Numpy matrix. This matrix can be stored in a *.binaryproto* file by calling the following function:

---

[5]Considering an RGB image as a three-dimensional matrix, mean image is computed by adding all images in the training set in element-wise fashion and dividing each element of the resulting matrix with the number of samples in the training set.

```
def write_mean_file(mean_npy, save_to):                                    1
                                                                           2
                                                                           3
    if mean_npy.ndim == 2:                                                 4
        mean_npy = mean_npy[np.newaxis, np.newaxis, ...]                   5
    else:                                                                  6
        mean_npy = np.transpose(mean_npy, (2, 0, 1))[np.newaxis, ...]      7
                                                                           8
    binaryproto_file = open(save_to, 'wb')                                 9
    binaryproto_file.write(caffe.io.array_to_blobproto(mean_npy).SerializeToString())  10
    binaryproto_file.close()                                              11
```

The GTSRB dataset is an imbalanced dataset. In this book, we have applied the upsampling technique for making this dataset balanced. Samples are picked randomly for being copied on minority classes. Finally, separate text files containing path of images and their corresponding class labels are created for training, validation, and test sets.

5.4 Analyzing Training/Validation Curves

Plotting accuracy of model on training and validation sets at different iterations during training phase provides diagnostic information about the model. Figure 5.6 shows three different scenarios that might happen during training. First, there is always an expected accuracy and we always try to achieve this accuracy. The plot on the left shows an acceptable scenario where the accuracy of model on both training and validation sets are close to the expected accuracy. In this case, the model can be thought appropriate and it might be applied on the test set.

The middle plot indicates a scenario where training and validation error are close to each other but the they are both far from the expected accuracy. In this case, we can conclude that the current model suffers from high bias. In this case, capacity of model can be increased by adding more neurons/layers to the model. Other solutions for this scenario is explained in Sect. 3.7.

Fig. 5.6 Accuracy of model on training and validation set tells us whether or not a model is acceptable or it suffers from high bias or high variance

The right plot illustrates a scenario where the accuracy on training set is very close to expected accuracy but the accuracy on validation set is far from the expected accuracy. This is a scenario where the model suffers from high variance. The quick solution for this issue is to reduce the model capacity of regularize it more. Other solutions for this problem are explained in Sect. 3.7. Also, this scenario may happen because training and validation sets are not drawn from the same distribution.

It is always is good practice to monitor training an validation accuracies during training. The vertical dashed line in the right plot shows the point in training where the model has started to overfit on data. This is a point where the training procedure can be stopped. This technique is called *early stopping* and it is an efficient way to save time in training a model and avoid overfitting.

5.5   ConvNets for Classification of Traffic Signs

In this section, we will explain different architectures for classification of traffic signs on the GTSRB dataset.[6] All architectures will be trained and validated on the same training and validation sets. For this reason, they will share the same ImageData layers for the training and validation ConvNets. These two layers are defined as follows:

```
def gtsrb_source(input_size=32):                                                       1
    shared_args = { 'is_color':True,                                                   2
                    'shuffle':True,                                                     3
                    'new_width':input_size,                                            4
                    'new_height':input_size,                                           5
                    'ntop':2,                                                          6
                    'transform_param':{'scale': 1. / 255}}                             7
    L = caffe.layers                                                                   8
    net_v = caffe.NetSpec()                                                            9
    net_v.data, net_v.label = L.ImageData(source='/home/pc/train.txt',               10
                                          batch_size=200,                             11
                                          include={'phase': caffe.TEST},              12
                                          **shared_args)                              13
    net_t = caffe.net_spec.NetSpec()                                                  14
    net_t.data, net_t.label = L.ImageData(source='/home/pc/validation.txt',          15
                                          batch_size=48,                              16
                                          include={'phase': caffe.TRAIN},             17
                                          **shared_args)                              18
    return net_t, net_v                                                               19
```

Later, this function will be used to define the architecture of ConvNets. Since the transformations applied on the training and validation samples are identical, we have created a dictionary of shared parameters and passed it as a keyword argument after unpacking it using the ** operator. Also, depending on the available memory of the graphic card you might need to reduce the batch size of the validation data and set it to a number smaller than 200. Finally, unless the above function is called by an integer argument the input images will be resized into 32 × 32 pixels by default.

---

[6]Implementations of the methods in this chapter are available at https://github.com/pcnn.

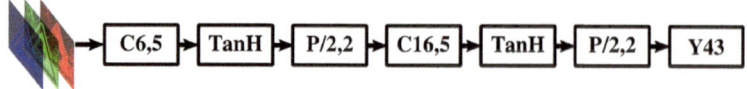

Fig. 5.7 A ConvNet consists of two *convolution-hyperbolic activation-pooling* blocks without fully connected layers. Ignoring the activation layers, this network is composed of five layers

We start with a very small ConvNet. The architecture of this ConvNet is illustrated in Fig. 5.7. This ConvNet has two blocks of *convolution-activation-pooling* layers where the Hyperbolic tangent function is used as activation of neurons. Without counting the activation layers, the depth of this ConvNet is 5. Also, the width of ConvNet (number of neurons in each layer) is not high. The last layer has 43 neurons one for each class in the GTSRB dataset.

This ConvNet is trained by minimizing the multiclass logistic loss function. In order to generate the definition file of this network and other networks in this chapter using Python, we can use the following auxiliary functions.

```
def conv_act(bottom, ks, nout, act='ReLU', stride=1, pad=0, group=1):           1
    L = caffe.layers                                                            2
    c = L.Convolution(bottom, kernel_size=ks, num_output=nout,                  3
                      stride=stride, pad=pad,                                   4
                      weight_filler={'type':'xavier'},                         5
                      bias_filler={'type':'constant',                          6
                                   'value':0},                                 7
                      param=[{'decay_mult':1},{'decay_mult':0}],               8
                      group=group)                                             9
    r = eval('L.{}(c)'.format(act))                                            10
    return c, r                                                                11
                                                                              12
def conv_act_pool(bottom, conv_ks, nout, conv_st=1, conv_p=0, pool_ks=2, pool_st=2, act='ReLU',  13
        group=1):
    L = caffe.layers                                                           14
    P = caffe.params                                                           15
    c,r = conv_act(bottom, conv_ks, nout,act,conv_st,conv_p, group=group)      16
    p = L.Pooling(r,                                                           17
    kernel_size=pool_ks,                                                       18
    stride=pool_st,                                                            19
    pool=P.Pooling.MAX)                                                        20
    return c,r,p                                                               21
                                                                              22
def fc_act_drop(bottom, nout, act='ReLU', drop_ratio=0.5):                     23
    L = caffe.layers                                                           24
    P = caffe.params                                                           25
    fc = L.InnerProduct(bottom, num_output=nout,                              26
                      weight_filler={'type':'xavier'},                        27
                      bias_filler={'type':'constant',                         28
                                   'value':0},                                29
                      param=[{'decay_mult': 1}, {'decay_mult': 0}])           30
    r = eval('L.{}(fc)'.format(act))                                          31
    d = L.Dropout(r, dropout_ratio=drop_ratio)                               32
    return fc, r, d                                                           33
                                                                              34
def fc(bottom, nout):                                                          35
    L = caffe.layers                                                          36
    return L.InnerProduct(bottom,                                            37
```

```
                    num_output=nout,                                            38
                    weight_filler={'type': 'xavier'},                           39
                    bias_filler={'type': 'constant',                            40
                    'value': 0})                                                41
```

The function *conv_act* creates a convolution layer and an activation layer and connects the activation layer to the convolution layer. The function *conv_act_pool* creates a convolution-activation layer and connects a pooling layer to the activation layer. The function *fc_act_drop* creates a fully connected layer and attaches an activation layer to it. It also connects a dropout layer to the activation layer. Finally, the function *fc* creates only a fully connected layer without an activation on top of it. The following Python code creates the network shown in Fig. 5.7 using the above functions:

```
def create_lightweight(save_to):                                                           1
    L = caffe.layers                                                                       2
    P = caffe.params                                                                       3
    n_tr , n_val = gtsrb_source(input_size=32, mean_file='/home/pc/gtsr_mean_32x32.binaryproto')  4
    n_tr.c1, n_tr.a1, n_tr.p1 = conv_act_pool(n_tr.data, 5, 6, act='TanH')                 5
    n_tr.c2, n_tr.a2, n_tr.p2 = conv_act_pool(n_tr.p1, 5, 16, act='TanH')                  6
    n_tr.f3_classifier = fc(n_tr.p2, 43)                                                   7
    n_tr.loss = L.SoftmaxWithLoss(n_tr.f3_classifier, n_tr.label)                          8
    n_tr.acc = L.Accuracy(n_tr.f3_classifier, n_tr.label)                                  9
                                                                                           10
    with open(save_to, 'w') as fs:                                                         11
        s_proto = str(n_val.to_proto()) + '\n' + str(n_tr.to_proto())                      12
        fs.write(s_proto)                                                                  13
        fs.flush()                                                                         14
```

The Accuracy layer computes the accuracy of predictions on the current mini-batch. It accepts actual label of samples in the mini-batch together with their predicted scores computed by the model and returns the fraction of samples that are correctly classified. Assuming that name of the solver definition file of this network is *solver_XXX.prototxt*, we can run the following script to train and validate the network and monitor the train/validation performance.

```
import caffe                                                                        1
import numpy as np                                                                  2
import matplotlib.pyplot as plt                                                     3
                                                                                    4
solver = caffe.get_solver(root + 'solver_{}.prototxt'.format(net_name))            5
                                                                                    6
train_hist_len = 5                                                                  7
test_interval = 250                                                                 8
test_iter = 100                                                                     9
max_iter = 5000                                                                     10
                                                                                    11
fig = plt.figure(1, figsize=(16,6), facecolor='w')                                 12
acc_hist_short = []                                                                 13
acc_hist_long = [0]*max_iter                                                        14
acc_valid_long = [0]                                                                15
acc_valid_long_x = [0]                                                             16
                                                                                    17
for i in xrange(max_iter):                                                          18
    solver.step(1)                                                                  19
                                                                                    20
    loss = solver.net.blobs['loss'].data.copy()                                     21
                                                                                    22
```

```
acc = solver.net.blobs['acc'].data.copy()                                         23
acc_hist_short.append(acc)                                                        24
if len(acc_hist_short) > train_hist_len:                                          25
    acc_hist_short.pop(0)                                                         26
acc_hist_long[i] = (np.asarray(acc_hist_short)).mean()*100                        27
                                                                                  28
if i > 0 and i % 10 == 0:                                                         29
    fig.clf()                                                                     30
    ax = fig.add_subplot(111)                                                     31
    a3 = ax.plot([0, i], [100, 100], color='k', label='Expected')                32
    a1 = ax.plot(acc_hist_long[:i], color='b', label='Training')                 33
    a2 = ax.plot(acc_valid_long_x, acc_valid_long, color='r', label='Validation')34
    plt.xlabel('iteration')                                                       35
    plt.ylabel('accuracy (%)')                                                    36
    plt.legend(loc='lower right')                                                 37
    plt.axis([0, i, 0, 105])                                                      38
    plt.draw()                                                                    39
    plt.show(block=False)                                                         40
    plt.pause(0.005)                                                              41
                                                                                  42
if i > 0 and i % test_interval == 0:                                              43
    acc_valid = [0]*test_iter                                                     44
    net = solver.test_nets[0]                                                     45
    net.share_with(solver.net)                                                    46
    for j in xrange(test_iter):                                                   47
        net.forward()                                                             48
        acc = net.blobs['acc'].data.copy()                                        49
        acc_valid[j] = acc                                                        50
    acc_valid_long.append(np.asarray(acc_valid).mean()*100)                       51
    acc_valid_long_x.append(i)                                                    52
    print 'Validation accuiracy:', acc_valid_long[-1]                             53
```

The above template can be used as a reference template for training and validating Caffe models in Python. Line 5 loads the information about the optimization algorithm as well as training and test networks. Depending on the value of the field *type* in the solver definition this function returns a different instance. For example, if value of *type* is set to "SGD", it will return an instance of *SGDSolver*. Likewise, if it is set to "RMSProp" it will return an instance of *RMSPropSolver*. All these objects are inherited from the same class and they share the same methods. Hence, regardless of type of solver, the method will be called in the above code to run the optimization algorithm.

The accuracy layer in the network always returns the accuracy of the current mini-batch. In order to compute the accuracy over more than one mini-batch, we have to compute the mean of accuracies of these mini-batches. In the above algorithm, the mean accuracy is computed over the last *train_hist_len* mini-batches.

To prevent Caffe from invoking the test network automatically, the field *test_interval* in the solver definition file must be set to a very large number. Also, the variable *test_iter* can be set to an arbitrary number. Its value does not have any effect on the optimization algorithm since the variable *test_interval* is set to a large number and the test phase will not be invoked by Caffe at all.

The variables in Lines 8–10 denote the test interval, number of mini-batches in the test interval, and maximum number of iterations in our algorithm. Also, the variables in Lines 13–16 will keep the mean accuracies of training samples and validation samples.

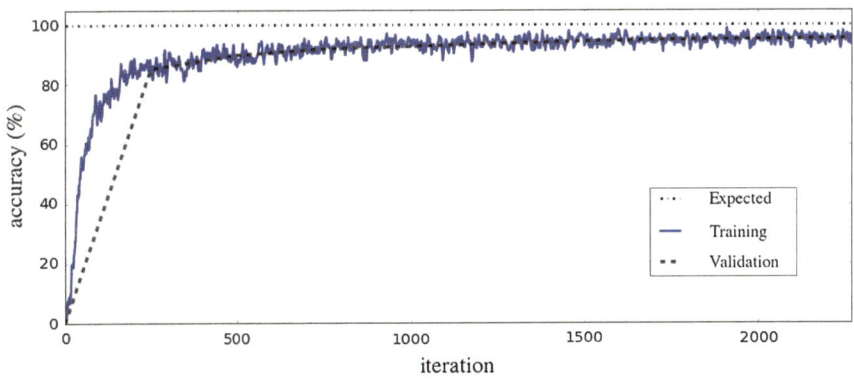

Fig. 5.8 Training, validation curve of the network illustrated in Fig. 5.7

The optimization loop starts in Line 18 and it is repeated *max_iter* times. First line in the loop runs the forward and backward steps on one mini-batch from the training set. Then, the loss of network on the current mini-batch is obtained in Line 21. Similarly, the accuracy of network on the current mini-batch is obtained on Line 23. Lines 24–27 stores accuracy of last *train_hist_len* mini-batches and updates the mean training accuracy of the current iteration.

Lines 29–41 draw the training, validation, and expected curves every 10 iterations. Most of time, it is a good practice to visually inspect these curves in order to stop the algorithm earlier if it is necessary.[7] Lines 43–53 validates the network every *test_interval* iterations using the validation set. Each time, it computes the mean accuracy over all mini-batches in the validation set.

Figure 5.8 shows the training/validation curve of the network in Fig. 5.7. According to the plot, the validation error is plateaued after 1000 iterations. Besides, the training error also is not reduced afterwards. In the case of traffic signs classification problem, it is expected to achieve 100% accuracy. Nonetheless, the training and validation error is much higher than the expected accuracy. The main reason is that the network in Fig. 5.7 has a very limited capacity. This is due to the fact that depth and width of network are low. Specifically, the number of neurons in each layer is very low. Also, the depth of network could be increased by adding more convolution-activation-pooling blocks and fully connected layers.

There was a competition for classification of traffic signs on the GTSRB dataset. The network in Ciresan et al. (2012a) won the competition and surpassed human accuracy on this dataset. The architecture of this network is illustrated in Fig. 5.9.

In order to be able to add one more convolution-pooling-activation block to the network in Fig. 5.7 the size of input image must be increased so that the spatial size of feature maps after the second convolution-activation-pooling block is big enough

---

[7]If the number of iterations is high, the above code should be changed slightly in order to always plot fixed number of points.

Fig. 5.9 Architecture of the network that won the GTSRB competition (Ciresan et al. 2012a)

to apply another convolution-activation-pooling block on this feature maps. For this reason, the size of input has increased from 32×32 pixels in Fig. 5.7 to 48×48 pixels in Fig. 5.9.

Also, the first layer has a bigger receptive field and it has 100 filters rather than 6 filters in the previous network. The second block has 150 filters of size 4×4 which yields a feature maps of size $150 \times 9 \times 9$. The third convolution-activation-pooling block consists of 250 filters of size 4×4. The output of this block is a $250 \times 3 \times 3$ feature map.

Another improvement on this network is utilizing a fully connected layer between the last pooling and the classification layer. Specifically, there is a fully connected layer with 300 neurons where each neuron is connect to $250 \times 3 \times 3$ neurons from the previous layer. This network can be define in Python as follows:

```
def create_net_jurgen(save_to):                                             1
    L = caffe.layers                                                        2
    P = caffe.params                                                        3
    net, net_valid = gtsrb_source(input_size=48,                            4
    mean_file='/home/hamed/Desktop/GTSRB/Training_CNN/gtsr_mean_48x48.binaryproto')   5
    net.conv1, net.act1, net.pool1 = conv_act_pool(net.data, 7, 100, act='TanH')      6
    net.conv2, net.act2, net.pool2 = conv_act_pool(net.pool1, 4, 150, act='TanH')     7
    net.conv3, net.act3, net.pool3 = conv_act_pool(net.pool2, 4, 250, act='TanH')     8
    net.fc1, net.fc_act, net.drop1 = fc_act_drop(net.pool3, 300, act='TanH')          9
    net.f3_classifier = fc(net.drop1, 43)                                   10
    net.loss = L.SoftmaxWithLoss(net.f3_classifier, net.label)              11
    net.acc = L.Accuracy(net.f3_classifier, net.label)                      12
                                                                            13
    with open(save_to, 'w') as fs:                                          14
    s_proto = str(net_valid.to_proto()) + '\n' + str(net.to_proto())        15
    fs.write(s_proto)                                                       16
    fs.flush()                                                              17
    print s_proto                                                           18
```

After creating the network, a solver must be created for this network. Then, it can be trained and validated using the script we mentioned earlier by loading the appropriate solver definition file. Figure 5.10 shows the training/validation curve of this network.

As it turns out from the training/validation curve, the above architecture is appropriate for classification of traffic signs in the GTSRB dataset. According to the curve, the training error is getting close to zero and if the network is trained longer, the training error might reduce more. In addition, the validation accuracy is ascending and with a longer optimization, the accuracy is likely to improve as well.

Size of the receptive field and number of filters in all layers are chosen properly in the above network. Also, flexibility (nonlinearity) of the network is enough for modeling a wide range of traffic signs. However, the number of parameters in this

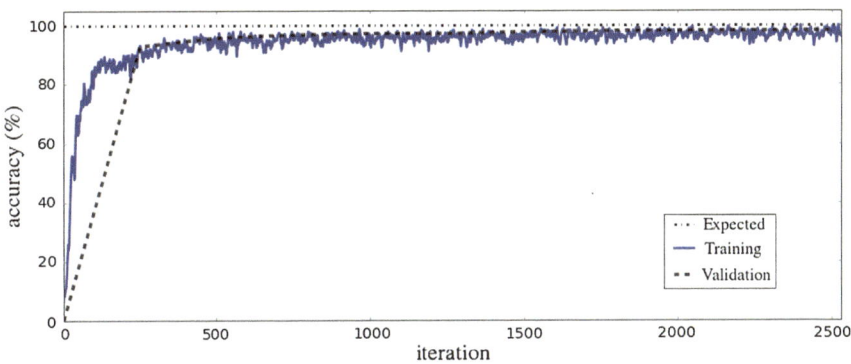

Fig. 5.10 Training/validation curve of the network illustrated in Fig. 5.9

network could be reduced in order to make it computationally and memory wise more efficient.

The above network utilizes the hyperbolic tangent function to compute neuron activations. The hyperbolic function is defined as $tanh(x) = \frac{e^x - e^{-x}}{e^x + e^{-x}} = \frac{e^{2x} - 1}{e^{2x} + 1}$. Even with an efficient implementation of exponentiation e^x, it still requires many multiplications. Note that x is a floating point number since it is the weighted sum of the input to the neuron. For this reason, e^x cannot be implemented using a lookup table.

An efficient way to calculate e^x is as follows: First, write $x = x_{int} + r$, where x_{int} is the nearest integer to x and $r \in [-0.5 \ldots 0.5]$ which gives $e^x = e^{x_{int}} \times e^r$. Second, multiply e by itself x_{int} times. The multiplication can be done quite efficiently. To further increase efficiency, various integer powers of e can be calculated and stored in a lookup table. Finally, e^r can be estimated using the polynomial $e^r = 1 + x + \frac{x^2}{2} + \frac{x^3}{6} + \frac{x^4}{24} + \frac{x^5}{120}$ with estimation error $+3e^{-5}$. Consequently, calculating $tanh(x)$ needs $[x] + 5$ multiplications and 5 divisions. We assuming that division and multiplication need the same amount of CPU cycles. Therefore, $tanh(x)$ can be computed using $[x] + 10$ multiplications. The simplest scenario is when $x \in [-0.5 \ldots 0.5]$. Then, $tanh(x)$ can be calculated using 10 multiplications. Based on this, the total number of multiplications of the network proposed in Ciresan et al. (2012a) is equal to 128,321,700. Since they build an ensemble of 25 networks, thus, the total number of the multiplications must be multiplied by 25 which is equal to 3,208,042,500 multiplications for making a prediction using an ensemble of 25 networks shown in Fig. 5.9.

Aghdam et al. (2016a) aimed to reduce the number of parameters together with the number of the arithmetic operations and increase the classification accuracy. To this end, they replaced the hyperbolic nonlinearities with the Leaky ReLU activation functions. Beside the favorable properties of ReLU activations, they are also computationally very efficient. To be more specific, a Leaky ReLU function needs only one multiplication in the worst case and if the input of the activation function is positive, it does not need any multiplication. Based on this idea, they designed the network illustrated in Fig. 5.11.

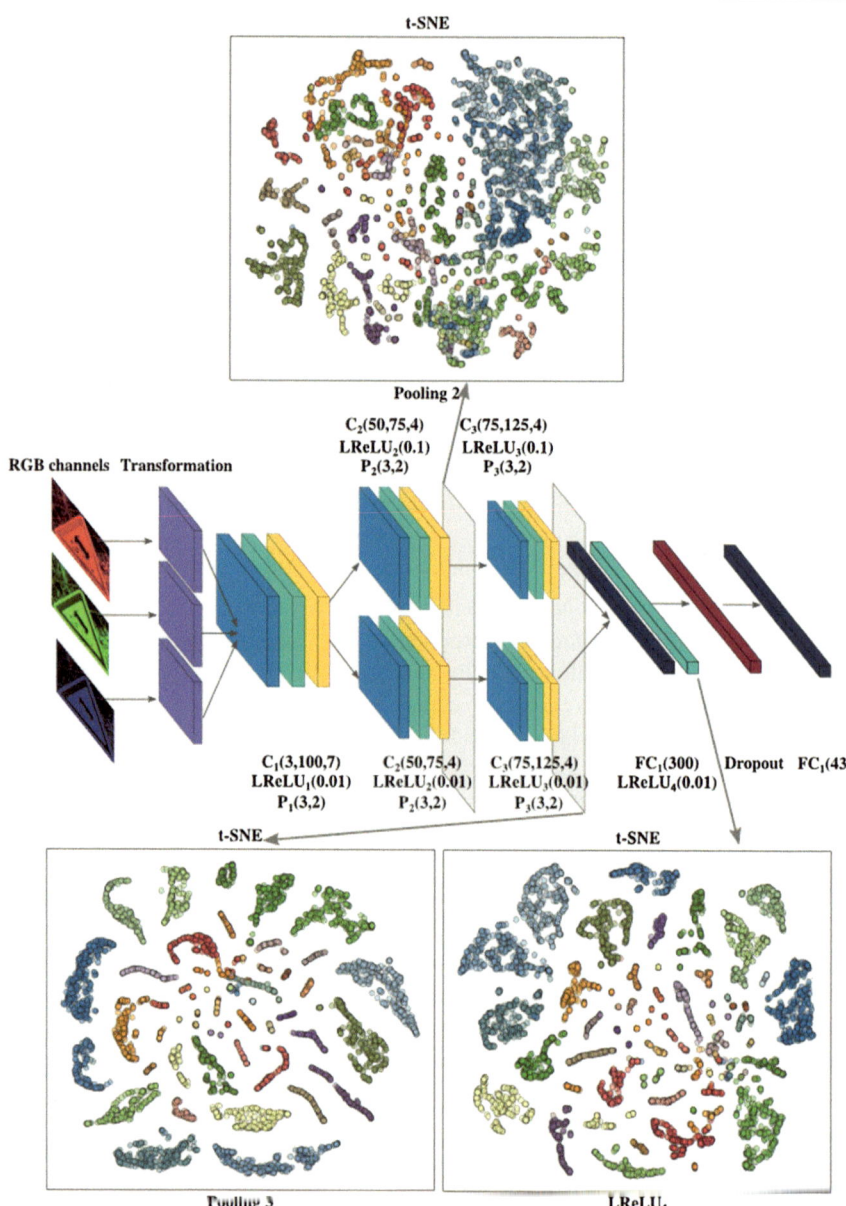

Fig. 5.11 Architecture of network in Aghdam et al. (2016a) along with visualization of the first fully connected layer as well as the last two pooling layers using the t-SNE method. *Light blue*, *green*, *yellow* and *dark blue* shapes indicate convolution, activation, pooling, and fully connected layers, respectively. In addition, each *purple* shape shows a linear transformation function. Each class is shown with a unique color in the scatter plots

Furthermore, the two middle convolution-pooling layers in the previous network is divided into two separate layers. There is also a layer connected to the input which applies linear transformation on each channel separately. Overall, this network consists of a transformation layer, three convolution-pooling layers and two fully connected layers with a dropout layer (Hinton 2014) in between. Finally, there is a Leaky ReLU layer after each convolution layer and after the first fully connected layer. The network accepts a 48×48 RGB image and classify it into one of the 43 traffic sign classes in the GTSRB dataset.

It is worth mentioning that number of the parameters is reduced by dividing the two middle convolution-pooling layers into two groups. More specifically, the transformation layer applies an element-wise linear transformation $f_c(x) = a_c x + b_c$ on c^{th} channel of the input image where a_c and b_c are trainable parameters and x is the input value. Note that each channel has a unique transformation function. Next, the image is processed using 100 filters of size 7×7. The notation $C(c, k, w)$ indicates a convolution layer with k filters of size $w \times w$ applied on the input with c channels. Then, the output of the convolution is passed through a Leaky ReLU layer and fetched into the pooling layer where a MAX-pooling operation is applied on the 3×3 window with stride 2.

In general, a $C(c, k, w)$ layer contains $c \times k \times w \times w$ parameters. In fact, the second convolution layer accepts a 100-channel input and applies 150 filters of size 4×4. Using this configuration, the number of the parameters in the second convolution layer would be 2,40,000. The number of the parameters in the second layer is halved by dividing the input channels into two equal parts and fetching each part into a layer including two separate $C(50, 75, 4)$ convolution-pooling units. Similarly, the third convolution-pooling layer halves the number of the parameters using two $C(75, 125, 4)$ units instead of one $C(150, 250, 4)$ unit. This architecture is collectively parametrized by 1,123,449 weights and biases which is 27, 22 and 3% reduction in the number of the parameters compared with the networks proposed in Ciresan et al. (2012a), Sermanet and Lecun (2011), and Jin et al. (2014), respectively. Compared with Jin et al. (2014), this network needs less arithmetic operations since Jin et al. (2014) uses a Local Response Normalization layer after each activation layer which needs a few multiplications per element in the resulting feature map from previous layer. The following script shows how to implement this network in Python using the Caffe library:

```
def create_net_ircv1(save_to):                                              1
    L = caffe.layers                                                        2
    P = caffe.params                                                        3
    net, net_valid = gtsrb_source(input_size=48,                            4
    mean_file='/home/pc/gtsr_mean_48x48.binaryproto')                       5
    net.tran = L.Convolution(net.data,                                      6
                    num_output=3,                                           7
                    group=3,                                                8
                    kernel_size=1,                                          9
                    weight_filler={'type':'constant',                       10
                            'value':1},                                     11
                    bias_filler={'type':'constant',                         12
                            'value'.0},                                     13
                    param=[{'decay_mult':1},{'decay_mult':0}])              14
    net.conv1, net.act1, net.pool1 = conv_act_pool(net.tran, 7, 100, act='ReLU')   15
```

```
net.conv2, net.act2, net.pool2 = conv_act_pool(net.pool1, 4, 150, act='ReLU', group=2)        16
net.conv3, net.act3, net.pool3 = conv_act_pool(net.pool2, 4, 250, act='TanH', group=2)        17
net.fc1, net.fc_act, net.drop1 = fc_act_drop(net.pool3, 300, act='ReLU')                      18
net.f3_classifier = fc(net.drop1, 43)                                                          19
net.loss = L.SoftmaxWithLoss(net.f3_classifier, net.label)                                     20
net.acc = L.Accuracy(net.f3_classifier, net.label)                                             21
                                                                                               22
with open(save_to, 'w') as fs:                                                                 23
s_proto = str(net_valid.to_proto()) + '\n' + str(net.to_proto())                               24
fs.write(s_proto)                                                                              25
fs.flush()                                                                                     26
print s_proto                                                                                  27
```

The above network is trained using the same training/validation procedure. Figure 5.12 shows the training/validation curve of this network. Although the number of parameters is reduced compared with Fig. 5.9, the network still accurately classifies the traffic signs in the GTSRB dataset. In general after finding a model which produces results very close to expected accuracy, it is likely to be able to reduce the model complexity by keeping the accuracy unaffected.

It is a common practice to inspect extracted features of a ConvNet using a visualization technique. The general procedure is to first train the network. Then, some samples are fed to the network and feature vectors extracted by a specific layer on all samples are collected. Assuming that each feature vector is a D dimensional vector, a feature generated by this layer will be a point in the D dimensional space. These D dimensional samples can be embedded into a two-dimensional space.

Embedding into two-dimensional space can be done using principal component analysis, self-organizing maps, isomaps, locally-linear embedding, and etc. One of the embedding methods which produces promising results is called *t-distributed stochastic neighbor embedding* (t-SNE) (Maaten and Hinton 2008). It nonlinearly embeds points into a lower dimensional (in particular two or three dimensional) space by preserving structure of neighbors as much as possible. This is an important property since it adequately reflects the neighborhood structure in the high dimensional space.

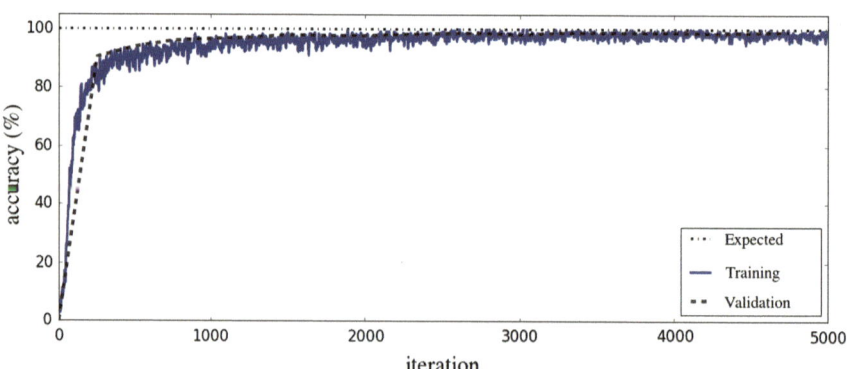

Fig. 5.12 Training/validation curve on the network illustrated in Fig. 5.11

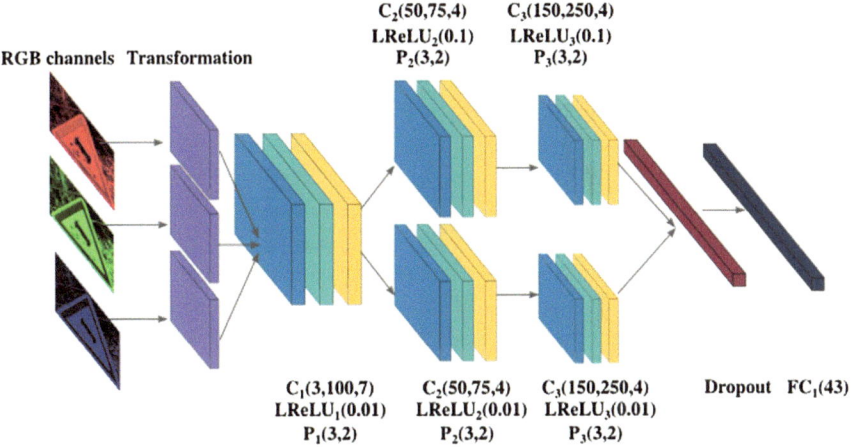

Fig. 5.13 Compact version of the network illustrated in Fig. 5.11 after dropping the first fully connected layer and the subsequent Leaky ReLU layer

Figure 5.11 shows the two-dimensional embedding of feature maps after the second and third pooling layers as well as the first fully connected layer. The embedding is done by using the t-SNE method. Samples of each class is represented using a different color. According to the embedding result, the classes are separated properly after the third pooling layer. This implies that the classification layer might be able to accurately discriminate the classes if we omit the first fully connected layer. According to the t-SNE visualization, the first fully connected layer does not increase the discrimination of the classes, considerably. Instead, it rearranges the classes in a lower dimensional space and it might mainly affect the interclass distribution of the samples.

Consequently, it is possible to discard the first fully connected layer and the subsequent Leaky ReLU layer from the network and connect the third pooling layer directly to the dropout layer. The more compact network is shown in Fig. 5.13. From optimization perspective, this decreases the number of the parameters from 1,123,449 to 531,999 which is 65, 63, and 54% reduction compared with Ciresan et al. (2012a), Sermanet and Lecun (2011), and Jin et al. (2014), respectively.

5.6 Ensemble of ConvNets

Given a training set $\mathcal{X} = \{x_1, \ldots, x_N\}$, we denote a model trained on this set using $\mathcal{M}$. Assuming that $\mathcal{M}$ is a model for classifying input x_i into one of K classes, $\mathcal{M}(x) \in \mathbb{R}^K$ returns the *per class score* (output of classification layer without applying the softmax function) of the model for input x. The main idea behind ensemble learning is to train L models $\mathcal{M}_1 \ldots \mathcal{M}_L$ on $\mathcal{X}$ and predict the class of sample $\mathbf{x}_i$ by combining the models using

$$\mathcal{L}(\mathcal{M}_1(\mathbf{x}_i), \ldots, \mathcal{M}_L(\mathbf{x}_i)). \tag{5.2}$$

In this equation, $\mathcal{L}$ is a functions which accepts classification scores of samples $\mathbf{x}_i$ predicted by L models and combines these scores in order to classify $\mathbf{x}_i$. Previous studies (Ciresan et al. 2012a; Jin et al. 2014; Sermanet et al. 2013; Aghdam et al. 2016a) show that creating an ensemble of ConvNets might increase the classification accuracy. In order to create an ensemble, we have to answer two questions. First, how to combine the predictions of made by models on sample $\mathbf{x}_i$? Second, how to train L different models on $\mathcal{X}$?

5.6.1 Combining Models

First step in creating an ensemble is to design a method for combining classification scores predicted by different models on the same sample. In this sections, we will explain some of these methods.

5.6.1.1 Voting
In this approach, first, class labels are predicted by each model. This way, each model votes for its class label. Then, the function $\mathcal{L}$ returns the class by combining all votes. One method for combining votes is to count the votes for each class and return the class with majority of votes. This technique is called *majority voting*. We may add more restrictions to this algorithm. For example, if the class with majority of votes does not have the minimum number of votes, the algorithm may return a value indicating that the samples cannot be classified with high confidence.

In the above approach, all models have the same impact in voting. It is also possible to give weight for each model so that votes of models are counted according to their weight. This method is called *weighted voting*. For example, if the weight of a model is equal to 3, its vote is counted three times. Then, majority of votes is returned taking into account the weights of each model. This technique is not widely used in neural networks.

5.6.1.2 Model Averaging
Model averaging is commonly used in creating ensemble with neural networks. In this approach, the function $\mathcal{L}$ is defined as:

$$\mathcal{L}(\mathcal{M}_1(\mathbf{x}_i), \ldots, \mathcal{M}_L(\mathbf{x}_i)) = \sum_{j=1}^{L} \alpha_j \mathcal{M}(\mathbf{x}_i) \tag{5.3}$$

where α_j is the weight of jth model. If we set all $\alpha_j = \frac{1}{L}$, $j = 1 \ldots L$ the above functions simply computes the average of classification scores. If each model is assigned with a different weight the above function will compute the weighted average of classification scores. Finally, the class of samples $\mathbf{x}_i$ is analogous to the index in $\mathscr{Z}$ with maximum value.

ConvNets can have low bias and high variance by increasing the number of layers and number of neurons in each layer. This means that the model has a higher chance to overfit on training data. However, the core idea behind model averaging is that computing average of many models with low bias and high variance represents a model with low bias and low variance on data which in turn increases the accuracy and generalization of the ensemble.

5.6.1.3 Stacking

Stacking is the generalized version of model averaging. In this method, $\mathscr{Z}$ is another function that learns how to combine the classification scores predicted by L models. This function can be a linear function such as weighted averaging or it can be a nonlinear function such as a neural network.

5.6.2 Training Different Models

The second step in creating an ensemble is to train L different models. The easiest way to achieve this goal is to sample the same model during the same training phase but in different iterations. For example, we can save the weights at 1000th, 5000th, $15,000^{th}$, and $40,000^{th}$ iterations in order to create four different models. Another way is to initialize the same model L times and execute the training procedure L times in order to obtain L models with different initializations. This method is more common than the former method. More general setting is to designs L networks with different architectures and train them on the training set. The previous two methods can be formulated as special cases of this method.

5.6.2.1 Bagging and Boosting

Bagging is a technique that can be used in training different models. In this technique, L random subsets of training set $\mathscr{X}$ are generated. Then, a model is trained on each subset independently. Clearly, some of samples might appear in more than one subset.

Boosting starts with assigning equal weights for each samples in the training set. Then, a model is trained taking into account the weight of each samples. After that, samples in the training set are classified using the model. The weights of correctly classified samples is reduced and the weights of incorrectly classified samples is increased. Then, second model is trained on the training set using the new weights. This procedure is repeated L times yielding L different models. Boosting using neural networks is not common and it is mainly used for creating ensemble using weak classifiers such as decision stumps.

5.6.3 Creating Ensemble

Works in Ciresan et al. (2012a), Jin et al. (2014) utilize the model averaging technique in which the average score of several ConvNets is computed. However, there are two limitations with this approach. First, sometimes, the classification accuracy of an ensemble might not be improved substantially compared with a single ConvNet in the ensemble. This is due to the fact that these ConvNets might have ended up in the same local minimum during the training process. As the result, their scores are very similar and their combination does not change the posterior of the classes. The second problem is that there might be some cases where adding a new ConvNet to the ensemble reduces the classification performance. One possibility is that the belief about a class posteriors of the new ConvNet is greatly different from the belief of the ensemble. Consequently, the new ConvNet changes the posterior of the ensemble dramatically which in turn reduces the classification performance.

To alleviate this problem, Ciresan et al. (2012a) and Jin et al. (2014) create ensembles consisting of many ConvNets. The idea is that the number of the ConvNets which contradicts the belief of the ensemble is less than the number of the ConvNets which increase the classification performance. Therefore, the overall performance of the ensemble increases as we add more ConvNets.

While the idea is generally correct but it posses a serious problem in practical applications. Concretely, an ensemble with many ConvNets needs more time to classify the input image. One solution to this problem is to formulate the ensemble construction as a LASSO regression (Tibshirani 1994) problem. Formally, given the classification score vector $\mathscr{L}_i^j$ of ith ConvNet computed on j^{th} image, our goal is to find coefficients a_i by minimizing the following error function:

$$E = \sum_{j=1}^{M} \|y_j - \sum_{i=1}^{N} a_i \mathscr{L}_i^j\| - \lambda \sum_{i=1}^{N} |a_i| \qquad (5.4)$$

where M is the total number of the test images, N is the number of the ConvNets in the ensemble, and λ is a user-defined value to determine the amount of sparseness. It is well-studied that L_1 norm regularization produces a sparse vector in which most of the coefficients a_i are zero. Thus, the ConvNets corresponding to these coefficients can be omitted from the ensemble. The remaining ConvNets are linearly combined according to their corresponding a_i value. Determining the correct value for λ is an empirical task and it might need many trials. More specifically, small values for λ retains most of the ConvNets in the ensemble. Conversely, increasing the value of λ drops more ConvNets from the ensemble.

Another method is to formulate the ensemble construction as the *optimal subset selection problem* by solving the following optimization problem (Aghdam et al. 2016a, b):

$$\arg\min_{I \subset \{1,...,N\}} \left[\frac{1}{M} \sum_{j=1}^{M} \delta \left(y_j - \arg\max \sum_{i \in I} \mathscr{L}_i^j \right) \right] - \lambda |I| \qquad (5.5)$$

where the arg max function returns the index of the maximum value in the classification score vector $\mathscr{L}_i^j$ and y_j is the actual class. The first term calculates the classification accuracy of the selected subset of ConvNets over the testing dataset and the second term penalizes the classification accuracy based on the cardinality of set I. In other words, we are looking for a subset of N ConvNets where classification accuracy is high and the number of the ConvNets is as few as possible. In contrast to the LASSO formulation, selecting the value of λ is straightforward. For example, assume two subsets I_1 and I_2 including 4 and 5 ConvNets, respectively. Moreover, consider that the classification accuracy of I_1 is 0.9888 and the classification accuracy of I_2 is 0.9890. If we set $\lambda = 3e^{-4}$ and calculate the score using (5.5), their score will be equal to 0.9876 and 0.9875. Thus, despite its higher accuracy, the subset I_2 is not better than the subset I_1 because adding an extra ConvNet to the ensemble improves the accuracy 0.02% which is discarded by the penalizing. However, if we choose $\lambda = 2e^{-4}$, the subset I_2 will have a higher score compared with the subset I_1. In sum, λ shows what is the expected minimum accuracy increase that a single ConvNet must cause in the ensemble. The above objective function can be optimized using an evolutionary algorithm such as Genetic algorithm.

In Aghdam et al. (2016b), genetic algorithms with population of 50 chromosomes is used for finding the optimal subset. Each chromosome in this method is encoded using the N-bit *binary coding* scheme. A gene with value 1 indicates the selection of the corresponding ConvNet in the ensemble. The *fitness* of each chromosome is computed by applying (5.5) on the *validation set*. The offspring is selected using the *tournament* selection operator with tournament size 3. The crossover operators are *single-point, two-point*, and *uniform* in which one of them is randomly applied in each iteration. The mutation operator flips the gene of a chromosome with probability $p = 0.05$. Finally, we also apply the *elitism* (with elite count = 1) to guarantee that the algorithm will not forget the best answer. Also, this can contribute for faster convergence by using the best individual so far during the selection process in the next iteration which may generate better answers during.

5.7 Evaluating Networks

We explained a few methods for evaluating classification models including ConvNets. In this section, we provide different techniques that can be used for analyzing ConvNets. To this end, we trained the network shown in Fig. 5.11 and its compact version 10 times and evaluated using the test set provided in the GTSRB dataset. Table 5.1 shows the results. The average classification accuracy of the 10 trials is 98.94 and 98.99% for the original ConvNet and its compact version, respectively, which are both above the average human performance reported in Stallkamp et al. (2012). In addition, the standard deviations of the classification accuracy is small which show that the proposed architecture trains the networks with very close accuracies. We argue that this stability is the result of reduction in the number of the

Table 5.1 Classification accuracy of the single network. Above The proposed network in Aghdam et al. Aghdam et al. (2016a) and below its compact version

Aghdam et al. (2016a) (original)

Trial	Top 1 acc. (%)	Top 2 acc. (%)
1	98.87	99.62
2	98.98	99.64
3	98.85	99.62
4	98.98	99.58
5	98.99	99.63
6	99.06	99.75
7	98.99	99.66
8	99.05	99.70
9	98.88	99.57
10	98.77	99.60
Average	98.94 ± 0.09	99.64 ± 0.05
Human	98.84	NA
Ciresan et al. (2012a)	98.52 ± 0.15	NA
Jin et al. (2014)	98.96 ± 0.20	NA

Aghdam et al. (2016a) (compact)

Trial	Top 1 acc. (%)	Top 2 acc. (%)
1	99.11	99.63
2	99.06	99.64
3	98.88	99.62
4	98.97	99.61
5	99.08	99.66
6	98.94	99.68
7	98.87	99.60
8	98.98	99.65
9	98.92	99.61
10	99.05	99.63
Average	98.99 ± 0.08	99.63 ± 0.02
Human	98.84	NA
Ciresan et al. (2012a)	98.52 ± 0.15	NA
Jin et al. (2014)	98.96 ± 0.20	NA

parameters and regularizing the network using a dropout layer. Moreover, we observe that the top-2[8] accuracies are very close in all trials and their standard deviations

---

[8]The percent of the samples which are always within the top 2 classification scores.

are 0.05 and 0.02 for the original ConvNet and its compact version, respectively. In other words, although the difference in the top-1 accuracies of the Trial 1 and the Trial 2 in the original network is 0.11%, notwithstanding, the same difference for top-2 accuracy is 0.02%. This implies that there are images that are classified correctly in Trial 1 and they are misclassified in Trial 2 (or vice versa) but they are always within the top-2 scores of both networks. As a consequence, if we fuse the scores of the two networks the classification accuracy might increase. The same argument is applied on the compact network, as well. Compared with the average accuracies of the single ConvNets proposed in Ciresan et al. (2012a) and Jin et al. (2014), the architecture in Aghdam et al. (2016a) and its compact version are more stable since their standard deviations are less than the standard deviations of these two ConvNets. In addition, despite the fact that the compact network has 52% fewer parameters than the original network, the accuracy of the compact network is more than the original network and the two other networks. This confirms the claim illustrated by t-SNE visualization in Fig. 5.11 that the fully connected layer in the original ConvNet does not increase the separability of the traffic signs. But, the fact remains that the compact network has less degree of freedom compared with the original network. Taking into account the Bias-Variance decomposition of the original ConvNet and the compact ConvNet, Aghdam et al. (2016a) claim that the compact ConvNet is more biased and its variance is less compared with the original ConvNet. To prove this, they created two different ensembles using the algorithm mentioned in Sect. 5.6.3. More specifically, one ensemble was created by selecting the optimal subset from a pool of 10 original ConvNets and the second pool was created in the same way but from a pool of 10 compact ConvNets. Furthermore, two other ensembles were created by utilizing the model averaging approach (Ciresan et al. 2012a; Jin et al. 2014). in which each ensemble contains 10 ConvNets. Tables 5.2 and 5.3 show the results and compare them with three other state-of-art ConvNets.

First, we observe that ensemble creating based on optimal subset selection method is more efficient than the model averaging approach. To be more specific, the ensemble created by selecting optimal subset of the ConvNets needs 50% less

Table 5.2 Comparing the classification performance of the ensembles created by model averaging and our proposed method on the pools of original and compact ConvNets proposed in Aghdam et al. (2016a) with three state-of-art ConvNets

Name	No. of ConvNets	Accuracy (%)	F_1-score
Ens. of original ConvNets	5	99.61	0.994
Ens. of original ConvNets (avg.)	10	99.56	0.993
Ens. of compact ConvNets	2	99.23	0.989
Ens. of compact ConvNets (avg.)	10	99.16	0.987
Ciresan et al. (2012a)	25	99.46	NA
Sermanet and Lecun (2011)	1	98.97	NA
Jin et al. (2014)	20	99.65	NA

Table 5.3 Comparing the run-time efficiency of the ensembles created by model averaging and optimal subset selection method on pools of original and compact ConvNets with three state-of-art ConvNets. Note that we have calculated the *worst* case by considering that every LReLU unit will perform one multiplications. In contrast, we have computed the *minimum* number of multiplications in Ciresan et al. (2012a) by assuming that the input of *tanh* function always falls in range $[-0.5, 0.5]$. Similarly, in the case of Jin et al. (2014), we have considered fast but inaccurate implementation of *pow(float, float)*

Name	No. of ConvNets	No. of parameters	No. of multiplications
Ens. of original ConvNets	5	1,123,449	382,699,560
Ens. of original ConvNets (avg.)	10	1,123,449	765,399,120
Ens. of compact ConvNets	2	531,999	151,896,924
Ens. of compact ConvNets (avg.)	10	531,999	759,484,620
Ciresan et al. (2012a)	25	1,543,443	3,208,042,500
Sermanet and Lecun (2011)	1	1,437,791	NA
Jin et al. (2014)	20	1,162,284	1,445,265,400

multiplications[9] and its accuracy is 0.05% higher compared with the ensemble created by averaging all the original ConvNets in the pool (10 ConvNets). Note that the number of ConvNets in the ensemble directly affects the number of the arithmetic operations required for making predictions. This means that the model averaging approach consumes double CPU cycles compared with optimal subset ensemble.

Moreover, the ensemble created by optimal subset criteria reduces the number of the multiplications 88 and 73% compared with the ensembles proposed in Ciresan et al. (2012a) and Jin et al. (2014), respectively. More importantly, the dramatic reduction in the number of the multiplications causes only five more misclassification (0.04% less accuracy) compared with the results obtained by the ensemble in Jin et al. (2014). We also observe that the ensemble in Ciresan et al. (2012a) makes 19 more mistakes (0.15% more misclassification) compared with optimal subset ensemble.

Besides, the number of the multiplications of the network proposed in Sermanet and Lecun (2011) is not accurately computable since its architecture is not clearly mentioned. However, the number of the parameters of this ConvNet is more than the

---

[9]We calculated the number of the multiplications of a ConvNet taking into account the number of the multiplications for convolving the filters of each layer with the N-channel input from the previous layer, number of the multiplications required for computing the activations of each layer and the number of the multiplications imposed by normalization layers. We previously explained that *tanh* function utilized in Ciresan et al. (2012a) can be efficiently computed using 10 multiplications. ReLU activation used in Jin et al. (2014) does not need any multiplications and Leaky ReLU units in Aghdam et al. (2016a) computes the results using only 1 multiplication. Finally, considering that *pow(float, float)* function needs only 1 multiplication and 64 shift operations (tinyurl.com/yehg932), the normalization layer in Jin et al. (2014) requires $k \times k + 3$ multiplications per each element in the feature map.

ConvNet in Aghdam et al. (2016a). In addition, it utilizes rectified sigmoid activation which needs 10 multiplications per each element in the feature maps. In sum, we can roughly conclude that the ConvNet in Sermanet and Lecun (2011) needs more multiplications. However, we observe that an ensemble of two compact ConvNets performs better than Sermanet and Lecun (2011) and, yet, it needs less multiplications and parameters.

Finally, although the single compact ConvNet performs better than single original ConvNet, nonetheless, the ensemble of compact ConvNets does not perform better. In fact, according to Table 5.2, an ensemble of two compact ConvNets shows a better performance compared with the ensemble of 10 compact ConvNets. This is due to the fact that the compact ConvNet is formulated with much fewer parameters and it is more biased compared with the original ConvNet. Consequently, their representation ability is more restricted. For this reason, adding more ConvNets to the ensemble does not increase the performance and it always vary around 99.20%. In contrary, the original network is able to model more complex nonlinearities so it is less biased about the data and its variance is more than the compact network. Hence, the ensemble of the original networks posses more discriminative representation which increases its classification performance. In sum, if run-time efficiency is more important than the accuracy, then, ensemble of two compact ConvNets is a good choice. However, if we need more accuracy and the computational burden imposed by more multiplications in the original network is negligible, then, the ensemble of the original ConvNets can be utilized.

It is worth mentioning that the time-to-completion (TTC) of ConvNets does not solely depend on the number of multiplications. Number of accesses to memory also affects the TTC of ConvNets. From the ConvNets illustrated in Table 5.3, a single ConvNet proposed in Jin et al. (2014) seems to have a better TTC since it needs less multiplications compared with Aghdam et al. (2016a) and its compact version. However, Jin et al. (2014, Table IX) shows that this ConvNets needs to pad the feature maps before each convolution layer and there are three local response normalization layers in this ConvNet. For this reason, it need more accesses to memory which can negatively affect the TTC of this ConvNets. To compute the TTC of these ConvNets in practice, we ran the ConvNets on both CPU (Inter Core i7-4960), and GPU (GeForce GTX 980). The hard disk was not involved in any other task and there were no running application or GPU demanding processes. The status of the hardware was fixed during the calculation of the TTC of ConvNets. Then, the average TTC of the forward-pass of every ConvNet was calculated by running each ConvNet 200 times. Table 5.4 shows the results in the scale of milliseconds for one forward-pass.

The results show that the TTC of single ConvNet proposed in Jin et al. (2014) is 12 and 37% more than Aghdam et al. (2016a) when it runs on CPU and GPU, respectively. This is consistent with our earlier discussion that the TTC of ConvNets does not solely depend on arithmetic operations. But, the number of memory accesses affects the TTC. Also, the TTC of the ensemble of Aghdam et al. (2016a) is 78 and 81% faster than the ensemble proposed in Jin et al. (2014).

Table 5.4 Benchmarking time-to-completion of Aghdam et al. (2016a) along with its compact ConvNet and Jin et al. (2014) obtained by running the forward-pass of each ConvNet 200 times and computing the average time for completing the forward-pass

	Aghdam et al. (2016a)	Aghdam et al. (2016a) (compact)	Jin et al. (2014)
CPU	12.96 ms	12.47 ms	14.47 ms
GPU	1.06 ms	1.03 ms	1.45 ms
	Aghdam et al. (2016a) ens.	Aghdam et al. (2016a) ens. (compact)	Jin et al. (2014) ens.
CPU	$5 \times 12.96 = 64.8$ ms	$2 \times 12.47 = 24.94$ ms	$20 \times 14.47 = 289.4$ ms
GPU	$5 \times 1.06 = 5.30$ ms	$2 \times 1.03 = 2.06$ ms	$20 \times 1.45 = 29.0$ ms

Table 5.5 Class-specific precision and recall obtained by the network in Aghdam et al. (2016a). Bottom images show corresponding class label of each traffic sign. The column support (sup) shows the number of the test images for each class

Class	precision	recall	sup	Class	precision	recall	sup	Class	precision	recall	sup
0	1.00	1.00	60	15	1.00	1.00	210	30	1.00	0.97	150
1	1.00	1.00	720	16	1.00	1.00	150	31	1.00	0.99	270
2	1.00	1.00	750	17	1.00	1.00	360	32	1.00	1.00	60
3	1.00	0.99	450	18	1.00	0.99	390	33	1.00	1.00	210
4	1.00	0.99	660	19	0.97	1.00	60	34	1.00	1.00	120
5	0.99	1.00	630	20	0.99	1.00	90	35	1.00	1.00	390
6	1.00	0.98	150	21	0.97	1.00	90	36	0.98	1.00	120
7	1.00	1.00	450	22	1.00	1.00	120	37	0.97	1.00	60
8	1.00	1.00	450	23	1.00	1.00	150	38	1.00	1.00	690
9	1.00	1.00	480	24	0.99	0.99	90	39	0.98	0.98	90
10	1.00	1.00	660	25	1.00	0.99	480	40	0.97	0.97	90
11	0.99	1.00	420	26	0.98	1.00	180	41	1.00	1.00	60
12	1.00	1.00	690	27	0.97	1.00	60	42	0.98	1.00	90
13	1.00	1.00	720	28	1.00	1.00	150				
14	1.00	1.00	270	29	1.00	1.00	90				

5.7.1   Misclassified Images

We computed the class-specific *precision* and *recall* (Table 5.5). Besides, Fig. 5.14 illustrates the incorrectly classified traffic signs. The blue and red numbers below each image show the actual and predicted class labels, respectively. For presentation purposes, all images were scaled to a fixed size. First, we observe that there are four cases where the images are incorrectly classified as class 11 while the true label is 30. Particularly, three of these cases are low-resolution images with poor illuminations. Moreover, class 30 is distinguishable from class 11 using the fine differences in the pictograph. However, rescaling a poorly illuminated low-resolution

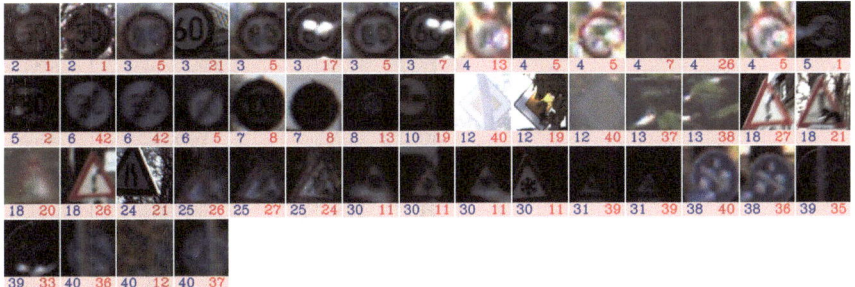

Fig. 5.14 Incorrectly classified images. The *blue* and *red* numbers below each image show the actual and predicted class labels, respectively. The traffic sign corresponding to each class label is illustrated in Table 5.5

image to 48 × 48 pixels causes some artifacts on the image. In addition, two of these images are inappropriately localized and their bounding boxes are inaccurately. As the result, the network is not able to discriminate these two classes on these images. In addition, by inspecting the rest of the misclassified images, we realize that the wrong classification is mainly due to *occlusion of pictograph* or *low-quality* of the images. However, there are a few cases where the main reason of the misclassification is due to inaccurate localization of the traffic sign in the detection stage (i.e., inaccurate bounding box).

5.7.2 Cross-Dataset Analysis and Transfer Learning

So far, we trained a ConvNet on the GTSRB dataset and achieved state-of-art results with much fewer arithmetic operations and memory accesses which led to a considerably faster approach for classification of traffic signs. In this section, we inspect how transferable is this ConvNet across different datasets. To this end, we first evaluate the cross-dataset performance of the network. To be more specific, we use the trained ConvNet to predict the class of the traffic signs in the Belgium traffic sign classification (BTSC) dataset (Radu Timofte 2011) (Fig. 5.15).

We inspected the dataset to make it consistent with the GTSRB. For instance, Class 32 in this dataset contains both signs "speed limit 50" and "speed limit 70". However, these are two distinct classes in the GTSRB dataset. Therefore, we separated the overlapping classes in the BTSC dataset according to the GTSRB dataset. Each image in the BTSC dataset contains one traffic sign and it totally consists of 4,672 color images for training and 2,550 color images for testing. Finally, we normalize the dataset using the mean image obtained from the GTSRB dataset and resize all the images to 48 × 48 pixels.

Among 73 classes in the BTSC dataset (after separating the overlapping classes), there are 23 common classes with the GTSRB dataset. We applied our ConvNet trained on the GTSRB dataset to classify these 23 classes inside both the *training* set

Fig. 5.15 Sample images from the BTSC dataset

and the *testing* set in the BTSC dataset. Table 5.6 shows the class-specific precision and recall.

In terms of accuracy, the trained network has correctly classified 92.12% of samples. However, precisions and recalls reveal that the classification of class 29 is worse than a random guess. To find out the reason, we inspect the misclassified images illustrated in Fig. 5.16.

Comparing the class 29 in the BTSC dataset with its corresponding class in the GTSRB (Table 5.5) shows that the pictograph of this class in the GTSRB dataset has significant differences with the pictograph of the same class in the BTSC dataset. In general, the misclassified images are mainly due to pictograph differences, perspective variation, rotation and blurriness of the images. We inspected the GTSRB dataset and found that perspective and rotation is more controlled than the BTSC dataset. As the result, the trained ConvNet has not properly captured the variations caused by different perspectives and rotations on the traffic signs. In other words, if we present adequate amount of data covering different combinations of perspective and rotation, the ConvNet might be able to accurately model the traffic signs in the BTSC dataset.

To prove that, we try to find out how transferable is the ConvNet. We follow the same procedure mentioned in Yosinski et al. (2014) and evaluate the degree of transferability of the ConvNet in different stages. Concretely, the original ConvNet is trained on the GTSRB dataset. The Softmax loss layer of this network consists of 43 neurons since there are only 43 classes in the GTSRB dataset. We can think of the transformation layer up to the $LReLU_4$ layer as a function which extracts the features of the input image. Thus, if this feature extraction algorithm performs accurately on the GTSRB dataset, it should also be able to model the traffic signs

Table 5.6 Cross-dataset evaluation of the trained ConvNet using the BTSC dataset. Class-specific precision and recall obtained by the network are shown. The column support (sup) shows the number of the test images for each class. Classes with support equal to zero do not have any test cases in the BTSC dataset

Class	Precision	Recall	Sup	Class	Precision	Recall	Sup	Class	Precision	Recall	Sup
0	NA	NA	0	15	0.91	0.86	167	30	NA	NA	0
1	NA	NA	0	16	1.00	0.78	45	31	NA	NA	0
2	NA	NA	0	17	1.00	0.93	404	32	NA	NA	0
3	NA	NA	0	18	0.99	0.93	125	33	NA	NA	0
4	1.00	0.93	481	19	1.00	0.90	21	34	NA	NA	0
5	NA	NA	0	20	0.93	0.96	27	35	0.92	1.00	96
6	NA	NA	0	21	0.92	0.92	13	36	1.00	0.83	18
7	NA	NA	0	22	0.72	1.00	21	37	NA	NA	0
8	NA	NA	0	23	1.00	0.95	19	38	NA	NA	0
9	0.94	0.94	141	24	0.66	1.00	21	39	NA	NA	0
10	NA	NA	0	25	0.90	1.00	47	40	0.99	0.87	125
11	0.88	0.91	67	26	0.75	0.86	7	41	NA	NA	0
12	0.97	0.95	382	27	NA	NA	0	42	NA	NA	0
13	0.97	0.99	380	28	0.89	0.91	241				
14	0.87	0.95	86	29	0.19	0.08	39				

Overall accuracy: 92.12%

Fig. 5.16 Incorrectly classified images from the BTSC dataset. The *blue* and *red* numbers below each image show the actual and predicted class labels, respectively. The traffic sign corresponding to each class label is illustrated in Table 5.5

in the BTSC dataset. To evaluate the generalization power of the ConvNet trained only on the GTSRB dataset, we replace the Softmax layer with a new Softmax layer including 73 neurons to classify the traffic signs in the BTSC dataset. Then, we freeze the weights of all the layers except the Sofmax layer and run the training algorithm on the BTSC dataset to learn the weights of the Softmax layer. Finally, we evaluate the performance of the network using the testing set in the BTSC dataset. This empirically computes how transferable is the network in Aghdam et al. (2016a) on other traffic signs datasets.

It is well studied that the first layer of a ConvNet is more general and the last layer is more class specific. This means that the $FC1$ layer in Fig. 5.11) is more specific than the $C3$ layer. In other words, the $FC1$ layer is adjusted to classify the 43 traffic signs in the GTSRB dataset. As the result, it might not be able to capture every aspects in the BTSC dataset. If this assumption is true, then we can adjust the weights in the FC1 layer beside the Softmax layer so it can model the BTSC dataset more accurately. Then, by evaluating the performance of the ConvNet on the testing set of the BTSC dataset we can find out to what extend the $C3$ layer is able to adjust on the BTSC dataset. We increasingly add more layers to be adjusted on the BTSC dataset and evaluate their classification performance. At the end, we have five different networks with the same configuration but different weight adjustment procedures on the BTSC dataset. Table 5.7 shows the weights which are fixed and adjusted in each network. We repeated the training 4 times for each row in this table. Figure 5.17 shows the results.

First, we observe that when we only adjust the softmax layer (layer 5) and freeze the previous layers, the accuracy drops dramatically compared with the results in the GTSRB dataset. In the one hand, layer 4 is adjusted such that the traffic signs in the GTSRB dataset become linearly separable and they can be discriminated using the linear classifier in the softmax layer. On the other hand, the number of the traffic signs in the BTSC dataset is increased 70% compared with GTSRB dataset. Therefore, layer 4 is not able to linearly differentiate fine details of the traffic signs in the BTSC dataset. This is observable from the t-SNE visualization of the $LReLU_4$ layer corresponding to $n = 5$ in Fig. 5.17. Consequently, the classification performance drops because of overlaps between the classes.

If the above argument is true, then, fine-tuning the layer 4 beside the layer 5 must increase the performance. Because, by this way, we let the $LReLU_4$ layer to be adjusted on the traffic signs included in the BTSC dataset. We see in the figure that adjusting the layer 4 ($n = 4$) and the layer 5 ($n = 5$) increases the classification

Table 5.7 Layers which are frozen and adjusted in each trial to evaluate the generality of each layer

ConvNet No.	Trans.	Conv1 layer 1	Conv2 layer 2	Conv3 layer 3	FC1 layer 4	Softmax layer 5
1	Fixed	Fixed	Fixed	Fixed	Fixed	Adjust
2	Fixed	Fixed	Fixed	Fixed	Adjust	Adjust
3	Fixed	Fixed	Fixed	Adjust	Adjust	Adjust
4	Fixed	Fixed	Adjust	Adjust	Adjust	Adjust
5	Fixed	Adjust	Adjust	Adjust	Adjust	Adjust

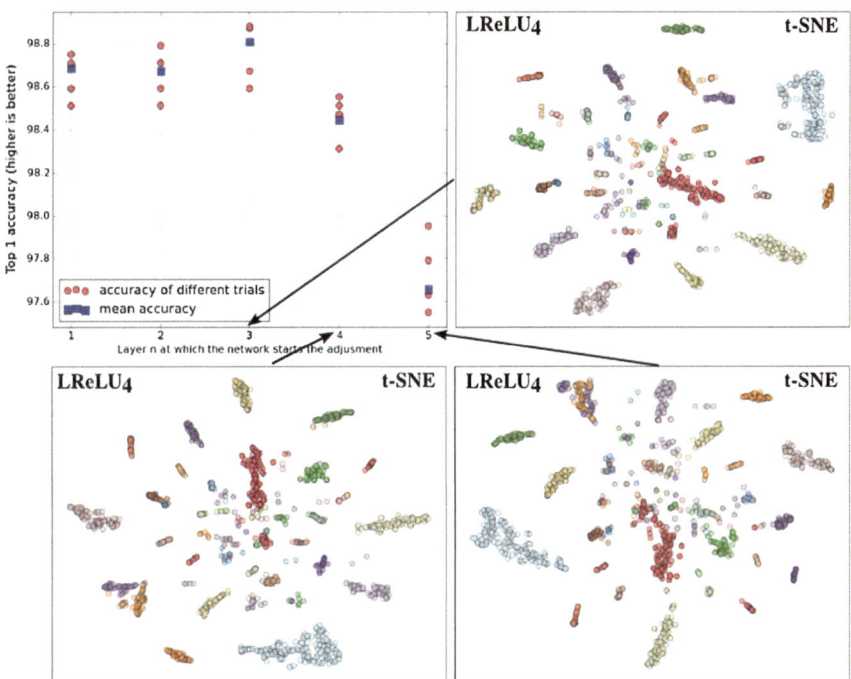

Fig. 5.17 The result of fine-tuning the ConvNet on the BTSC dataset that is trained using GTSRB dataset. Horizontal axis shows the layer n at which the network starts the weight adjustment. In other words, weights of the layers before the layer n are fixed (frozen). The weights of layer n and all layers after layer n are adjusted on the BTSC dataset. We repeated the fine-tuning procedure 4 times for each $n \in \{1, \ldots, 5\}$, separately. *Red circles* show the accuracy of each trial and *blue* squares illustrate the mean accuracy. The t-SNE visualizations of the best network for $n = 3, 4, 5$ are also illustrated. The t-SNE visualization is computed on the $LReLU_4$ layer

accuracy from 97.65 to 98.44%. Moreover, the t-SNE visualization corresponding to $n = 4$ reveals that the traffic signs classes are more separable compared with the result from $n = 5$. Thus, adjusting both $LReLU_4$ and Softmax layers make the network more accurate for the reason we mentioned above.

Recall from Fig. 5.11 that $LReLU_4$ was not mainly responsible for increasing the separability of the classes. Instead, we saw that this layer mainly increases the variance of the ConvNet and improves the performance of the ensemble. In fact, we showed that traffic signs are chiefly separated using the last convolution layer. To further inspect this hypothesis, we fine-tuned the ConvNet on the BTSC dataset starting from layer 3 (i.e., the last convolution layer). Figure 5.17 illustrate an increase up to 98.80% in the classification accuracy. This can be also seen on the t-SNE visualization corresponding to the layer 3 where traffic signs of the BTSC dataset become more separable when the ConvNet is fine-tuned starting from the layer 3.

Interestingly, we observe a performance reduction when the weights adjustment starts from layer 2 or layer 1. Specifically, the mean classification accuracy drops from 98.80% in layer 3 to 98.67 and 98.68% in layer 2 and layer 1, respectively. This is due to the fact that the first two layers are more general and they do not significantly change from the GTSRB to the BTSC dataset. In fact, these two layers are trained to detect blobs and oriented edges. However, because the number of data is very few in the BTSC dataset compared with the number of the parameters in the ConvNet, hence, it adversely modifies the general filters in the first two layers which consequently affects the weight of the subsequent layers. As the result, the ConvNet overfits on data and does not generalize well on the test set. For this reason, the accuracy of the network drops when we fine-tune the network starting from layer 1 or layer 2.

Finally, it should be noted that 98.80 accuracy is obtained using only a single network. As we showed earlier, creating an ensemble of these networks could improve the classification performance. In sum, the results obtained from cross-dataset analysis and transferability evaluation reveals that the network is able to model a wide range of traffic signs and in the case of new datasets it only needs to be fine-tuned starting from the last convolution layer.

5.7.3 Stability of ConvNet

A ConvNet is a nonlinear function that transforms a D-dimensional vector into a K-dimensional vector in the layer before the classification layer. Ideally, small changes in the input should produce small changes in the output. In other words, if image $f \in \mathcal{R}^{M \times N}$ is correctly classified as c using the ConvNet, then, the image $g = f + r$ obtained by adding a small degradation $r \in \mathcal{R}^{M \times N}$ to f must also be classified as c.

However, f is strongly degraded as $\|r\|$(norm of r) increases. Therefore, at a certain point, the degraded image g is not longer recognizable. We are interested in finding r with minimum $\|r\|$ that causes the g and f are classified differently. Szegedy et al. (2014b) investigated this problem and they proposed to minimize the following objective function with respect to r:

$$minimize \ \ \lambda|r| + score(f + r, l)$$
$$s.t \ \ \ f + r \in [0, 1]^{M \times N} \tag{5.6}$$

where l is the actual class label, λ is the regularizing weight, and $score(f + r, l)$ returns the score of the degraded image $f + r$ given the actual class of image f. In the ConvNet, the classification score vector is 43 dimensional since there are only 43 classes in the GTSRB dataset. Denoting the classification score vector by $\mathcal{L} \in [0, 1]^{43}$, $\mathcal{L}[k]$ returns the score of the input image for class k. The image is classified correctly if $c = \arg \max \mathcal{L} = l$ where c is the index of the maximum value in the score vector $\mathcal{L}$. If $max(\mathcal{L}) = 0.9$, the ConvNet is 90% confident that the input image belongs to class c. However, there might be an image where $max(\mathcal{L}) = 0.3$. This means that the image belongs to class c with probability 0.3. If we manually inspect the scores of other classes we might realize that $\mathcal{L}[c_2] = 0.2$, $\mathcal{L}[c_3] =$

0.2, $\mathscr{L}[c_4] = 0.2$, and $\mathscr{L}[c_5] = 0.1$ where c_i depicts the ith maximum in the score vector $\mathscr{L}$.

Conversely, assume two images that are misclassified by the ConvNet. In the first image, $\mathscr{L}[l] = 0.1$ and $\mathscr{L}[c] = 0.9$ meaning that the ConvNet believes the input image belongs to class l and class c with probabilities 0.1 and 0.9, respectively. But, in the second image, the beliefs of the ConvNet are $\mathscr{L}[l] = 0.49$ and $\mathscr{L}[c] = 0.51$. Even tough in both cases the images are misclassified, the degrees of misclassification are different.

One problem with the objective function (5.6) is that it finds r such that $score(f + r, l)$ approaches to zero. In other words, it finds r such that $\mathscr{L}[l] = \varepsilon$ and $\mathscr{L}[c] = 1 - \varepsilon$. Assume the current state of the optimization function is r_t where $\mathscr{L}[l] = 0.3$ and $\mathscr{L}[c] = 0.7$. In other words, the input image f is misclassified using the current degradation r_t. Yet, the goal of the objective function (5.6) is to settle in a point where $score(f + r_t, l) = \varepsilon$. As the result, it might change r_t which results in a greater $\|r_t\|$. Consequently, the degradation found by minimizing the objective function (5.6) might not be optimal. To address this problem, we propose the following objective function to find the degradation r:

$$minimize \quad \psi(\mathscr{L}, l) + \lambda \|\mathscr{L}\|_1 \tag{5.7}$$

$$\psi(\mathscr{L}, l) = \begin{cases} \beta \times \mathscr{L}[l] & \arg\max_c \mathscr{L} = l \\ \max(\mathscr{L}) - \mathscr{L}[l] & otherwise \end{cases} \tag{5.8}$$

In this equation, λ is the regularizing weight, β is a multiplier to penalize those values of r that do not properly degrade the image so it is not misclassified by the ConvNet and $\|.\|_1$ is the sparsity inducing term that forces r to be sparse. The above objective function finds the value r such that degrading the input image f using r causes the image to be classified incorrectly and the difference between the highest score in $\mathscr{L}$ and the true label of f is minimum. This guarantees that $f + r$ will be outside the decision boundary of actual class l but it will be as close as possible to this decision boundary.

We minimize the objective function (5.7) using *genetic algorithms*. To this end, we use real-value encoding scheme for representing the population. The size of each chromosome in the population is equal to the number of the elements in r. Each chromosome, represents a solution for r. We use *tournament* method with tour size 5 for selecting the offspring. Then, a new offspring is generated using *arithmetic*, *intermediate* or *uniform* crossover operators. Finally, the offspring is mutated by adding a small number in range $[-10, 10]$ on some of the genes in the population. Finally, we use *elitism* to always keep the best solution in the population. We applied the optimization procedure for one image from each traffic sign classes. Figure 5.18 shows the results.

Inspecting all the images in this figure, we realize that the ConvNet can easily make mistakes even for noises which are not perceivable by human eye. This conclusion is also made by Szegedy et al. (2014b). This suggests that the function presenting by the ConvNet is highly nonlinear where small changes in the input may cause a

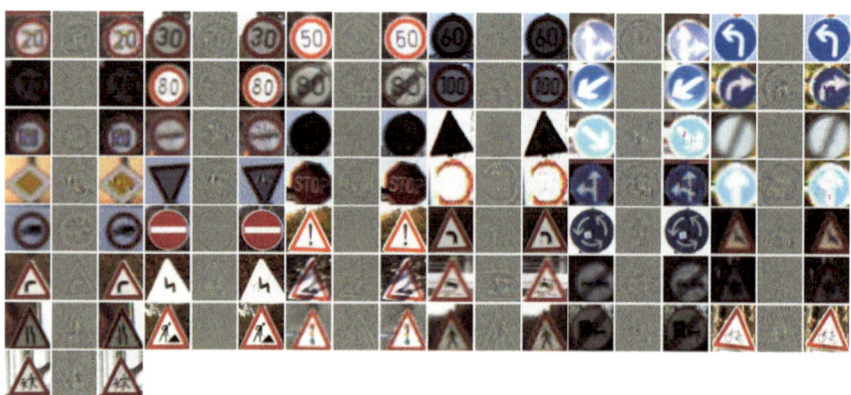

Fig. 5.18 Minimum additive noise which causes the traffic sign to be misclassified by the minimum different compared with the highest score

significant change in the output. When the output changes dramatically, it might fall in a wrong class in the feature space. Hence, the image is incorrectly classified. Note that, because of the proposed objective function, the difference between the wrongly predicted class and the true class is positive but it is very close the decision boundary of the two classes. We repeated the above procedure on 15 different images and calculated the mean Signal-to-Noise Ratio (SNR) of each class, separately. Figure 5.19 shows the results. First, we observe that classes 4 and 30 have the lowest SNR values. In other words, the images from these two classes are more tolerant against noise. In addition, class 15 has the highest SNR values which shows it is more prone to be misclassified with small changes. Finally, most of the classes are tolerant against noise with approximately the same degree of tolerance since they have close mean SNR values. One simple solution to increase the tolerance of the ConvNet is to augment

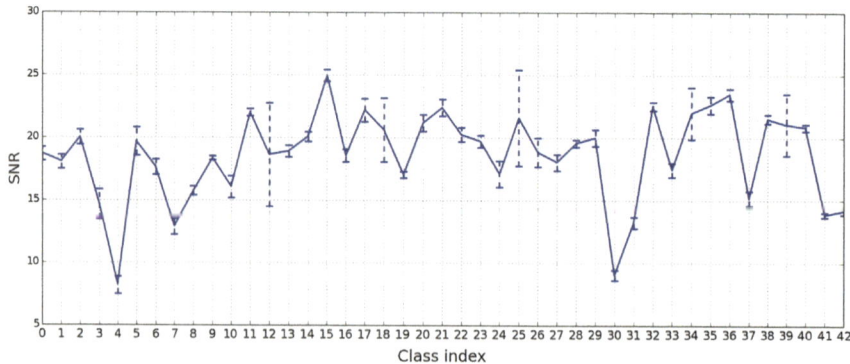

Fig. 5.19 Plot of the SNRs of the noisy images found by optimizing (5.7). The mean SNR and its variance are illustrated

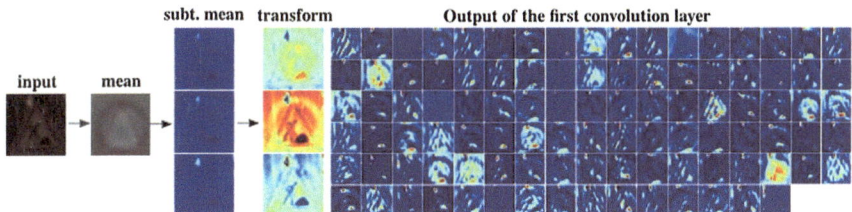

Fig. 5.20 Visualization of the transformation and the first convolution layers

noisy images with various SNR values so the network can learn how to handle small changes.

5.7.3.1 Effect of Linear Transformation

We manually inspected the database and realized that there are images with poor illumination. In fact, the transformation layer enhances the illumination of the input image by multiplying the each channel with different constant factors and adding different intercepts to the result. Note that there is a unique transformation function per each channel. This is different from applying the same linear transformation function on all channels in which it does not have any effect on the results of convolution filters in the next layer (unless the transformation causes the intensity of the pixels exceed their limits). In this ConvNet, applying a different transformation function on each channel affects the output of the subsequent convolution layer. By this way, the transformation layer learns the parameters of the linear transformation such that it increases the classification accuracy. Figure 5.20 illustrates the output of the transformation and the first convolution layers. We observe that the input image suffers from a poor illumination. However, applying the linear transformation on the image enhances the illumination of each channel differently and, consequently, the subsequent layers represent the image properly so it is classified correctly.

5.7.4 Analyzing by Visualization

Visualizing ConvNets helps to understand them under different circumstances. In this section, we propose a new method for assessing the stability of the network and, then, conduct various visualization techniques to analyze different aspects of the proposed ConvNet.

5.8 Analyzing by Visualizing

Understanding the underlying process of ConvNets is not trivial. To be more specific, it is not easy to mathematically analyze a particular layer/neuron and determine what the layer/neuron exactly does on the input and what is extracted using the layer/neuron. Visualizing is a technique that helps to better understand underlying process of a ConvNet. There are several ways for visualizing a ConvNet. In this section, we will explain a few techniques that can be utilized in practical applications.

5.8.1 Visualizing Sensitivity

Assume we are given a pure image which is classified correctly by the ConvNet. We might be interested in localizing those areas on the image where degrading one of these areas by noise causes the image to be misclassified. This helps us to identify the sensitive regions of each traffic sign. To this end, we start from a window size equal to 20% of the image size and slide this window on the image. At each location, the region under the window is degraded by noise and the classification score of the image is computed. By this way, we obtain a score matrix H^c where element $H^c(m, n)$ is the score of the image belonging to class c when a small region of the image starting from (m, n) is degraded by noise (i.e., (m, n) is the top-left corner of the window not its center). We computed the matrix $H_i^c, i \in 1 \ldots 20$ for 20 different instances of the same class and calculated the average matrix $\bar{H}^c = \frac{\sum_{i=1}^{20} H_i}{20}$ as well as the average image. Figure 5.21 illustrates the heat map of $\bar{H}$. First, we observe that the ConvNet is mainly sensitive to small portion of the pictographs in the traffic signs. For example, in the speed limits signs related to speeds less than 100, it is clear that the ConvNet is mainly sensitive to some part of the first digit. Conversely, the score is affected by whole three digits in the "speed limit 100" sign. In addition, the score of the "speed limit 120" sign mainly depends on the second digit. These are all reasonable choices made by the ConvNet since the best way to classify two-digit speed limit signs is to compare their first digit. In addition, the "speed limit 100" is differentiable from "speed limit 120" sign through only the middle digit.

Furthermore, there are traffic signs such as the "give way" and the "no entry" signs in which the ConvNet is sensitive in almost every location on the image. In other

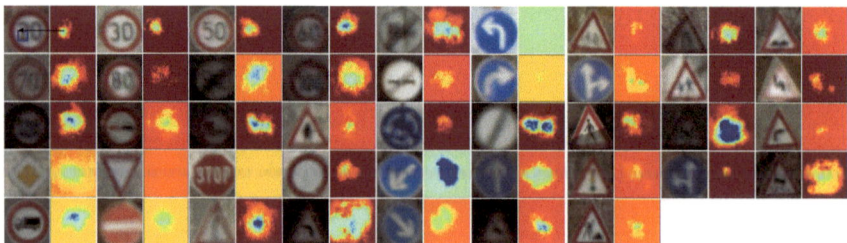

Fig. 5.21 Classification score of traffic signs averaged over 20 instances per each traffic sign. The *warmer color* indicates a higher score and the colder color shows a lower score. The corresponding window of element (m, n) in the score matrix is shown for one instance. It should be noted that the (m, n) is the *top-left* corner of the window not its center and the size of the window is 20% of the image size in all the results

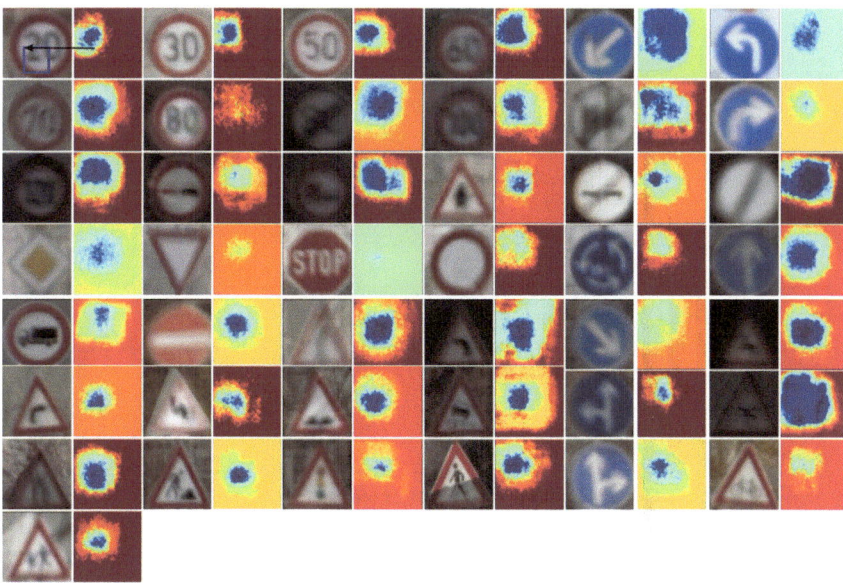

Fig. 5.22 Classification score of traffic signs averaged over 20 instances per each traffic sign. The warmer color indicates a higher score. The corresponding window of element (m, n) in the score matrix is shown for one instance. It should be noted that the (m, n) is the *top-left* corner of the window not its center and the size of the window is 40% of the image size in all the results

words, the score of the ConvNet is affected regardless of the position of the degraded region when the size of the degradation window is 20% of the image. We increased the size of the window to 40% and repeated the above procedure. Figure 5.22 shows the result. We still see that all analysis mentioned for window size 20% hold true for window size 40%, as well. In particular, we observe that the most sensitive regions of the "mandatory turn left" and the "mandatory turn right" traffic signs emerge by increasing the window size. Notwithstanding, degradation affects the classification score regardless of its location in these two signs.

5.8.2   Visualizing the Minimum Perception

Classifying traffic signs at night is difficult because perception of the traffic signs is very limited. In particular, the situation is much worse in interurban areas at which the only lightening source is the headlights of the car. Unless the car is very close to the signs, it is highly probable that the traffic signs are partially perceived by the camera. In other words, most part of the perceived image might be dark. Hence, this question arises that "what is the minimum area to be perceived by the camera to successfully classify the traffic signs".

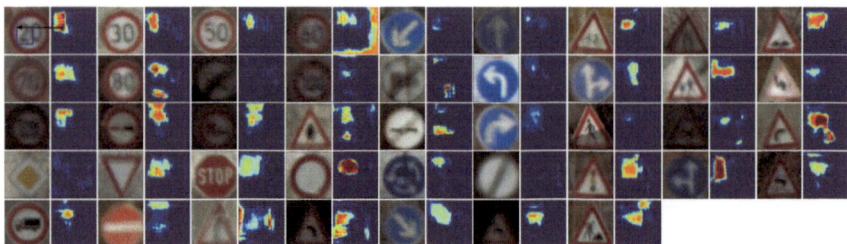

Fig. 5.23 Classification score of traffic signs averaged over 20 instances per each traffic sign. The warmer color indicates a higher score. The corresponding window of element (m, n) in the score matrix is shown for one instance. It should be noted that the (m, n) is the *top-left* corner of the window not its center and the size of the window is 40% of the image size in all the results

To answer this question, we start from a window size equal to 40% of the image size and slide this window on the image. At each location, we keep the pixels under the window untouched and zero out the rest of the pixels. Then, the image is entered into the ConvNet and the classification score is computed. By this way, we obtain a score matrix H where element $H(m, n)$ is the score of the traffic sign when only a small region of the image starting from (m, n) is perceived by the camera. As before, we computed the average score matrix $\bar{H}$ using 20 instances for each traffic sign. Figure 5.23 illustrates the heat map plot of $\bar{H}$ obtained by sliding a window which its size is 40% of the image size. Based on this figure, we realize that in most of the traffic signs, the pictograph is the region with highest response. In particular, some parts of the pictograph have the greatest importance to successfully identify the traffic signs. However, there are signs such as the "priority road" sign which are not recognizable using 40% of the image. It seems instead of pictograph, the ConvNet learns to detect color blobs as well as the shape information of the sign to recognize these traffic signs. We also computed the results obtained by increasing the window size to 60%. Nonetheless, since the same analysis applies on these results we do not show them in this section to avoid redundancy of figures. But, these results are illustrated in the supplementary material.

5.8.3   Visualizing Activations

We can think of the value of the activation functions as the amount of excitement of a neuron to the input image. Since the output of the neuron is linearly combined using the neuron in the next layer, then, as the level of excitement increases, it also changes the output of the subsequent neurons in the next layer. So, it is a common practice to inspect which images significantly excite a particular neuron.

To this, we enter all the images in the test set of the GTSRB dataset into the ConvNet and keep record of the activation of neuron (k, m, n) in the last pooling layer where m and n depict the coordinates of the neuron in channel k of the last pooling result. According to Fig. 5.11, there are 250 channels in the last pooling

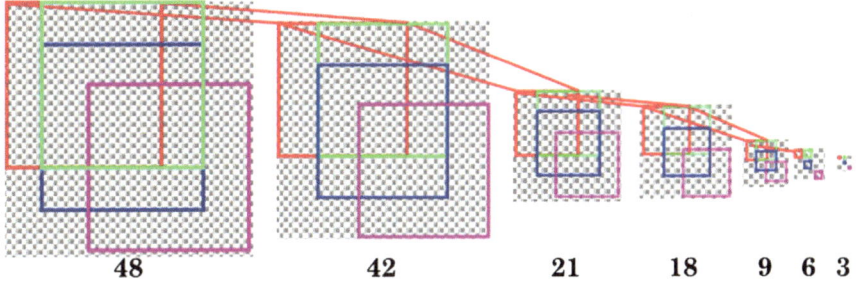

Fig. 5.24 Receptive field of some neurons in the last pooling layer

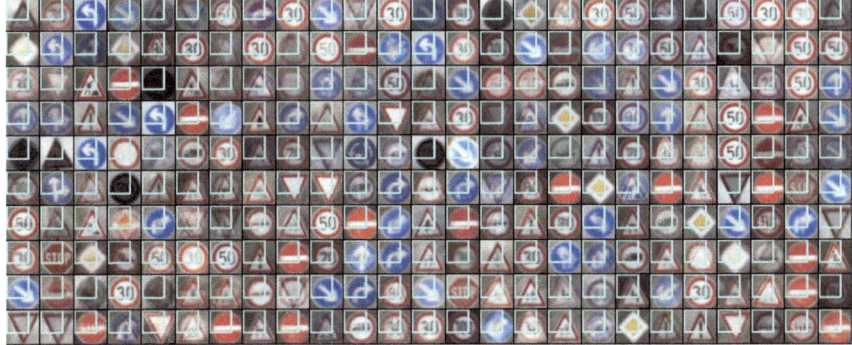

Fig. 5.25 Average image computed over each of 250 channels using the 100 images with highest value in position $(0, 0)$ of the last pooling layer. The corresponding receptive field of this position is shown using a cyan rectangle

layer and each channel is a 3×3 matrix. Then, the images are sorted in descending order according to their value in position (k, m, n) of the last pooling layer and the average of the first 100 images is computed. It should be noted that each location (m, n) in the pooling layer has a corresponding receptive field in the input image. To compute the receptive field of each position we must back project the results from the last pooling layer to the input layer. Figure 5.24 shows the receptive field of some neurons in the last pooling layer.

We computed the average image of each neuron in the last pooling layer as we mentioned above. This is shown in Fig. 5.25 where each image $i\,m_i$ depicts the receptive field of the neuron $(0, 0)$ from i^{th} channel in the last pooling layer. According to these figures, most of the neurons in the last pooling layer are mainly activated by a specific traffic sign. There are also some neurons which are highly activated by more than one traffic signs. To be more specific, these neurons are mainly sensitive to 2–4 traffic signs. By entering an image of a traffic sign to the ConvNet, some of these neurons are highly activated while other neurons are deactivated (they are usually close to zero or negative). The pattern of highly activated neurons are different for each traffic sign and this is the reason that the classes become linearly separable in the last pooling layer.

5.9 More Accurate ConvNet

Visualizing the ConvNet in Fig. 5.11 showed that classification of traffic signs is mainly done using the shape and the pictograph of traffic signs. Therefore, it is possible to discard color information and use only gray-scale pixels to learn a representation by the ConvNet.

In this section, we will train a more accurate and less computational ConvNet for classification of traffic signs. To this end, Habibi Aghdam et al. (2016) computed the layerwise time-to-completion of the network in Fig. 5.11 using the command line tools in the Caffe library. More specifically, executing the *caffe* command with parameter *time* will analyze the run-time of the given network and return a layer-wise summary. Table 5.8 shows the results in milliseconds. We observe that the two middle layers with 4×4 kernels consume most of the GPU time. Moreover, the fully connected layer does not significantly affect the overall time-to-completion. Likewise, the first convolution layer can be optimized by reducing the size of the kernel and number of the input channels.

From accuracy perspective, the aim is to reach a accuracy higher than the previously trained network. The basic idea behind ConvNets is to learn a representation which makes objects linearly separable in the last layer. Fully connected layers facilitates this by learning a nonlinear transformation to project the representation into another space. We can increase the degree of freedom of the ConvNet by adding more fully connected layers to it. This may help to learn a better linearly separable representation. Based these ideas, Habibi Aghdam et al. (2016) proposed the ConvNet illustrated in Fig. 5.26.

First, the color image is replaced with gray-scale image in this ConvNet. In addition, because a gray-scale image is a single-channel input, the linear transformation layer must be also discarded. Second, we have utilized Parametric Rectified Linear Units (PReLU) to learn separate α_i for each feature map in a layer where α_i depicts the value of leaking parameter in LReLU. Third, we have added another fully connected layer to the network to increase its flexibility. Fourth, the size of the first kernel and the middle kernels have been reduced to 5×5 and 3×3, respectively.

Table 5.8 Per layer time-to-completion (milliseconds) of the previous classification ConvNet

Layer	Data	c1 × 1	c7 × 7	pool1	c4 × 4	pool2	c4 × 4	pool3	fc	Class
Time (ms)	0.032	0.078	0.082	0.025	0.162	0.013	0.230	0.013	0.062	0.032

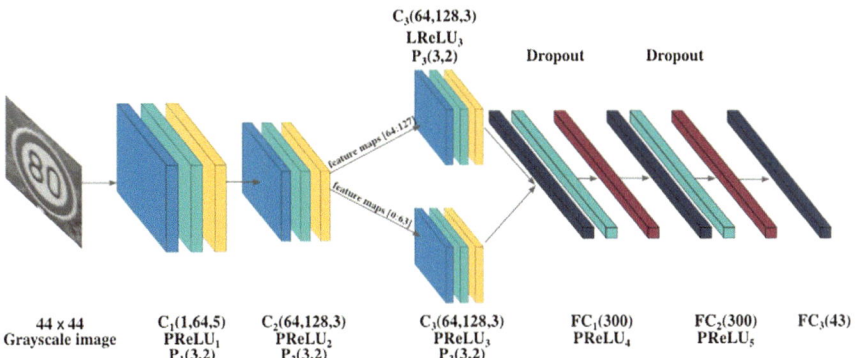

Fig. 5.26 The modified ConvNet architecture compare with Fig. 5.11

Table 5.9 Per layer time-to-completion (milliseconds) of the classification ConvNet in Habibi Aghdam et al. (2016)

Layer	Data	c5 × 5	pool1	c3 × 3	pool2	c3 × 3	pool3	fc1	fc2	Class
Time (ms)	0.036	0.076	0.0166	0.149	0.0180	0.159	0.0128	0.071	0.037	0.032

Last but not the least, the size of the input image is reduced to 44×44 pixels to reduce the dimensionality of the feature vector in the last convolution layer. Table 5.9 shows the layer-wise time-to-completion of the ConvNet illustrated in Fig. 5.26.

According to this table, time-to-completion of the middle layers has been reduced. Especially, time-to-completion of the last convolution layer has been reduced substantially. In addition, the ConvNet has saved 0.078 ms by removing $c1 \times 1$ layer in the previous architecture. It is worth mentioning that the overhead caused by the second fully connected layer is slight. In sum, the overall time-to-completion of the above ConvNet is less than the ConvNet in Fig. 5.11. Finally, we investigated the effect of batch size on the time-to-completion of ConvNet. Figure 5.27 illustrates the relation between the batch size of the classification ConvNet and its time-to-completion.

According to the figure, while processing 1 image takes approximately 0.7 ms using the classification, processing 50 images takes approximately 3.5 ms using the same ConvNet (due to parallel architecture on the GPU). In other words, if the detection ConvNet generates 10 samples, we do not need to enter the samples one by one to the ConvNet. This will take $0.7 \times 10 = 7$ ms to complete. Instead, we can fetch a batch of 10 samples to the network and process them in approximately 1 ms. By this way, we can save more GPU time.

Fig. 5.27 Relation between the batch size and time-to-completion of the ConvNet

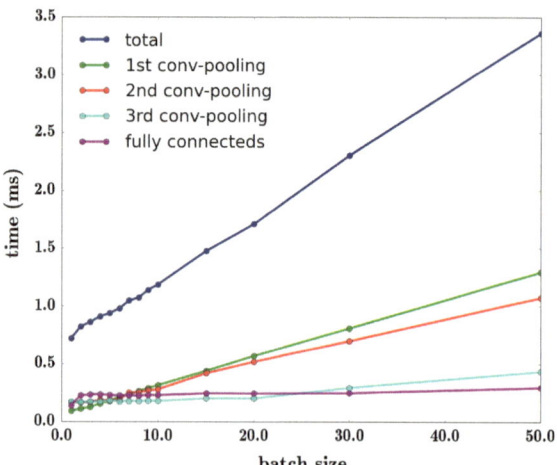

5.9.1 Evaluation

The classification ConvNet is also trained using the mini-batch stochastic gradient descent (batch size $= 50$) with exponential learning rate annealing. We fix the learning rate to 0.02, momentum to 0.9, L2 regularization to 10^{-5}, annealing parameter to 0.99996, dropout ratio to 0.5, and initial value of leaking parameters to 0.01. The network is trained 12 times and their classification accuracies on the test set are calculated. It is worth mentioning that Stallkamp et al. (2012) have only reported the classification accuracy and it is the only way to compare the following results with Ciresan et al. (2012b) and Sermanet and Lecun (2011).

Table 5.10 shows the result of training 10 ConvNets with the same architecture and different initializations. The average classification accuracy of the 10 trials is 99.34% which is higher than the average human performance reported in Stallkamp et al. (2012). In addition, the standard deviation of the classification accuracy is small which shows that the proposed architecture trains the networks with very close

Table 5.10 Classification accuracy of the single network

Trial	top-1 acc. (%)	top-2 acc. (%)	Trial	Top 1 acc. (%)	Top 2 acc. (%)
1	99.21	99.77	6	99.54	99.78
2	99.38	99.73	7	99.25	99.70
3	99.55	99.83	8	99.21	99.73
4	99.16	99.72	9	99.53	99.82
5	99.35	99.75	10	99.24	99.64
Average top-1		99.34 ± 0.02	Average top-2		99.75 ± 0.002
Human top-1		98.84	Human top-2		NA

Table 5.11 Comparing the results with ConvNets in Ciresan et al. (2012a,b), Stallkamp et al. (2012), and Sermanet and Lecun (2011)

ConvNet	Accuracy(%)
Single ConvNet (best) (Ciresan et al. 2012a, b)	98.80
Single ConvNet (avg.) (Ciresan et al. 2012a, b)	98.52
Multi-scale ConvNet (official) (Stallkamp et al. 2012)	98.31
Multi-scale ConvNet (best) (Sermanet and Lecun 2011)	98.97
Proposed ConvNet (best)	99.55
Proposed ConvNet (avg.)	99.34
Committee of 25 ConvNets (Ciresan et al. 2012a,b; Stallkamp et al. 2012)	99.46
Ensemble of 3 proposed ConvNets	**99.70**

accuracies. We argue that this stability is the result of reduction in number of the parameters and regularizing the network using a dropout layer. Moreover, we observe that the top-2^{10} accuracy is very close in all trials and their standard deviation is 0.002. In other words, although the difference in top-1 accuracy of the Trial 5 and the Trial 6 is 0.19%, notwithstanding, the same difference for top-2 accuracy is 0.03%. This implies that some cases are always within the top-2 results. In other words, there are images that have been classified correctly in Trial 5 but they have been misclassified in Trial 6 (or vice versa). As a consequence, if we fuse the scores of two networks the classification accuracy might increase.

Based on this observation, an ensemble was created using the optimal subset selection method. The created ensemble consists of three ConvNets (ConvNets 5, 6, and 9 in Table 5.1). As it is shown in Table 5.11, the overall accuracy of the network increases to 99.70% by this way. Furthermore, the proposed method has established a new record compared with the winner network reported in the competition Stallkamp et al. (2012). Besides, we observe that the results of the single network has outperformed the two other ConvNets. Depending on the application, one can use the single ConvNet instead of ensemble since it already outperforms state-of-art methods as well as human performance with much less time-to-completion.

Misclassified images: We computed the class-specific *precision* and *recall* (Table 5.12). Besides, Fig. 5.28 illustrates the incorrectly classified traffic signs. The number below each image shows the predicted class label. For presentation purposes, all images were scaled to a fixed size. First, we observe that there are 4 cases where the images are incorrectly classified as class 5 while the true label is 3. We note that all these cases are degraded. Moreover, class 3 is distinguishable from class 5 using the fine differences in the first digit of the sign. However, because of degradation the ConvNet is not able to recognize the first digit correctly. In addition, by inspecting the rest of the misclassified images, we realize that the wrong classification is mainly

---

[10]Percent of the samples which are always within the top 2 classification scores.

Table 5.12 Class specific precision and recall obtained by the network in Habibi Aghdam et al. (2016). Bottom images show corresponding class number of each traffic sign

class	precision	recall	class	precision	recall	class	precision	recall
0	1.00	1.00	15	1.00	1.00	30	1.00	0.98
1	1.00	1.00	16	1.00	1.00	31	1.00	1.00
2	1.00	1.00	17	0.99	1.00	32	1.00	1.00
3	1.00	0.99	18	0.99	1.00	33	1.00	1.00
4	1.00	1.00	19	0.98	1.00	34	1.00	1.00
5	0.99	1.00	20	1.00	1.00	35	0.99	1.00
6	1.00	1.00	21	1.00	1.00	36	0.99	1.00
7	1.00	1.00	22	1.00	1.00	37	0.97	1.00
8	1.00	1.00	23	1.00	1.00	38	1.00	0.99
9	1.00	1.00	24	1.00	0.96	39	1.00	0.97
10	1.00	1.00	25	1.00	1.00	40	0.97	0.97
11	0.99	1.00	26	0.97	1.00	41	1.00	1.00
12	1.00	0.99	27	0.98	1.00	42	1.00	1.00
13	1.00	1.00	28	1.00	1.00			
14	1.00	1.00	29	1.00	1.00			

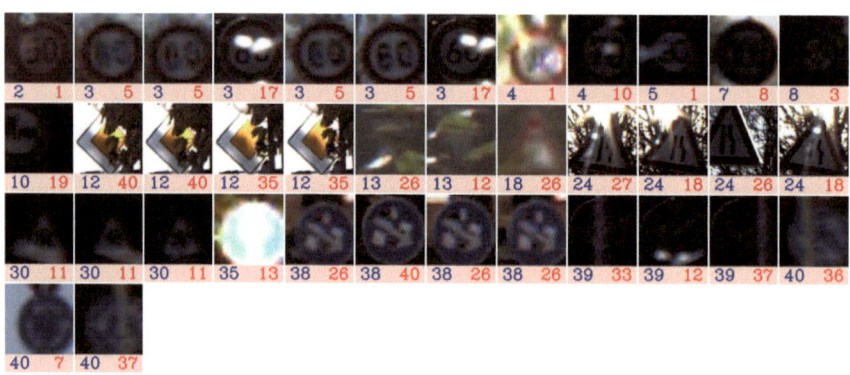

Fig. 5.28 Misclassified traffic sings. The *blue* and the *red* number indicate the actual and predicted class labels, respectively

due to occlusion of the signs and blurry or degraded images. In addition, the class specific precision and recall show that the ConvNet is very accurate in classifying the traffic signs in all the classes.

5.9.2 Stability Against Noise

In real applications, it is necessary to study stability of the ConvNet against image degradations. To empirically study the stability of the classification ConvNet against Gaussian noise, the following procedure is conducted. First, we pick the test

Table 5.13 Accuracy of the ConvNets obtained by degrading the *correctly classified test images* in the original datasets using a Gaussian noise with various values of σ

Accuracy (%) for different values of σ										
	1	2	4	8	10	15	20	25	30	40
Single	99.4	99.4	99.3	98.9	98.3	96.3	93.2	89.7	86.0	78.7
Ensemble	99.5	99.5	99.4	99.2	98.8	97.1	94.4	91.4	88.0	81.4
Correctly classified samples										
Single	99.95	99.94	99.9	99.3	98.8	96.9	93.8	90.3	86.6	79.3
Ensemble	99.94	99.93	99.9	99.5	99.1	97.5	95.0	92.0	88.7	82.1

images from the original datasets. Then, 100 noisy images are generated for each $\sigma \in \{1, 2, 4, 8, 10, 15, 20, 25, 30, 40\}$. In other words, 1000 noisy images are generated for each test image in the original dataset. Next, each noisy image is entered to the ConvNet and its class label is computed. Table 5.13 reports the accuracy of the single ConvNet and the ensemble of three ConvNets per each value of σ. It is divided into two sections. In the first section, the accuracies are calculated on the all images. In the second section, we have only considered the noisy images whose *clean* version are correctly classified by the single model and the ensemble model. Our aim in the second section is to study how noise may affect a sample which is originally classified correctly by our models.

According to this table, there are cases in which adding a Gaussian noise with $\sigma = 1$ on the images causes the models to incorrectly classify the noisy image. Note that, a Gaussian noise with $\sigma = 1$ is not easily perceivable by human eye. However, it may alter the classification result. Furthermore, there are also a few clean images that have been correctly classified by both models but they are misclassified after adding a Gaussian noise with $\sigma = 1$. Notwithstanding, we observe that both models generate admissible results when $\sigma < 10$.

This phenomena is partially studied by Szegedy et al. (2014b) and Aghdam et al. (2016c). The above behavior is mainly due to two reasons. First, the interclass margins might be very small in some regions in the feature space where a sample may fall into another class using a slight change in the feature space. Second, ConvNets are highly nonlinear functions where a small change in the input may cause a significant change in the output (feature vector) where samples may fall into a region representing to another class. To investigate nonlinearity of the ConvNet, we computed the Lipschitz constant of the ConvNet *locally*. Denoting the transformation from the input layer up to layer fc_2 by $\mathscr{C}_{fc_2}(x)$ where $m \in \mathbb{R}^{W \times H}$ is a gray-scale image, we compute the Lipschitz constant for every noisy image $x + \mathcal{N}(0, \sigma)$ using the following equation:

$$d(x, x + \mathcal{N}(0, \sigma)) \leq K d(\mathscr{C}_{fc_2}(x), \mathscr{C}_{fc_2}(x + \mathcal{N}(0, \sigma))) \tag{5.9}$$

where K is the Lipschitz constant and $d(a, b)$ computes the Euclidean distance between a and b. For each traffic sign category in the GTSRB dataset, we pick

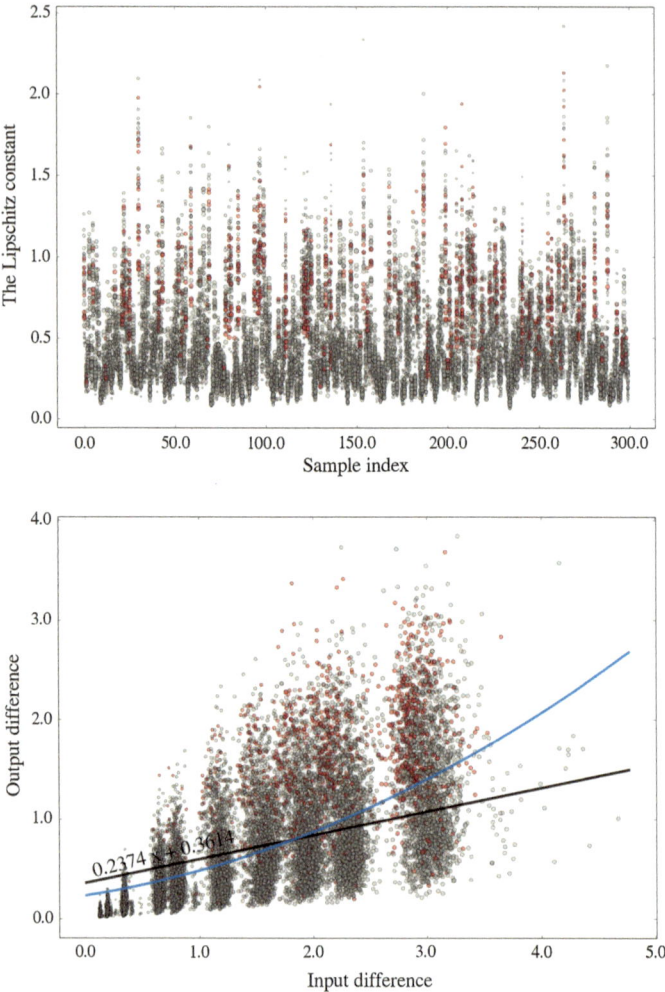

Fig. 5.29 Lipschitz constant (*top*) and the correlation between $d(x, x + \mathcal{N}(0, \sigma))$ and $d(\mathscr{C}_{fc_2}(x), \mathscr{C}_{fc_2}(x + \mathcal{N}(0, \sigma)))$ (*bottom*) computed on 100 samples from every category in the GTSRB dataset. The *red circles* are the noisy instances that are incorrectly classified. The size of each circle is associated with the values of σ in the Gaussian noise

100 correctly classified samples and compute the Lipschitz constant between the clean images and their noisy versions. The top graph in Fig. 5.29 illustrates the Lipschitz constant for each sample, separately. Besides, the bottom graph shows the $d(x, x + \mathcal{N}(0, \sigma))$ and $d(\mathscr{C}_{fc_2}(x), \mathscr{C}_{fc_2}(x + \mathcal{N}(0, \sigma)))$. The black and blue lines are the linear regression and second order polynomial fitted on the point. The size of circles in the figure is associated with the value of σ in the Gaussian noise. A sample with a bigger σ appears bigger on the plot. In addition, the red circles shows the samples which are incorrectly classified after adding a noise to them.

There are some important founding in this figure. First, $\mathscr{C}_{fc_2}$ is *locally contraction* in some regions since there are instances in which their Lipschitz constant is $0 \leq K \leq 1$ regardless of the value of σ. Also, $K \in [\varepsilon, 2.5)$ which means that the ConvNet is very nonlinear in some regions. Besides, we also see that there are some instances which their Lipschitz constants are small but they are incorrectly classified. This could be due to the first reason that we mentioned above. Interestingly, we also observe that there are some cases where the image is degraded using a low magnitude noise (very small dots in the plot) but its Lipschitz constant is very large meaning that in that particular region, the ConvNet is very nonlinear along a specific direction. Finally, we also found out that misclassification can happen regardless of value of the Lipschitz constant.

5.9.3   Visualization

As we mentioned earlier, an effective way to examine each layer is to nonlinearly map the feature vector of a specific layer into a two-dimensional space using the t-SNE method (Maaten and Hinton 2008). This visualization is important since it shows how discriminating are different layers and how a layer changes the behavior of the previous layer. Although there are other techniques such as Local Linear Embedding, Isomap and Laplacian Eigenmaps, the t-SNE method usually provides better results given high dimensional vectors. We applied this method on the fully connected layer before the classification layer as well as the last pooling layer on the detection and the classification ConvNets individually. Figure 5.30 illustrates the results for the classification ConvNets.

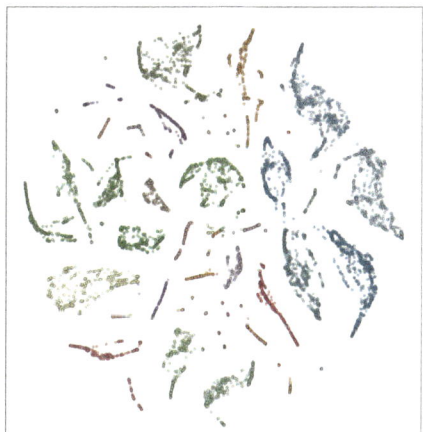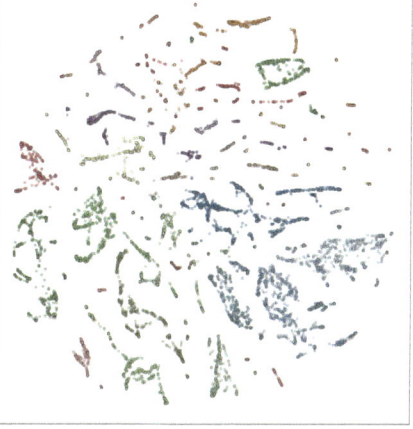

Fig. 5.30 Visualizing the *relu4* (*left*) and the *pooling3* (*right*) layers in the classification ConvNet using the t-SNE method. Each class is shown using a different color

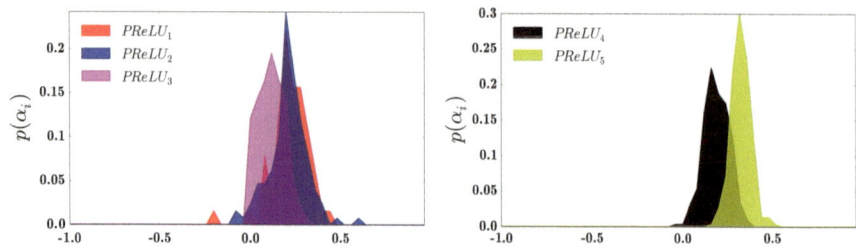

Fig. 5.31 Histogram of leaking parameters

Comparing the results of the classification ConvNet show that although the traffic sign classes are fairly separated in the last pooling layer, the fully connected layers increase the separability of the classes and make them linearly separable. Moreover, we observe that the two classes in the detection ConvNet are not separable in the pooling layer. However, the fully connected layer makes these two classes to be effectively separable. These results also explain why the accuracy of the above ConvNets are high. This is due to the fact that both ConvNets are able to accurately disperse the classes using the fully connected layers before the classification layers.

Leaking parameters: We initialize the leaking parameters of all PReLU units in the classification ConvNet to 0.01. In practice, applying PReLU activations takes slightly more time compared with LReLU activations in Caffe framework. It is important to study the distribution of the leaking parameters to see if we can replace them with LReLU parameters. To this end, we computed the histogram of leaking parameters for each layer, separately. Figure 5.31 shows the results.

According to the histograms, mean of leaking parameters for each layer is different except for the first and the second layers. In addition, variance of each layer is different. One can replace the PReLU activations with ReLU activations and set the leaking parameter of each layer to the mean of leaking parameter in this figure. By this way, time-to-completion of ConvNet will be reduced. However, it is not clear if it will have a negative impact on the accuracy. In the future work, we will investigate this setting.

5.10 Summary

This chapter started with reviewing related work in the field of traffic sign classification. Then, it explained the necessity of splitting data and some of methods for splitting data into training, validation and test sets. A network should be constantly assessed during training in order to diagnose it if it is necessary. For this reason, we showed how to train a network using Python interface of Caffe and evaluate it constantly using training-validation curve. We also explained different scenarios that may happen during training together with their causes and remedies. Then, some of the successful architectures that are proposed in literature for classification of traffic signs were introduced. We implemented and trained these architectures and analyzed their training-validation plots.

Creating ensemble is a method to increase classification accuracy. We mentioned various methods that can be used for creating ensemble of models. Then, a method based on optimal subset selection using genetic algorithms were discussed. This way, we create ensembles with minimum number of models that together they increase the classification accuracy.

After that, we showed how to interpret and analyze quantitative results such as precision, recall, and accuracy on a real dataset of traffic signs. We also explained how to understand behavior of convolutional neural networks using data-driven visualization techniques and nonlinear embedding methods such as t-SNE.

Finally, we finished the chapter by implementing a more accurate and computationally efficient network that is proposed in literature. The performance of this network was also analyzed using various metrics and from different perspective.

5.11 Exercises

5.1 Why if test set and training sets are not drawn from the same distribution, the model might not be accurate enough on the test set?

5.2 When splitting the dataset into training, validation, and test set, each sample in the dataset is always assigned to one of these sets. Why it is not correct to assign a sample to more than one set?

5.3 Computer the number of multiplications of the network in Fig. 5.9.

5.4 Change the pooling size from 3 × 3 to 2 × 2 and trained the networks again. Does that affect the accuracy? Can we generalize the result to any other datasets?

5.5 Replace the leaky ReLU with ReLU in the network illustrated in Fig. 5.11 and train the network again? Does it have any impact on optimization algorithm or accuracy?

5.6 Change the regularization coefficient to 0.01 and trained the network. Explain the results.

References

Aghdam HH, Heravi EJ, Puig D (2015) A unified framework for coarse-to-fine recognition of traffic signs using bayesian network and visual attributes. In: Proceedings of the 10th international conference on computer vision theory and applications, pp 87–96. doi:10.5220/0005303500870096

Aghdam HH, Heravi EJ, Puig D (2016a) A practical and highly optimized convolutional neural network for classifying traffic signs in real-time. Int J Comput Vis 1–24. doi:10.1007/s11263-016-0955-9

Aghdam HH, Heravi EJ, Puig D (2016b) Analyzing the stability of convolutional neural networks against image degradation. In: Proceedings of the 11th international conference on computer vision theory and applications, vol 4(Visigrapp), pp 370–382. doi:10.5220/0005720703700382

Aghdam HH, Heravi EJ, Puig D (2016c) Computer vision ECCV 2016. Workshops 9913:178–191. doi:10.1007/978-3-319-46604-0

Baró X, Escalera S, Vitrià J, Pujol O, Radeva P (2009) Traffic sign recognition using evolutionary adaboost detection and forest-ECOC classification. IEEE Trans Intell Transp Syst 10(1):113–126. doi:10.1109/TITS.2008.2011702

Ciresan D, Meier U, Schmidhuber J (2012a) Multi-column deep neural networks for image classification. In: 2012 IEEE conference on computer vision and pattern recognition, IEEE, pp 3642–3649. doi:10.1109/CVPR.2012.6248110, arXiv:1202.2745v1

Ciresan D, Meier U, Masci J, Schmidhuber J (2012b) Multi-column deep neural network for traffic sign classification. Neural Netw 32:333–338. doi:10.1016/j.neunet.2012.02.023

Coates A, Ng AY (2012) Learning feature representations with K-means. Lecture notes in computer science (Lecture notes in artificial intelligence and lecture notes in bioinformatics), vol 7700. LECTU:561–580. doi:10.1007/978-3-642-35289-8-30

Fleyeh H, Davami E (2011) Eigen-based traffic sign recognition. IET Intell Transp Syst 5(3):190. doi:10.1049/iet-its.2010.0159

Gao XW, Podladchikova L, Shaposhnikov D, Hong K, Shevtsova N (2006) Recognition of traffic signs based on their colour and shape features extracted using human vision models. J V Commun Image Represent 17(4):675–685. doi:10.1016/j.jvcir.2005.10.003

Greenhalgh J, Mirmehdi M (2012) Real-time detection and recognition of road traffic signs. IEEE Trans Intell Transp Syst 13(4):1498–1506. doi:10.1109/tits.2012.2208909

Habibi Aghdam H, Jahani Heravi E, Puig D (2016) A practical approach for detection and classification of traffic signs using convolutional neural networks. Robot Auton Syst 84:97–112. doi:10.1016/j.robot.2016.07.003

He K, Zhang X, Ren S, Sun J (2015) Delving deep into rectifiers: surpassing human-level performance on imagenet classification. arXiv:1502.01852

Hinton G (2014) Dropout: a simple way to prevent neural networks from overfitting. J Mach Learn Res (JMLR) 15:1929–1958

Hsu SH, Huang CL (2001) Road sign detection and recognition using matching pursuit method. Image Vis Comput 19(3):119–129. doi:10.1016/S0262-8856(00)00050-0

Huang Gb, Mao KZ, Siew Ck, Huang Ds (2013) A hierarchical method for traffic sign classification with support vector machines. In: The 2013 international joint conference on neural networks (IJCNN), IEEE, pp 1–6. doi:10.1109/IJCNN.2013.6706803

Jin J, Fu K, Zhang C (2014) Traffic sign recognition with hinge loss trained convolutional neural networks. IEEE Trans Intell Transp Syst 15(5):1991–2000. doi:10.1109/TITS.2014.2308281

Krizhevsky A, Sutskever I, Hinton G (2012) Imagenet classification with deep convolutional neural networks. In: Advances in neural information processing systems. Curran Associates, Inc., Red Hook, pp 1097–1105

Larsson F, Felsberg M (2011) Using fourier descriptors and spatial models for traffic sign recognition. In: Image analysis lecture notes in computer science, vol 6688. Springer, Berlin, pp 238–249. doi:10.1007/978-3-642-21227-7_23

Liu H, Liu Y, Sun F (2014) Traffic sign recognition using group sparse coding. Inf Sci 266:75–89. doi:10.1016/j.ins.2014.01.010

Lu K, Ding Z, Ge S (2012) Sparse-representation-based graph embedding for traffic sign recognition. IEEE Trans Intell Transp Syst 13(4):1515–1524. doi:10.1109/TITS.2012.2220965

Maaten LVD, Hinton G (2008) Visualizing data using t-SNE. J Mach Learn Res 9:2579–2605. doi:10.1007/s10479-011-0841-3

Maldonado-Bascon S, Lafuente-Arroyo S, Gil-Jimenez P, Gomez-Moreno H, Lopez-Ferreras F (2007) Road-sign detection and recognition based on support vector machines. IEEE Trans Intell Transp Syst 8(2):264–278. doi:10.1109/TITS.2007.895311

Maldonado Bascón S, Acevedo Rodríguez J, Lafuente Arroyo S, Fernndez Caballero A, López-Ferreras F (2010) An optimization on pictogram identification for the road-sign recognition task using SVMs. Comput Vis Image Underst 114(3):373–383. doi:10.1016/j.cviu.2009.12.002

Mathias M, Timofte R, Benenson R, Van Gool L (2013) Traffic sign recognition – how far are we from the solution? Proc Int Jt Conf Neural Netw. doi:10.1109/IJCNN.2013.6707049

Møgelmose A, Trivedi MM, Moeslund TB (2012) Vision-based traffic sign detection and analysis for intelligent driver assistance systems: perspectives and survey. IEEE Trans Intell Transp Syst 13(4):1484–1497. doi:10.1109/TITS.2012.2209421

Moiseev B, Konev A, Chigorin A, Konushin A (2013) Evaluation of traffic sign recognition methods trained on synthetically generated data. In: 15th international conference on advanced concepts for intelligent vision systems (ACIVS). Springer, Poznań, pp 576–583

Paclík P, Novovičová J, Pudil P, Somol P (2000) Road sign classification using Laplace kernel classifier. Pattern Recognit Lett 21(13–14):1165–1173. doi:10.1016/S0167-8655(00)00078-7

Piccioli G, De Micheli E, Parodi P, Campani M (1996) Robust method for road sign detection and recognition. Image Vis Comput 14(3):209–223. doi:10.1016/0262-8856(95)01057-2

Radu Timofte LVG (2011) Sparse representation based projections. In: 22nd British machine vision conference, BMVA Press, pp 61.1–61.12. doi:10.5244/C.25.61

Ruta A, Li Y, Liu X (2010) Robust class similarity measure for traffic sign recognition. IEEE Trans Intell Transp Syst 11(4):846–855. doi:10.1109/TITS.2010.2051427

Sermanet P, Lecun Y (2011) Traffic sign recognition with multi-scale convoltional networks. In: Proceedings of the international joint conference on neural networks, pp 2809–2813. doi:10.1109/IJCNN.2011.6033589

Sermanet P, Eigen D, Zhang X, Mathieu M, Fergus R, LeCun Y (2013) OverFeat: integrated recognition, localization and detection using convolutional networks, pp 1–15. arXiv:1312.6229

Simonyan K, Zisserman A (2015) Very deep convolutional networks for large-scale image recognition. In: International conference on learning representation (ICLR), pp 1–13. arXiv:1409.1556v5

Stallkamp J, Schlipsing M, Salmen J, Igel C (2012) Man vs. computer: benchmarking machine learning algorithms for traffic sign recognition. Neural Netw 32:323–332. doi:10.1016/j.neunet.2012.02.016

Sun ZL, Wang H, Lau WS, Seet G, Wang D (2014) Application of BW-ELM model on traffic sign recognition. Neurocomputing 128:153–159. doi:10.1016/j.neucom.2012.11.057

Szegedy C, Reed S, Sermanet P, Vanhoucke V, Rabinovich A (2014a) Going deeper with convolutions, pp 1–12. arXiv:1409.4842

Szegedy C, Zaremba W, Sutskever I (2014b) Intriguing properties of neural networks. arXiv:1312.6199v4

Tibshirani R (1994) Regression selection and shrinkage via the Lasso. doi:10.2307/2346178

Timofte R, Zimmermann K, Van Gool L (2011) Multi-view traffic sign detection, recognition, and 3D localisation. Mach Vis Appl 1–15. doi:10.1007/s00138-011-0391-3

Wang J, Yang J, Yu K, Lv F, Huang T, Gong Y (2010) Locality-constrained linear coding for image classification. In: 2010 IEEE computer society conference on computer vision and pattern recognition, IEEE, pp 3360–3367. doi:10.1109/CVPR.2010.5540018

Yosinski J, Clune J, Bengio Y, Lipson H (2014) How transferable are features in deep neural networks? Nips'14 vol 27. arXiv:1411.1792v1

Yuan X, Hao X, Chen H, Wei X (2014) Robust traffic sign recognition based on color global and local oriented edge magnitude patterns. IEEE Trans Intell Transp Syst 15(4):1466–1474. doi:10.1109/TITS.2014.2298912

Zaklouta F, Stanciulescu B (2012) Real-time traffic-sign recognition using tree classifiers. IEEE Trans Intell Transp Syst 13(4):1507–1514. doi:10.1109/TITS.2012.2225618

Zaklouta F, Stanciulescu B (2014) Real-time traffic sign recognition in three stages. Robot Auton Syst 62(1):16–24. doi:10.1016/j.robot.2012.07.019

Zaklouta F, Stanciulescu B, Hamdoun O (2011) Traffic sign classification using K-d trees and random forests. In: Proceedings of the international joint conference on neural networks, pp 2151–2155. doi:10.1109/IJCNN.2011.6033494

Zeng Y, Xu X, Fang Y, Zhao K (2015) Traffic sign recognition using deep convolutional networks and extreme learning machine. In: Intelligence science and big data engineering. Image and video data engineering (IScIDE). Springer, Berlin, pp 272–280

Detecting Traffic Signs

6.1 Introduction

Recognizing traffic signs is mainly done in two stages including *detection* and *classification*. The detection module performs a multi-scale analysis on the image in order to locate the patches consisting only of one traffic sign. Next, the classification module analyzes each patch individually and classifies them into classes of traffic signs.

The ConvNets explained in the previous chapter are only suitable for the *classification* module and they cannot be directly used in the task of *detection*. This is due to the fact that applying these ConvNets on high-resolution images is not computationally feasible. On the other hand, accuracy of the *classification* module also depends on the *detection* module. In other words, any false-positive results produced by the detection module will be entered into the classification module and it will be classified as one of traffic signs. Ideally, the false-positive rate of the detection module must be zero and its true-positive rate must be 1. Achieving this goal usually requires more complex image representation and classification models. However, as the complexity of these models increases, the detection module needs more time to complete its task.

Sermanet et al. (2013) proposed a method for implementing a multi-scale sliding window approach within a ConvNet. Szegedy et al. (2013) formulated the object detection problem as a regression problem to object bounding boxes. Girshick et al. (2014) proposed a method so-called Region with ConvNets in which they apply ConvNet to bottom-up region proposals for detecting the domain-specific objects. Recently, Ouyang et al. (2015) developed a new pooling technique called *deformation constrained pooling* to model the deformation of object parts with geometric constraint.

© Springer International Publishing AG 2017
H. Habibi Aghdam and E. Jahani Heravi, *Guide to Convolutional Neural Networks*, DOI 10.1007/978-3-319-57550-6_6

6.2 ConvNet for Detecting Traffic Signs

In contrast to offline applications, an ADAS requires algorithms that are able to perform their task in real time. On the other hand, the detection module consumes more time compared with the classification module especially when it is applied on a high-resolution image. For this reason, the detection module must be able to locate traffic signs in real time. However, the main barrier in achieving this speed is that the detection module must analyze high-resolution images in order to be able to locate traffic signs that are in a distance up to 25 m. This is illustrated in Fig. 6.1 in which the width of the image is 1020 pixels.

Assuming that the length of the bus is approximately 12.5 m, we can estimate that the distance between the traffic sign indicated by the arrow and the camera is approximately 20 m. Although the distance is not large, the bounding box of the sign is nearly 20 × 20 pixels. Consequently, it is impractical to apply the detection algorithm on low-resolution images since signs that are located in 20 m of distance from the camera might not be recognizable. Moreover, considering that the speed of the car is 80 km/h in an interurban road, it will travel 22 m in one second. For this reason, the detection module must be able to analyze more than one frame per second in order to be able to deal with high speeds motions of a car.

In practice, a car might be equipped with a stereo camera. In that case, the detection module can be applied much faster since most of the non-traffic sign pixels can be discarded using the distance information. In addition, the detection module can be calibrated on a specific car and use the calibration information to ignore non-traffic sign pixels. In this work, we propose a more general approach by considering that

Fig. 6.1 The detection module must be applied on a high-resolution image

there is only one color camera that can be mounted in front of any car. In other words, the detection module must analyze all the patches on the image in order to identify the traffic signs.

We trained separate traffic sign detectors using HOG and LBP features together with a linear classifier and a random forest classifier. However, previous studies showed that the detectors based on these features suffer from low precision and recall values. More importantly, applying these detectors on a high-resolution image using a CPU is impractical since it takes a long time to process the whole image. Besides, it is not trivial to implement the whole scanning window approach using these detectors on a GPU in order to speed up the detection process. For this reason, we developed a lightweight but accurate ConvNet for detecting traffic signs. Figure 6.2 illustrates the architecture of the ConvNet.

The above ConvNet is inspired by Gabor feature extraction algorithm. In this method, a bank of convolution kernels is applied on image and the output of each kernel is individually aggregated. Then, the final feature is obtained by concatenating the aggregated values. Instead of handcrafted Gabor filters, the proposed ConvNet learns a bank of 60 convolution filters each with $9 \times 9 \times 3$ coefficients. The output of the first convolution layer is a $12 \times 12 \times 60$ tensor where each slice is a feature map (i.e., 60 feature maps). Then, the aggregation is done by spatially dividing each feature map into four equal regions and finding the maximum response in each region. Finally, the extracted feature vector is nonlinearly transformed into a 300-dimensional space where the ConvNet tries to linearly separate the two classes (traffic sign versus non-traffic sign) in this space.

One may argue that we could attach a fully connected network to HOG features (or other handcrafted features) and train an accurate detection model. Nonetheless, there are two important issues with this approach. First, it is not trivial to implement a sliding window detector using these kind of features on a GPU. Second, as we will

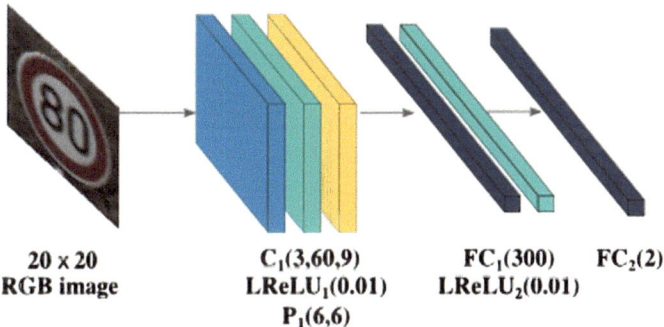

$$20 \times 20 \qquad C_1(3,60,9) \qquad FC_1(300) \qquad FC_2(2)$$
$$RGB\ image \qquad LReLU_1(0.01) \qquad LReLU_2(0.01)$$
$$P_1(6,6)$$

Fig. 6.2 The ConvNet for detecting traffic signs. The *blue*, *green*, and *yellow* color indicate a convolution, LReLU and pooling layer, respectively. $C(c, n, k)$ denotes n convolution kernel of size $k \times k \times c$ and $P(k, s)$ denotes a max-pooling layer with pooling size $k \times k$ and stride s. Finally, the number in the LReLU units indicate the leak coefficient of the activation function

Fig. 6.3 Applying the trained ConvNet for hard-negative mining

show in experiments, their representation power is limited and they produce more false-positive results compared with this ConvNet.

To train the ConvNet, we collect the positive samples and pick some image patches randomly in each image as the negative samples. After training the ConvNet using this dataset, it is applied on each image in the training set using the *multi-scale sliding window* technique in order to detect traffic signs. Figure 6.3 illustrates the result of applying the detection ConvNet on an image.

The red, the blue, and the green rectangles show the false-positive, ground-truth, and the true-positive patches. We observe that the ConvNet is not very accurate and it produces some false-positive results. Although the false-positive rate can be reduced by increasing the threshold value of the classification score, the aim is to increase the overall accuracy of the ConvNet.

There are mainly two solutions to improve the accuracy. Either to increase the complexity of the ConvNet or refine the current model using more appropriate data. Increasing the complexity of the ConvNet is not practical since it will also increase its time to completion. However, it is possible to refine the model using more appropriate data. Here we utilize the *hard-negative mining* method.

In this method, the current ConvNet is applied on all the training images. Next, all the patches that are classified as positive (the red and the green boxes) are compared with the ground-truth bounding boxes and those which do not align well are selected as the new negative image patches (the red rectangles). They are called *hard-negative* patches. Having the all hard-negative patches collected from all the training images, the ConvNet is fine-tuned on the new dataset. Mining hard-negative data and fine-tuning the ConvNet can be done repeatedly until the accuracy of the ConvNet converges.

6.3   Implementing Sliding Window Within the ConvNet

The detection procedure starts with sliding a 20×20 mask over the image and classifying the patch under the mask using the detection ConvNet. After all the pixels are scanned, the image is downsampled and the procedure repeats on the smaller image. The downsampling can be done several times to ensure that the closer objects will be also detected. Applying this simple procedure on a high-resolution image may take several minutes (even on GPU because of redundancy in computation and transferring data between the main memory and GPU). For this reason, we need to find an efficient way for running the above procedure in real time.

Currently, advanced embedded platforms such as NVIDIA Drive Px[1] come with a dedicated GPU module. This makes it possible to execute highly computational models in real time on these platforms. Therefore, we consider a similar platform for running the tasks of ADAS. There are two main computational bottlenecks in naive implementation of the sliding window detector. On the one hand, the input image patches are very small and they may use small fraction of GPU cores to complete a forward pass in the ConvNet. In other words, two or more image patches can be simultaneously processed depending on the number of GPU cores. However, the aforementioned approach considers the two consecutive patches are independent and applies the convolution kernels on the each patches separately. On the other hand, transferring overlapping image patches between the main memory and GPU is done thousands of time which adversely affects the time-to-completion. To address these two problems, we propose the following approach for implementing the sliding window method on a GPU. Figure 6.4 shows the intuition behind this implementation.

Normally, the input of the ConvNet is a $20 \times 20 \times 3$ image and the output of the pooling layer is a $2 \times 2 \times 60$ tensor. Also, each neuron in the first fully connected layer is connected to $2 \times 2 \times 60$ neurons in the previous layer. In this paper, traffic signs are detected in a 1020×600 image. Basically, sliding window approach scans every pixel in the image to detect the objects.[2] In other words, for each pixel in the image, the first step is to crop a 20×20 patch and, then, to apply the bank of the convolution filters in the ConvNet on the patch. Next, the same procedure is repeated for the pixel next to the current pixel. Note that 82% of the pixels are common between two consecutive 20×20 patches. As the result, transferring the common pixels to GPU memory is redundant. The solution is that the whole high-resolution image is transferred to the GPU memory and the convolution kernels are applied on different patches simultaneously.

The next step in the ConvNet is to aggregate the pixels in the output of the convolution layer using the max-pooling layer. When the ConvNet is provided by a 20×20 image, the convolution layer generates a 12×12 feature map for each kernel. Then, the pooling layer computes the maximum values in 6×6 regions. The distance between each region is 6 pixels (stride $= 6$). Therefore, the output of the pooling layer

---

[1] www.nvidia.com/object/drive-px.html.
[2] We may set the stride of scanning to two or more for computational purposes.

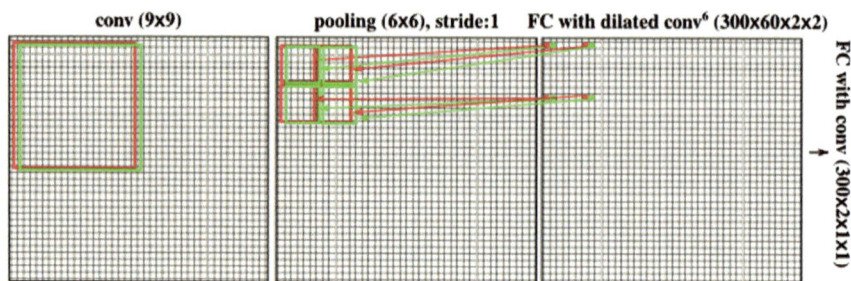

Fig. 6.4 Implementing the sliding window detector within the ConvNet

on single feature map is a 2×2 feature map. Our goal is to implement the sliding window approach within the ConvNet. Assume the two consecutive patches indicated by the red and green rectangles in the convolution layer (Fig. 6.4). The pooling layer will compute the maximum value of 6×6 regions. Based on the original ConvNet, the output of the pooling layer for the red rectangle must be computed using the 4 small red rectangles illustrated in the middle figure.

In addition, we also want to aggregate the values inside the green region in the next step. Its corresponding 6×6 regions are illustrated using 4 small green rectangles. Since we need to apply the pooling layer consecutively, we must change the stride of the pooling layer to 1. With this formulation, the pooling result of the red and green regions will not be consecutive. Rather, there will be 6 pixels gap between the two consecutive nonoverlapping 6×6 regions. The pooling results of the red rectangle are shown using 4 small filled squares in the figure. Recall from the above discussion that each neuron in the first fully connected layer is connected to $2 \times 2 \times 60$ regions in the output of the pooling layer.

Based on the above discussion, we can implement the fully connected layer using $2 \times 2 \times 60$ *dilated convolution filters* with dilation factor 6. Formally, a $W \times H$ convolution kernel with dilation factor τ is applied using the following equation:

$$(f(m, n) * g)^{\tau} = \sum_{h=-\frac{H}{2}}^{\frac{H}{2}} \sum_{w=-\frac{W}{2}}^{\frac{W}{2}} f(m + \tau h, n + \tau w) g(h, w). \qquad (6.1)$$

Note that the number of the arithmetic operations on a normal convolution and its dilated version is identical. In other words, dilated convolution does not change the computational complexity of the ConvNet. Likewise, we can implement the last fully connected layer using $1 \times 1 \times 2$ filters. Using this formulation, we are able to implement the sliding window method in terms of convolution layers. The architecture of the sliding window ConvNet is shown in Fig. 6.5.

The output of the fully convolutional ConvNet is a $y' = \mathbb{R}^{1012 \times 592 \times 2}$ where the patch at location (m, n) is a traffic sign if $y'(m, n, 0) > y'(m, n, 1)$. It is a common practice in the sliding window method to process patches every 2 pixels. This is easily implementable by changing the stride of the pooling layer of the fully

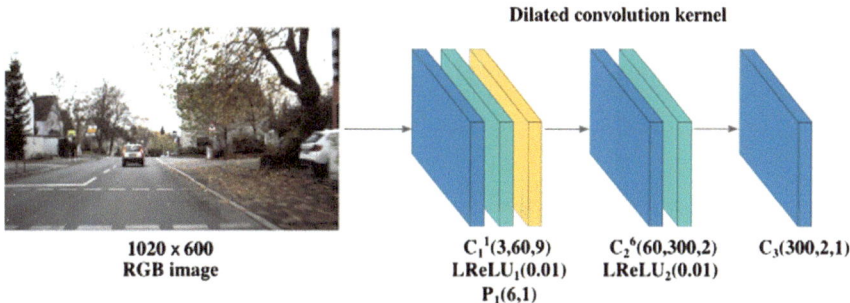

Fig. 6.5   Architecture of the sliding window ConvNet

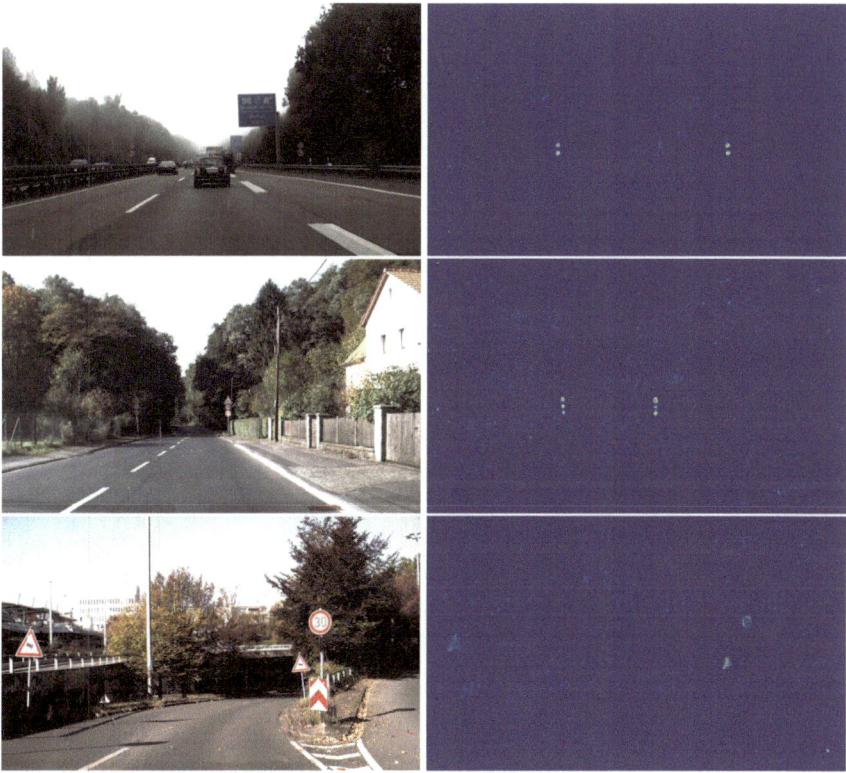

Fig. 6.6   Detection score computed by applying the fully convolutional sliding network to 5 scales of the high-resolution image

convolutional ConvNet to 2 and adjusting the dilation factor of convolution kernel in the first fully connected layer accordingly (i.e., $s = 3$). Finally, to implement the multi-scale sliding window, we only need to create different scales of the original image and apply the sliding window ConvNet on it. Figure 6.6 shows the detection score computed by the detection ConvNet on high-resolution images.

Table 6.1 Time-to-completion (milliseconds) of the sliding ConvNet computed on 5 different image scales

Layer name	Per layer time to completion in milliseconds for different scales				
	1020×600	816×480	612×360	480×240	204×120
Data	0.167	0.116	0.093	0.055	0.035
Conv	7.077	4.525	2.772	1.364	0.321
Relu	1.744	1.115	0.621	0.330	0.067
Pooling	6.225	3.877	2.157	1.184	0.244
Fully connected	8.594	5.788	3.041	1.606	0.365
Relu	2.126	1.379	0.746	0.384	0.081
Classify	1.523	0.893	0.525	0.285	0.101
Total	27.656	17.803	10.103	5.365	1.336

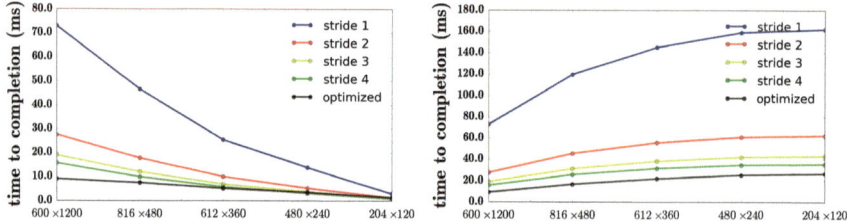

Fig. 6.7 Time to completion of the sliding ConvNet for different strides. *Left* time to completion per resolution and *Right* cumulative time to completion

Detecting traffic signs in high-resolution images is the most time-consuming part of the processing pipeline. For this reason, we executed the sliding ConvNet on a GeForce GTX 980 card and computed the time to completion of each layer separately. To be more specific, each ConvNet repeats the forward pass 100 times and the average time to completion of each layer is computed. The condition of the system is fixed during all calculations. Table 6.1 shows the results of the sliding ConvNet on 5 different scales. Recall from our previous discussion that the stride of the pooling layer is set to 2 in the sliding ConvNet.

We observe that applying the sliding ConvNet on 5 different scales of a high-resolution image takes 62.266 ms in total which is equal to processing of 19.13 frames per second. We also computed time to completion of the sliding ConvNet by changing the stride of the pooling layer from 1 to 4. Figure 6.7 illustrates time to completion per image resolution (left) as well as the cumulative time to completion.

The results reveal that it is not practical to set the stride to 1 since it takes 160 ms to detect traffic sign on an image (6.25 frames per second). In addition, it consumes a considerable amount of GPU memory. However, it is possible to process 19 frames per second by setting the stride to 2. In addition, the reduction in the processing time between stride 3 and stride 4 is negligible. But, stride 3 is preferable compared with

Fig. 6.8 Distribution of traffic signs in different scales computed using the training data

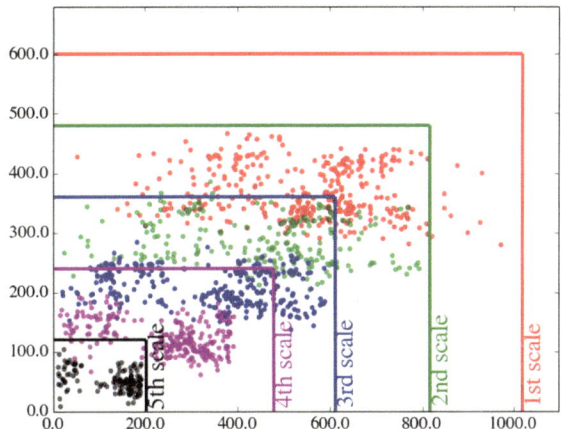

stride 4 since it produces a denser detection score. Last but not least, it is possible to apply a combination of stride 2 and stride 3 in various scales to improve the overall time to completion. For instance, we can set the stride to 3 for first scale and set it to 2 for rest of the image scales. By this way, we can save about 10 ms per image. The execution time can be further improved by analyzing the database statistics.

More specifically, traffic signs bounded in 20×20 regions will be detected in the first scale. Similarly, signs bounded in 50×50 regions will be detected in the 4th scale. Based on this fact, we divided traffic signs in training set into 5 groups according to the image scale they will be detected. Figure 6.8 illustrates the distribution of traffic signs in each scale.

According to this distribution, we must expect to detect 20×20 traffic signs inside a small region in the first scale. That said, the region between row 267 and row 476 must be analyzed to detect 20×20 signs rather than whole 600 rows in the first scale. Based on the information depicted in the distribution of signs, we process only fetch the 945×210, 800×205, 600×180 and 400×190 pixels in the first 4 scales to the sliding ConvNet. As it is illustrated by a black line in Fig. 6.7, this reduces the time to completion of the ConvNet with stride 2 to 26.506 ms which is equal to processing 37.72 high-resolution frames per second.

6.4   Evaluation

The detection ConvNet is trained using the mini-batch stochastic gradient descent (batch size $= 50$) with learning rate annealing. We fix the learning rate to 0.02, momentum to 0.9, L2 regularization to 10^{-5}, step size of annealing to 10^4, annealing rate to 0.8, the negative slope of the LReLU to 0.01 and the maximum number of iterations to 150,000. The ConvNet is first trained using the ground-truth bounding boxes (the blue boxes) and the negative samples collected randomly from image.

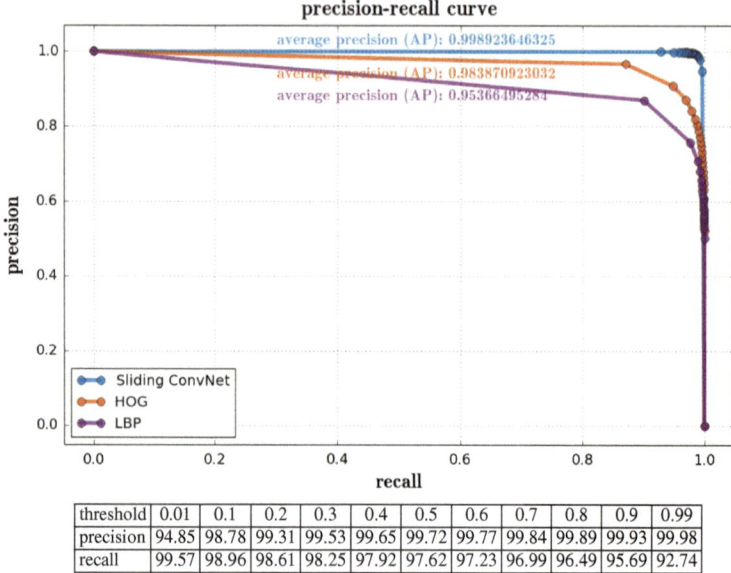

Fig. 6.9 *Top* precision-recall curve of the detection ConvNet along with models obtained by HOG and LBP features. *Bottom* Numerical values (%) of precision and recall for the detection ConvNet

Then, a hard-negative mining is performed on the training set in order to collect more organized negative samples and the ConvNet is trained again. To compare with handcrafted features, we also trained detection models using HOG and LBP features and Random Forests by following the same procedure. Figure 6.9 illustrates the precision-recall plot of these models.

Average precision (AP) of the sliding ConvNet is 99.89% which indicates a nearly perfect detector. In addition, average precision of the models based on HOG and LBP features are 98.39 and 95.37%, respectively. Besides, precision of the sliding ConvNet is considerably higher than HOG and LBP features. In other words, the number of the false-positive samples in the sliding ConvNet is less than HOG and LBP. It should be noted that false-positive results will be directly fetched into the classification ConvNet where they will be classified into one of traffic sign classes. This may produce dangerous situations in the case of autonomous cars. For example, consider a false-positive result produced by detection module of an autonomous car is classified as "speed limit 100" in an educational zone. Clearly, the autonomous car may increase the speed according to the wrongly detected sign. This may have vital consequences in real world. Even though average precision on the sliding ConvNet and HOG models are *numerically* comparable, using the sliding ConvNet is certainly safer and more applicable than the HOG model.

Post-processing bounding boxes: One solution to deal with the *false-positive* results of the detection ConvNet is to post-process the bounding boxes. The idea is that if a bounding box is classified positive, all the bounding boxes in distance

Fig. 6.10 Output of the detection ConvNet before and after post-processing the bounding boxes. A *darker* bounding box indicate that it is detected in a lower scale image

of $\{-1, 0, -1\} \times \{-1, 0, -1\}$ must be also classified positive. In other words, if a region of the image consists of a traffic sign, there must be at least 10 bounding boxes over that region in which the detection ConvNet classifies them positive. By only applying this technique, the false-positive rate can be considerably reduces. Figure 6.10 illustrates the results of the detection ConvNet on a few images before and after post-processing the bounding boxes.

In general, the detection ConvNet is able to locate traffic signs with high precision and recall. Furthermore, post-processing the bounding boxes is able to effectively discard the false-positive results. However, a few false-positive bounding boxes may still exist in the result. In practice, we can create a second step verification by creating a ConvNet with more complexity and apply it on the results from the detection ConvNet in order to remove all the false-positive results.

6.5 Summary

Object detection is one of the hard problems in computer vision. It gets even harder in time-demanding tasks such as ADAS. In this chapter, we explained a convolutional neural network that is able to analyze high-resolution images in real time and it accurately finds traffic signs. We showed how to quantitatively analyze the networks and visualize it using an embedding approach.

6.6 Exercises

6.1 Read the documentation of *dilation* from caffe.proto file and implement the architecture mentioned in this chapter.

6.2 Tweak the number of filters and neurons in the fully connected layer and train the networks. Is there a more compact architecture that can be used for accurately detecting traffic sign?

References

Girshick R, Donahue J, Darrell T, Berkeley UC, Malik J (2014) Rich feature hierarchies for accurate object detection and semantic segmentation. doi:10.1109/CVPR.2014.81, arXiv:abs/1311.2524
Ouyang W, Wang X, Zeng X, Qiu S, Luo P, Tian Y, Li H, Yang S, Wang Z, Loy CC, Tang X (2015) DeepID-Net: deformable deep convolutional neural networks for object detection. In: Computer vision and pattern recognition. arXiv:1412.5661
Sermanet P, Eigen D, Zhang X, Mathieu M, Fergus R, LeCun Y (2013) OverFeat: integrated recognition, localization and detection using convolutional networks, pp 1–15. arXiv:1312.6229
Szegedy C, Toshev A, Erhan D (2013) Deep neural networks for object detection. In: Advances in neural information processing systems (NIPS). IEEE, pp 2553–2561. http://ieeexplore.ieee.org/stamp/stamp.jsp?tp=&arnumber=6909673

Visualizing Neural Networks

7.1 Introduction

A neural network is a method that transforms input data into a feature space through a highly nonlinear function. When a neural network is trained to classify input patterns, it learns a transformation function from input space to the feature space such that patterns from different classes become linearly separable. Then, it trains a linear classifier on this feature space in order to classify input patterns. The beauty of neural networks is that they simultaneously learn a feature transformation function as well as a linear classifier.

Another method is to design the feature transformation function by hand and train a linear or nonlinear classifiers to differentiate patterns in this space. Feature transformation functions such as histogram of oriented gradients and local binary pattern histograms are two of commonly used feature transformation functions. Understanding the underlying process of these functions is more trivial than a transformation function represented by a neural network.

For example, in the case of histogram of oriented gradients, if there are many strong vertical edges in an image we know that the bin related to vertical edges is going to be significantly bigger than other bins in the histogram. If a linear classifier is trained on top of these histograms and if the magnitude of weight of linear classifier related to the vertical bin is high, we can imply that vertical edges have a great impact on the classification score.

As it turns out from the above example, figuring out that how a pattern is classified using a linear classifier trained on top of histogram of oriented gradients is doable. Also, if an interpretable nonlinear classifier such as decision trees or random forest is trained on the histogram, it is still possible to explain how a pattern is classified using these methods.

© Springer International Publishing AG 2017
H. Habibi Aghdam and E. Jahani Heravi, *Guide to Convolutional Neural Networks*, DOI 10.1007/978-3-319-57550-6_7

The problem with deep neural networks is that it is hard or even impossible to inspect weights of neural networks and understand how the feature transformation function works. In other words, it is not trivial to know how a pattern with many strong vertical edges will be transformed into the feature space. Also, in contrast to histogram of oriented gradients where each axis in the feature spaces has an easy-to-understand meaning for human, axes of feature spaces represented by a neural network are not easily interpretable.

For these reasons, diagnosing neural networks and understanding the underlying process of a neural network are not possible. Visualization is a way to make sense of complex models such as neural networks. In Chap. 5, we showed a few data-oriented techniques for understanding the feature transformation and classification process of neural networks. In this chapter, we will briefly review these techniques again and introduce gradient-based visualization techniques.

7.2 Data-Oriented Techniques

In general, data-oriented visualization methods work by feeding images to a network and collecting information from desired neurons.

7.2.1 Tracking Activation

In this method, N images are fed into the network and the activation (i.e., output of neuron after applying the activation function) of a specific neuron on each of these images is stored in an array. This way, we will obtain an array containing N real numbers in which each real number shows the activation of a specific neuron. Then, $K \ll N$ images with the highest activations are selected (Girshick et al. 2014). This method shows that what information about objects in the receptive field of the neuron increases the activation of the neuron. In Chap. 5, we visualized the classification network trained on the GTSRB dataset using this method.

7.2.2 Covering Mask

Assume that image $\mathbf{x}$ is correctly classified by a neural network with a probability close to 1.0. In order to understand which parts of the image have a greater impact on the score, we can run a multi-scale scanning window approach. In scale s and at location (m, n) on $\mathbf{x}$, $x(m, n)$ and all pixels in its neighborhood are set to zero. The size of neighborhood depends on s. This is equivalent zeroing the inputs to the network. In other words, information in this particular part of the image is missing. If the classification score highly depends on the information centered at (m, n), the score must be dropped significantly by zeroing the pixels in this region. If the above procedure is repeated for different scales and on all the locations in the image, we will end up with a map for each scale where the value of map will be close to 1 if zeroing its analogous region does not have any effect on the score. In contrast, the value of map will be close to zero if zeroing its corresponding region has a great impact on score. This method is previously used in Chap. 5 on the classification network trained on GTSRB dataset. One problem with this method is that it could be very time consuming to apply the above method on many samples for each class to figure out which regions are important in the final classification score.

7.2.3 Embedding

Embedding is another technique which provides important information about feature space. Basically, given a set of feature vectors $\mathscr{X} = \{\Phi(\mathbf{x}_1), \Phi(\mathbf{x}_2), \ldots, \Phi(\mathbf{x}_N)\}$, where $\Phi : \mathbb{R}^{H \times W \times 3} \rightarrow \mathbb{R}^d$ is the feature transformation function, the goal of embedding is to find the mapping $\Psi : \mathbb{R}^d \rightarrow \mathbb{R}^{\hat{d}}$ to project the d-dimensional feature vector into a $\hat{d}$-dimensional space. Usually, $\hat{d}$ is set to 2 or 3 since inspecting vectors visually in this spaces can be easily done using scatter plots.

There are different methods for finding the mapping Ψ. However, there is a specific mapping which is particularly used for mapping into two-dimensional space in the field of neural network. This mapping is called *t-distributed stochastic neighbor embedding* (t-SNE). It is a structure preserving mapping meaning that it tries to preserve the structure of neighbors in the $\hat{d}$-dimensional space as similar as possible to the structure of neighbors in d-dimensional space. This is an important property since it shows that how separable are patterns from different classes in the original feature space.

Denoting the feature transformation function up to layer L in a network by $\Phi_L(\mathbf{x})$, we collect the set $\mathscr{X}_L = \{\Phi_L(\mathbf{x}_1), \Phi_L(\mathbf{x}_2), \ldots, \Phi_L(\mathbf{x}_N)\}$ by feeding many images from different classes to the network and collecting $\Phi_L(\mathbf{x}_N)$ for each image. Then, the t-SNE algorithm is applied on $\mathscr{X}_L$ in order to find a mapping into the two-dimensional space. The mapped points can be plotted using scatter plots. This technique was used for analyzing networks in Chaps. 5 and 6.

7.3 Gradient-Based Techniques

Gradient-based methods explain neural networks in terms of their gradient with respect to the input image $\mathbf{x}$ (Simonyan et al. 2013). Depending on how the gradients are interpreted, a neural network can be studied from different perspectives.[1]

---

[1]Implementations of the methods in this chapter are available at *github.com/pcnn/*.

7.3.1 Activation Maximization

Denoting the classification score of $\mathbf{x}$ on class c with $S_c(\mathbf{x})$, we can find an input $\hat{\mathbf{x}}$ by maximizing the following objective function:

$$S_c(\hat{\mathbf{x}}) - \lambda \|\hat{\mathbf{x}}\|_2^2, \tag{7.1}$$

where λ is the regularization parameter defined by user. In other words, we are looking for an input image $\hat{\mathbf{x}}$ that maximizes the classification score on class c and it is always within n-sphere defined by the second term in the above function. This loss can be implemented using a Python layer in the Caffe library. Specifically, the layer accepts a parameter indicating the class of interest. Then, it will return the score of class of interest during forward pass. In addition, in the backward pass derivative of all classes except the class of interest will be set to zero. Obviously, any change in the inputs of layer other than class of interest does not change the output. Consequently, derivative of the loss with respect to these inputs will be equal to zero. In contrast, derivative of loss with respect to class of interest will be equal to 1 since it just passes the value from class of interest to the output. One can think of this loss as a multiplexer which directs inputs according to its address.

The derivative of the second term in the objective function with respect to classification scores is always zero. However, derivative of the second term with respect to input x_i is equal to $2\lambda x_i$. In order to formulate the above objective function as a minimization problem, we can simply multiply the function with -1. In that case, derivative of the first term with respect to the class of interest will be equal to -1. Putting all this together, the Python layer for the above loss function can be defined as follows:

```
class score_loss(caffe.Layer):                                              1
    def setup(self, bottom, top):                                           2
        params = eval(self.param_str)                                       3
        self.class_ind = params['class_ind']                               4
        self.decay_lambda = params['decay_lambda' ] if params.has_key('decay_lambda') else 0   5
                                                                            6
    def reshape(self, bottom, top):                                        7
        top[0].reshape(bottom[0].data.shape[0], 1)                         8
                                                                            9
    def forward(self, bottom, top):                                        10
        top[0].data[...] = 0                                               11
        top[0].data[:, 0] = bottom[0].data[:, self.class_ind]              12
                                                                            13
    def backward(self, top, propagate_down, bottom):                      14
        bottom[0].diff[...] = np.zeros(bottom[0].data.shape)               15
        bottom[0].diff[:, self.class_ind] = −1                             16
                                                                            17
        if len(bottom) == 2 and self.decay_lambda > 0:                    18
            bottom[1].diff[...] = self.decay_lambda * bottom[1].data[...]  19
```

After designing the loss layer, it has to be connected to the *trained* network. The following Python script shows how to do this.

```
def create_net(save_to, class_ind):                                              1
    L = caffe.layers                                                             2
    P = caffe.params                                                             3
    net = caffe.NetSpec()                                                        4
    net.data = L.Input(shape=[{'dim':[1,3,48,48]}])                              5
    net.tran = L.Convolution(net.data,                                           6
                    num_output=3,                                                7
                    group=3,                                                     8
                    kernel_size=1,                                               9
                    weight_filler={'type':'constant',                          10
                            'value':1},                                         11
                    bias_filler={'type':'constant',                            12
                            'value':0},                                         13
                    param=[{'decay_mult':1},{'decay_mult':0}],                 14
                    propagate_down=True)                                        15
    net.conv1, net.act1, net.pool1 = conv_act_pool(net.tran, 7, 100, act='ReLU')  16
    net.conv2, net.act2, net.pool2 = conv_act_pool(net.pool1, 4, 150, act='ReLU', group=2)  17
    net.conv3, net.act3, net.pool3 = conv_act_pool(net.pool2, 4, 250, act='ReLU', group=2)  18
    net.fc1, net.fc_act, net.drop1 = fc_act_drop(net.pool3, 300, act='ReLU')    19
    net.f3_classifier = fc(net.drop1, 43)                                       20
    net.loss = L.Python(net.f3_classifier, net.data, module='py_loss', layer='score_loss',  21
        param_str="{'class_ind':%d, 'decay_lambda':5}" %class_ind)              22
    with open(save_to, 'w') as fs:                                              23
        s_proto = 'force_backward:true\n' + str(net.to_proto())                 24
        fs.write(s_proto)                                                       25
        fs.flush()                                                              26
        print s_proto                                                           27
```

Recall from Chap. 4 that the Python file has to be placed next to the network definition file. We also set *force_backward* to *true* in order to force Caffe to always perform the backward pass down to the data layer. Finally, the image $\hat{\mathbf{x}}$ can be found by running the following momentum-based gradient descend algorithm.

```
caffe.set_mode_gpu()                                                             1
root = '/home/pc/'                                                                2
net_name = 'ircv1'                                                                3
save_to = root + 'cnn_{}.prototxt'.format(net_name)                              4
class_ind = 1                                                                     5
create_net(save_to, class_ind)                                                    6
                                                                                 7
net = caffe.Net(save_to, caffe.TEST)                                             8
net.copy_from('/home/pc/cnn.caffemodel')                                         9
                                                                                10
im_mean = read_mean_file('/home/pc/gtsr_mean_48x48.binaryproto')               11
im_res = read_mean_file('/home/pc/gtsr_mean_48x48.binaryproto')                12
im_res = im_res[np.newaxis,...]/255.                                            13
                                                                                14
alpha = 0.0001                                                                  15
momentum = 0.9                                                                  16
momentum_vec = 0                                                                17
                                                                                18
for i in xrange(4000):                                                          19
    net.blobs['data'].data[...] = im_res[np.newaxis, ...]                       20
    net.forward()                                                              21
    net.backward()                                                             22
    momentum_vec = momentum * momentum_vec + alpha * net.blobs['data'].diff    23
    im_res = im_res  − momentum_vec                                            24
    im_res = np.clip(im_res, −1, 1)                                            25
                                                                                26
                                                                                27
fig1 = plt.figure(1, figsize=(6, 6), facecolor='w')                            28
plt.clf()                                                                      29
res = np.transpose(im_res[0].copy()*255+im_mean, [1, 2, 0])[:,:,[2,1,0]]       30
res = np.divide(res − res.min(), res.max()−res.min())                          31
plt.imshow(res)                                                                32
```

Lines 1–9 create a network with the Python layer connected to this network and loads weights of the trained network into the memory. Line 11 loads the mean image into memory. The variable in this line will be used for applying the backward transformation on the result for illustration purposes. Lines 12 and 13 initialize the optimization algorithm by setting it to the mean image.

Lines 15–17 configure the optimization algorithm. Lines 19–25 perform the momentum-based gradient descend algorithm. Line 18 executes the forward pass and the next line performs the backward pass and computes derivative of loss function with respect to the input data. Finally, the commands after the loop show the obtained image. Figure 7.1 illustrates the result of running the above script on each of classes, separately.

It turns out that classification score of each class mainly depends on pictograph inside of each sign. Furthermore, shape of each sign has impact on the classification score as well. Finally, we observe that the network does a great job in eliminating the background of traffic sign.

It is worth mentioning that the optimization is directly done on the classification scores rather than output of softmax function. The reason is that maximizing the output of softmax may not necessarily maximize the score of class of interest. Instead, it may try to reduce the score of other classes.

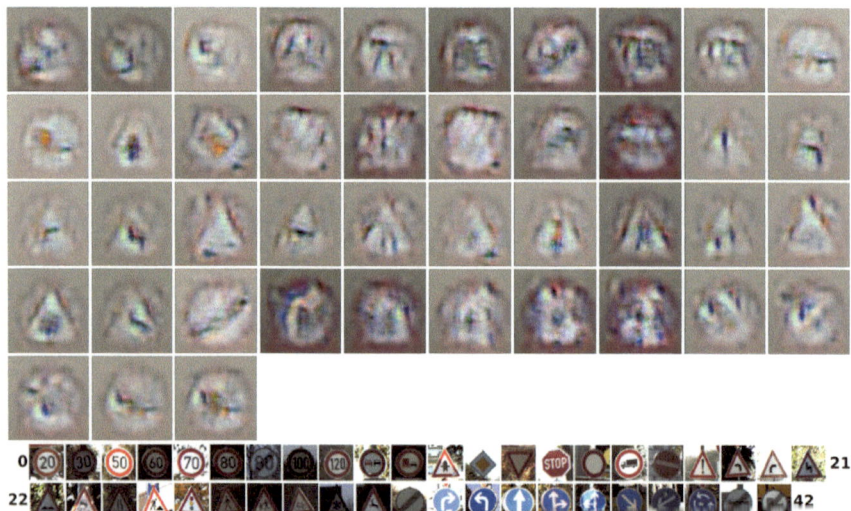

Fig. 7.1 Visualizing classes of traffic signs by maximizing the classification score on each class. The *top-left* image corresponds to class 0. The class labels increase from *left* to *right* and *top* to *bottom*

7.3.2 Activation Saliency

Another way for visualizing neural networks is to asses how sensitive is a classification score with respect to every pixel on the input image. This is equivalent to computing gradient of the classification score with respect to the input image. Formally, given the image $\mathbf{x} \in \mathbb{R}^{H \times W \times 3}$ belonging to class c, we can compute:

$$\nabla \mathbf{x}_{mnk} = \frac{\delta S_c(\mathbf{x})}{\mathbf{x}_{mnk}}, \qquad m = 0, \ldots, H, n = 0, \ldots, W, k = 0, 1, 2. \qquad (7.2)$$

In this equation, $\nabla \mathbf{x}_{mnk} \mathbb{R}^{H \times W \times 3}$ stores the gradient of classification score with respect to every pixel in $\mathbf{x}$. If $\mathbf{x}$ is a grayscale image the output will only have one channel. Then, the output can be illustrated by mapping each gradient to a color. In the case that $\mathbf{x}$ is a color image, maximum of $\nabla \mathbf{x}$ is computed across channels.

$$\nabla \mathbf{x}'_{mn} = \max_{k=0,1,2} \nabla \mathbf{x}_{mnk}. \qquad (7.3)$$

Then, $\nabla \mathbf{x}'_{mn}$ is illustrated by mapping each element in this matrix to a color. This roughly shows saliency of each pixel in $\mathbf{x}$. Figure 7.2 visualizes the class saliency of a random sample from each class.

In general, we see that the pictograph region in each image has a great effect on the classification score. Besides, in a few cases, we also observe that background pixels have impact on the classification score. However, this might not be generalized to all images in the same class. In order to understand expected saliency of pixel, we can compute $\nabla \mathbf{x}'$ for many samples from the same class and compute their average. Figure 7.3 shows expected class saliency obtained by computing the average of class saliency of 100 samples coming from the same class.

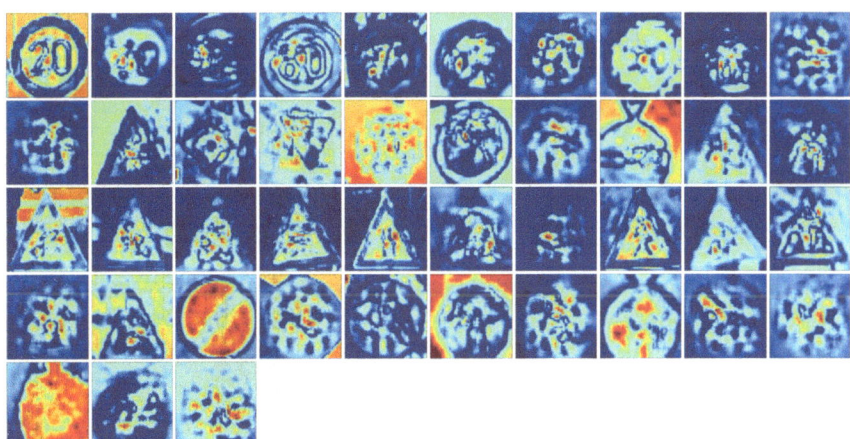

Fig. 7.2 Visualizing class saliency using a random sample from each class. The order of images is similar Fig. 7.1

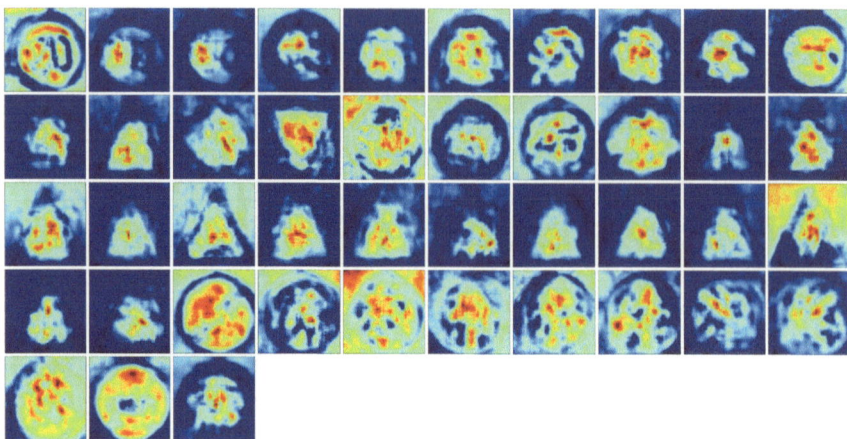

Fig. 7.3 Visualizing expected class saliency using 100 samples from each class. The order of images is similar to Fig. 7.1

The expected saliency reveals that the classification score mainly depends on pictograph region. In other words, slight changes in this region may dramatically change the classification score which in turn may alter the class of image.

7.4 Inverting Representation

Inverting a neural network (Mahendran and Vedaldi 2015) is a way to roughly know what information is retained by a specific layer in a neural network. Denoting the representation produced by L^{th} layer in a ConvNet for the input image $\mathbf{x}$ with $\Phi(\mathbf{x})_L$, inverting a ConvNet can be done by minimizing

$$\hat{\mathbf{x}} = \underset{\mathbf{x}' \in \mathbb{R}^{H \times W \times 3}}{\arg\min} \ \|\Phi(\mathbf{x}') - \Phi(\mathbf{x})\|_2 + \lambda \|\mathbf{x}'\|_p^p, \qquad (7.4)$$

where the first term computes the Euclidean distance between the representations of the source image $\mathbf{x}$ and reconstructed image $\mathbf{x}'$ and the second term regularizes the cost by the *p-norm* of the reconstructed image.

If the regularizing term is omitted, it is possible to design a network using available layers in Caffe which accepts the representation of an image and tries to find the reconstructed image $\hat{\mathbf{x}}$. However, it is not possible to implement the above cost function including the second term using available layers in Caffe. For this reason, a Python layer has to be implemented for computing the loss and its gradient with respect to its bottoms. This layer could be implemented as follows:

```
class euc_loss(caffe.Layer):                                                          1
    def setup(self, bottom, top):                                                     2
        params = eval(self.param_str)                                                 3
        self.decay_lambda = params['decay_lambda'] if params.has_key('decay_lambda') else 0   4
        self.p = params['p'] if params.has_key('p') else 2                            5
                                                                                      6
    def reshape(self, bottom, top):                                                   7
        top[0].reshape(bottom[0].data.shape[0], 1)                                    8
                                                                                      9
    def forward(self, bottom, top):                                                   10
                                                                                      11
        if bottom[0].data.ndim == 4:                                                  12
            top[0].data[:, 0] = np.sum(np.power(bottom[0].data−bottom[1].data,2), axis=(1,2,3))   13
        elif bottom[0].data.ndim == 2:                                                14
            top[0].data[:, 0] = np.sum(np.power(bottom[0].data − bottom[1].data, 2), axis=1)   15
                                                                                      16
        if len(bottom) == 3:                                                          17
            top[0].data[:,0] += np.sum(np.power(bottom[2].data,2))                    18
                                                                                      19
    def backward(self, top, propagate_down, bottom):                                 20
        bottom[0].diff[...] = bottom[0].data − bottom[1].data                         21
        if len(bottom) == 3:                                                          22
            bottom[2].diff[...] = self.decay_lambda *self.p* np.multiply(bottom[2].data[...], np.power(np.abs(bottom   23
            [2].data[...]),self.p−2))
```

Then, the above loss layer is connected to the network trained on the GTSRB dataset.

```
def create_net_ircv1_vis(save_to):                                                    1
    L = caffe.layers                                                                  2
    P = caffe.params                                                                  3
    net = caffe.NetSpec()                                                             4
    net.data = L.Input(shape=[{'dim':[1,3,48,48]}])                                   5
    net.rep = L.Input(shape=[{'dim': [1, 250, 6, 6]}])  #output shape of conv3        6
                                                                                      7
    net.tran = L.Convolution(net.data,                                                8
                    num_output=3,                                                     9
                    group=3,                                                          10
                    kernel_size=1,                                                    11
                    weight_filler={'type':'constant',                                 12
                                'value':1},                                           13
                    bias_filler={'type':'constant',                                   14
                                'value':0},                                           15
                    param=[{'decay_mult':1},{'decay_mult':0}],                        16
                    propagate_down=True)                                              17
    net.conv1, net.act1, net.pool1 = conv_act_pool(net.tran, 7, 100, act='ReLU')      18
    net.conv2, net.act2, net.pool2 = conv_act_pool(net.pool1, 4, 150, act='ReLU', group=2)   19
    net.conv3, net.act3, net.pool3 = conv_act_pool(net.pool2, 4, 250, act='ReLU', group=2)   20
    net.fc1, net.fc_act, net.drop1 = fc_act_drop(net.pool3, 300, act='ReLU')          21
    net.f3_classifier = fc(net.drop1, 43)                                             22
    net.loss = L.Python(net.act3, net.rep, net.data, module='py_loss', layer='euc_loss',   23
    param_str="{'decay_lambda':10,'p':6}")                                            24
```

The network accepts two inputs. The first input shows the reconstructed image and the second input indicates the representation of the source image. In the above network, our goal is to reconstruct the image using representation produced by the activation of the third convolution layer. The output shape of the third convolution layer is $250 \times 3 \times 3$. Hence, the shape of second input in the network is set to $1 \times 250 \times 6 \times 6$. Moreover, as it is proposed in Mahendran and Vedaldi (2015), we set the value of p

in the above network to 6. Having the network created, we can execute the following momentum-based gradient descend for finding $\hat{x}$.

```
im_mean = read_mean_file('/home/pc/gtsr_mean_48x48.binaryproto')              1
im_mean = np.transpose(im_mean, [1, 2, 0])                                    2
                                                                              3
im = cv2.imread('/home/pc/GTSRB/Training_CNN/00016/crop_00001_00029.ppm')     4
im = cv2.resize(im, (48,48))                                                  5
im_net = (im.astype('float32')-im_mean)/255.                                  6
net.blobs['data'].data[...] = np.transpose(im_net, [2, 0, 1])[np.newaxis, ...]  7
                                                                              8
net.forward()                                                                 9
rep = net.blobs['act3'].data.copy()                                          10
                                                                             11
                                                                             12
im_res = im*0                                                                13
im_res = np.transpose(im_res, [2,0,1])                                       14
                                                                             15
alpha = 0.000001                                                             16
momentum = 0.9                                                               17
momentum_vec = 0                                                             18
                                                                             19
for i in xrange(10000):                                                      20
    net.blobs['data'].data[...] = im_res[np.newaxis, ...]                    21
    net.blobs['rep'].data[...] = rep[...]                                    22
                                                                             23
    net.forward()                                                           24
    net.backward()                                                          25
                                                                             26
    momentum_vec = momentum * momentum_vec - alpha * net.blobs['data'].diff  27
                                                                             28
    im_res = im_res + momentum_vec                                          29
    im_res = np.clip(im_res, -1, 1)                                         30
                                                                             31
plt.figure(1)                                                               32
plt.clf()                                                                   33
res = np.transpose(im_res[0].copy(), [1, 2, 0])                             34
res = np.clip(res*255 + im_mean, 0, 255)                                    35
res = np.divide(res - res.min(), res.max()-res.min())                       36
plt.imshow(res[:,:, [2,1,0]])                                               37
plt.show()                                                                  38
```

In the above code, the source image is first fed to the network and the output of the third convolution layer is copied into memory. Then, the optimization is done in 10,000 iterations. At each iteration, the reconstructed image is entered to the network and the backward pass is computed down to the input layer. This way, gradient of the loss function is obtained with respect to the input. Finally, the reconstructed image is updated using the momentum gradient descend rule. Figure 7.4 shows the result of inverting the classification network from different layers. We see that the first convolution layer keeps photo-realistic information. For this reason, the reconstructed image is very similar to the source image. Starting from the second convolution layer, photo-realistic information starts to vanish and they are replaced with parts of image which is important to the layer. For example, the fully connected layer mainly depends on the specific part of pictograph on the sign and it ignores background information.

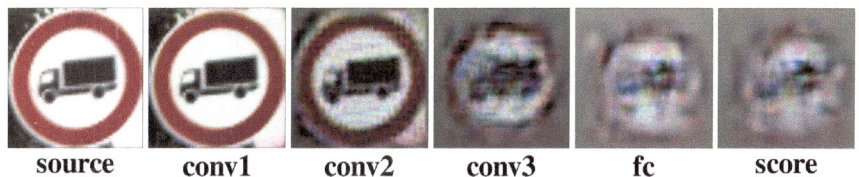

| source | conv1 | conv2 | conv3 | fc | score |

Fig. 7.4 Reconstructing a traffic sign using representation of different layers

7.5 Summary

Understanding behavior of neural networks is necessary in order to better analyze and diagnose them. Quantitative metrics such as classification accuracy and F1 score just give us numbers indicating how good is the classifier in our problem. They do not tell us how a neural network achieves this result. Visualization is a set of techniques that are commonly used for understanding structure of high-dimensional vectors.

In this chapter, we briefly reviewed data-driven techniques for visualization and showed that how to apply them on neural networks. Then, we focused on techniques that visualize neural networks by minimizing an objective function. Among them, we explained three different methods.

In the first method, we defined a loss function and found an image that maximizes the classification score of a particular class. In order to generate more interpretable images, the objective function was regularized using L_2 norm of the image. In the second method, gradient of a particular neuron was computed with respect to the input image and it is illustrated by computing its magnitude.

The third method formulated the visualizing problem as an image reconstruction problem. To be more specific, we explained a method that tries to find an image in which the representation of this image is very close to the representation of the original image. This technique usually tells us what information is usually discarded by a particular layer.

7.6 Exercises

7.1 Visualizing a ConvNet can be done by maximizing the softmax score of a specific class. However, this may not exactly generate an image that maximizes the classification score. Explain the reason taking into account the softmax score.

7.2 Try embed a feature extracted by neural network using local linear embedding method.

7.3 Use Isomap to embed features into two-dimensional space.

7.4 Assume an image of traffic signs belonging to class c which is correctly classified by the ConvNet. Instead of maximizing $S_c(\mathbf{x})$, try to minimize directly $S_c(\mathbf{x})$ such that $\mathbf{x}$ is no longer classified correctly by ConvNets but it is still easily recognizable for human.

References

Girshick R, Donahue J, Darrell T, Berkeley UC, Malik J (2014) Rich feature hierarchies for accurate object detection and semantic segmentation. doi:10.1109/CVPR.2014.81, arXiv:abs/1311.2524

Mahendran A, Vedaldi A (2015) Understanding deep image representations by inverting them. In: Computer vision and pattern recognition. IEEE, Boston, pp 5188–5196. doi:10.1109/CVPR.2015.7299155, arXiv:abs/1412.0035

Simonyan K, Vedaldi A, Zisserman A (2013) Deep inside convolutional networks: visualising image classification models and saliency maps, pp 1–8. arXiv:13126034

Appendix A
Gradient Descend

Any classification model such as neural networks is trained using an objective function. The goal of an objective function is to compute a scaler based on training data and current configuration of parameters of the model. The scaler shows how good is the model in classifying training samples. Assume that the range of objective function is in internal [0, inf) where it returns 0 for a model that classifies training samples perfectly. As the error of model increases, the objective function returns a larger positive number.

Let $\Phi(\mathbf{x}; \theta)$ denotes a model which classifies the sample $\mathbf{x} \in \mathbb{R}^d$. The model is defined using its parameter vector $\theta \in \mathbb{R}^q$. Based on that, a training algorithm aims to find $\hat{\theta}$ such that the objective function returns a number close to zero given the model $\Phi(.; \hat{\theta})$. In other words, we are looking for a parameter vector $\hat{\theta}$ that minimizes the objective function.

Depending on the objective function, there are different ways to find $\hat{\theta}$. Assume that the objective function is differentiable everywhere. Closed-form solution for finding minimum of the objective function is to set its derivative to zero and solve the equation. However, in practice, finding a closed-form solution for this equation is impossible. The objective function that we use for training a classifier is multivariate which means that we have to set the norm of gradient to zero in order to find the minimum of objective function. In a neural network with 1 million parameters, the gradient of objective function will be a 1-million-dimensional vector. Finding a closed-form solution for this equation is almost impossible.

For this reason, we always use a numerical method for finding the (local) minimum of objective function. Like many numerical methods, this is also an iterative process. The general algorithm for this purpose is as follows. The algorithm always starts from an initial point. Then, it iteratively updates the initial solution using vector δ. The only unknown in the above algorithm is the vector δ. A randomized hill climbing

H. Habibi Aghdam and E. Jahani Heravi, *Guide to Convolutional Neural Networks*, DOI 10.1007/978-3-319-57550-6

Algorithm 1 Numerical optimization

$x' \leftarrow random\ vector$
while stopping condition **do**
   $x' = x' + \delta$
return x'

---

algorithm sets δ to a random vector.[1] However, it is not guaranteed that the objective function will be constantly minimized by this way. Hence, its convergence speed might not be fast especially when the dimensionality of the parameter vector θ is high. For this reason, we need a better heuristic for finding δ.

Assume that you are in the middle of hill and you want to go down as quickly as possible. There might be many paths that you can choose as your next step. Some of them will get you closer to the bottom of the hill. There is only one move that will get you much more closer than all other moves. It is the move exactly along steepest descend. In the above algorithm, the hill is the objective function. Your current location on the hill is analogous to x' in the above algorithm. Steepest descend is also related to δ.

From mathematical perspective, steepest descend is related to gradient of function in the current location. Gradient descend is an iterative optimization algorithm to find a *local minimum* of a function. It sets δ proportional to the negative of gradient. The following table shows the pseudocode of the gradient descend algorithm:

Algorithm 2 Gradient descend algorithm

$x' \leftarrow random\ vector$
while stopping condition **do**
   $x' = x' - \alpha \nabla J(\theta; x)$
return x'

In this algorithm, α is the learning rate and ∇ denotes the gradient of the objective function $J(.)$ with respect to θ. The learning rate α shows how big should be the next step in the direction of steepest descend. In addition, the stopping condition might be implemented by assuming a maximum number of iterations. Also, the loop can be stopped if the changes of θ are less than a threshold. Let us explain this using an example. Assume that our model is defined as follows.

$$\Phi(x; \theta) = \theta_1 x_1 + \theta_2 x_2, \tag{A.1}$$

---

[1]A randomize hill climbing algorithm accepts $x' + \delta$ if it reduces the objective function. Otherwise, it rejects the current δ and generates new δ. The above process is repeated until the stopping criteria are reached.

where $x, \theta \in \mathbb{R}^2$ are two-dimensional vectors. We are given a dataset $\mathcal{X} = \{x^1, x^2, \ldots, x^n\}$ and our goal is to minimize the following objective function:

$$J(\theta^t) = \frac{1}{2n} \sum_{i=1}^{n} \left[\Phi(x^i; \theta^t) \right]^2 \tag{A.2}$$

In order to minimize $J(.)$ using the gradient descend algorithm, we need to compute its gradient vector which is given by

$$\bigtriangledown J(\theta^t) = \left[\frac{\delta J(\theta^t)}{\delta \theta_1}, \frac{\delta J(\theta^t)}{\delta \theta_2} \right]$$
$$\left[\frac{1}{n} \sum_{i=1}^{n} x_1 (\theta_1 x_1 + \theta_2 x_2), \frac{1}{n} \sum_{i=1}^{n} x_2 (\theta_1 x_1 + \theta_2 x_2) \right]. \tag{A.3}$$

Since $J(.)$ is a two-dimensional function it can be illustrated using filled contour plots. To be more specific, the dataset $\mathcal{X}$ is fixed and the variables of this function are θ_1 and θ_2. Therefore, we can evaluate J for different values of θ and show it using a contour plot. The following Python script plots this function:

```
def J(x, w):                                                                            1
    e = (np.dot(x, w.transpose())) ** 2                                                 2
    return np.mean(e, axis = 0)                                                          3
                                                                                        4
def dJ(x, w):                                                                           5
    return np.mean(x*(x* w.transpose()), axis = 0)                                       6
                                                                                        7
x1, x2 = np.meshgrid(np.linspace(-5, 5, 100),np.linspace(-5, 5, 100), indexing='ij')    8
x1x2 = np.stack((x1.flatten(), x2.flatten()), axis = 1)                                  9
                                                                                        10
w1,w2 = np.meshgrid(np.linspace(-0.9, 0.9, 50),np.linspace(-0.9, 0.9, 50), indexing = 'ij') 11
w = np.stack((w1.flatten(), w2.flatten()), axis = 1)                                     12
                                                                                        13
e = J(x1x2, w)                                                                          14
                                                                                        15
plt.figure(1, figsize=(9,8), facecolor ='w')                                            16
plt.contourf(w1,w2, np.reshape(e, w1.shape),50)                                          17
plt.colorbar()                                                                          18
plt.show()                                                                             19
```

If we execute the above code, we will obtain the result illustrated in Fig. A.1.

Since we know $\bigtriangledown J(\theta^t)$, we can plug it into the gradient descend algorithm and find the minimum of J by constantly updating θ^t until the algorithm converges. The following Python script shows how to do this:

```
w_sol = np.asarray([0.55, 0.50])                                                        1
batch_size = x1x2.shape[0]                                                               2
for _ in xrange(50):                                                                     3
    x = x1x2                                                                            4
                                                                                        5
    e = J(x, w)                                                                         6
    de = dJ(x, w_sol)                                                                    7
                                                                                        8
    w_sol = w_sol - alpha * de                                                           9
```

Fig. A.1 Surface of the error function in (A.2)

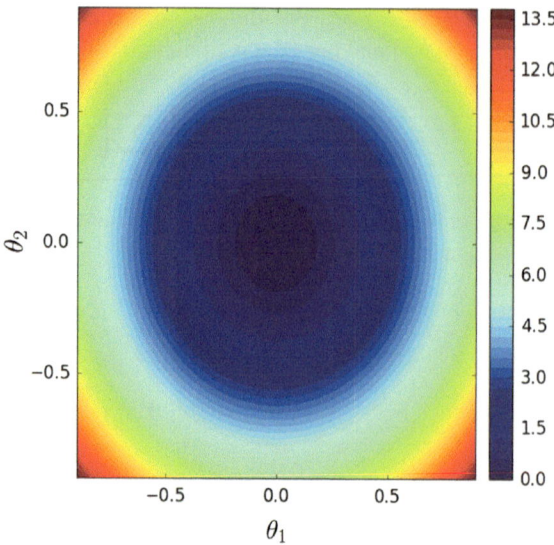

The algorithm starts with initializing θ to $[0.55, 0.50]$. Then, it executes the gradient calculation and parameters update for 50 iterations. At each iteration, the variable *w_sol* slightly changes and moves toward minimum of J. Figure A.2 shows the trajectory of parameters in 50 iterations. The function is more steep at initial location. Hence, the parameters are updated using bigger steps. However, as the parameters approach minimum of the function, the gradient becomes smaller. For this reason, the parameters are updated using smaller steps. Assume we change $\Phi(x; \theta)$ to

Fig. A.2 Trajectory of parameters obtained using the gradient descend algorithm

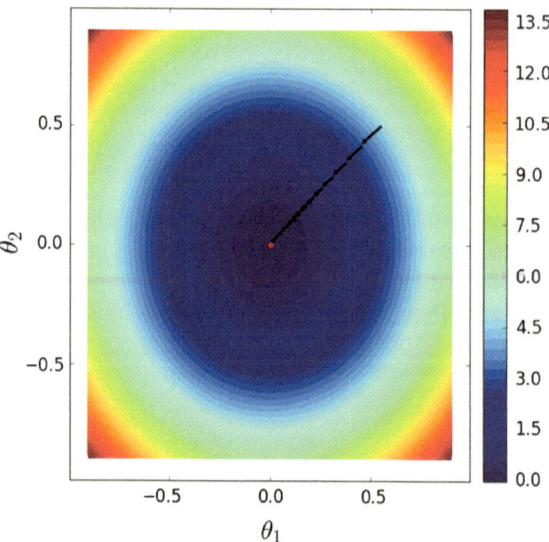

Fig. A.3 Surface of $J(.)$ using Φ in (A.4)

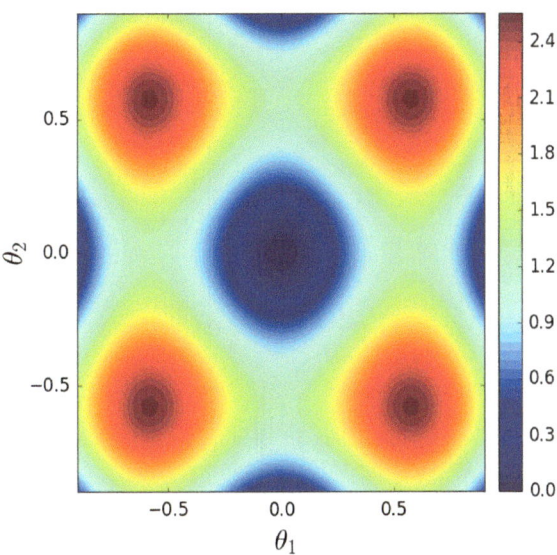

$$\Phi(x; \theta) = \theta_1 x_1 + \theta_2 x_2 - \theta_1^3 x_1 + \theta_2^3 x_2. \tag{A.4}$$

Figure A.3 illustrates the surface of $J(.)$ using the above definition for $\Phi(.)$. In contrast to the previous definition of Φ, the surface of $J(.)$ with the new definition of Φ is multi-modal. In other words, the surface is not convex anymore.

An immediate conclusion from a non-convex function is that there are more than one local minimum in the function. Consequently, depending on the initial location on the surface of $J(.)$, trajectory of the algorithm might be different. This property is illustrated in Fig. A.4.

As it is clear from the figure, although initial solutions are very close to each other, their trajectory are completely different and they have converged to distinct local minimums. Sensitivity of the gradient descend algorithm to the initial solutions is an inevitable issue. For a linear classifier, $J()$ is a convex function of the parameter vector θ. However, for models such as multilayer feed-forward neural networks, $J(.)$ is a non-convex function. Therefore, depending on the initial value of θ, the gradient descend algorithm is likely to converge to different local minimums.

Regardless of the definition of Φ, the gradient descend algorithm applied on the above definition of $J()$ is called *vanilla gradient descend* or *batch gradient descend*. In general, the objective function J can be defined in three different ways:

$$J(\theta) = \frac{1}{n} \sum_{i=1}^{n} \mathscr{L}\left(\Phi(x^i; \theta)\right) \tag{A.5}$$

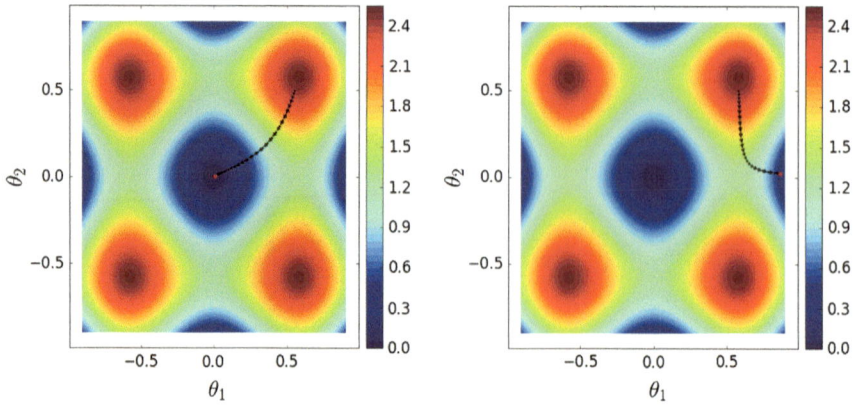

Fig. A.4 Depending on the location of the initial solution, the gradient descend algorithm may converge toward different local minimums

$$J(\theta) = \sum_{i=m}^{m} \mathscr{L}\big(\Phi(x^i; \theta)\big) \quad m = \{1, \ldots, n-k\} \tag{A.6}$$

$$J(\theta) = \frac{1}{k} \sum_{i=m}^{m+k} \mathscr{L}\big(\Phi(x^i; \theta)\big) \quad k \ll n, m = \{1, \ldots, n-k\}. \tag{A.7}$$

In the above equations, $\mathscr{L}$ is a loss function which computes the loss of Φ given the vector x^i. We have explained different loss functions in Chap. 2 that can be used in the task of classifications. The only difference in the above definitions is the number of iterations in the summation operation. The definition in (A.5) sums the loss over *all n* samples in training set. This is why it is called *batch gradient descend*. As the result, $\frac{\delta J}{\delta \theta_j}$ is also computed over all the samples in the training set. Assume we want to train a neural network with $10M$ parameters on a dataset containing $20M$ samples. Suppose that computing the gradient on one sample takes 0.002 s. This means that it will take $20M \times 0.002 = 40{,}000$ s (11 h) in order to compute (A.5) and do a single update on parameters. Parameters of a neural network may require thousands of updates before converging to a local minimum. However, this is impractical to do using (A.5).

The formulation of J in (A.6) is called *stochastic gradient descend* and it computes the gradient only on one sample and updates the parameter vector θ using the gradient over a single sample. Using this formulation, it is possible to update parameters thousand times in a tractable period. The biggest issue with this formulation is the fact that only one sample may not represent the error surface with an acceptable precision. Let us explain this on an example. Assume the formulation of Φ in (A.4). We showed previously the surface of error function (A.5) in Fig. A.3. Now, we

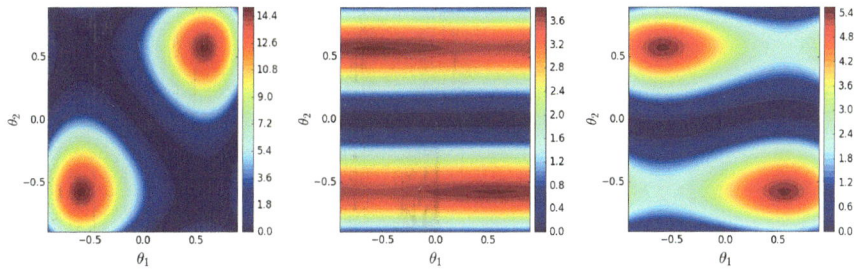

Fig. A.5 Contour plot of (A.6) computed using three different samples

compute surface of (A.6) using only three samples in the training set rather than all samples. Figure A.5 illustrates the contour plots associated with each sample.

As we expected, a single sample is not able to accurately represent the error surface. As the result, $\frac{\delta J}{\theta_j}$ might be different if it is computed on two different samples. Therefore, the magnitude and direction of parameter update will highly depend on the sample at current iteration. For this reason, we expect that the trajectory of parameters update to be jittery. Figure A.6 shows the trajectory of the stochastic gradient descend algorithm.

Compared with the trajectory of the vanilla gradient descend, trajectory of the stochastic gradient descend is jittery. From statistical point of view, if we take into account the gradients of J along its trajectory, the gradient vector of the stochastic gradient descend method has a higher variance compared with vanilla gradient descend. In highly nonlinear functions such as neural networks, unless the learning

Fig. A.6 Trajectory of stochastic gradient descend

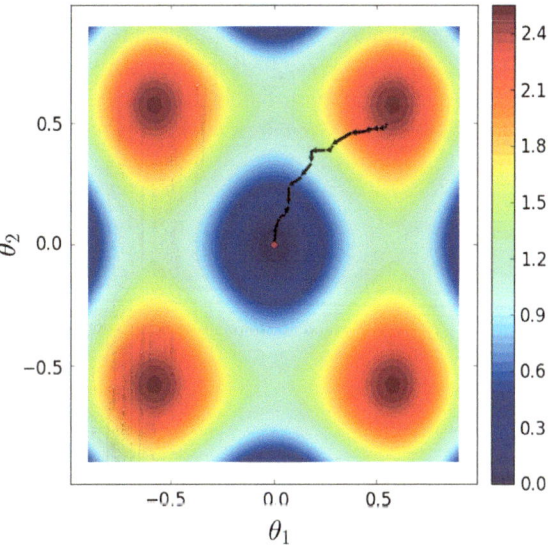

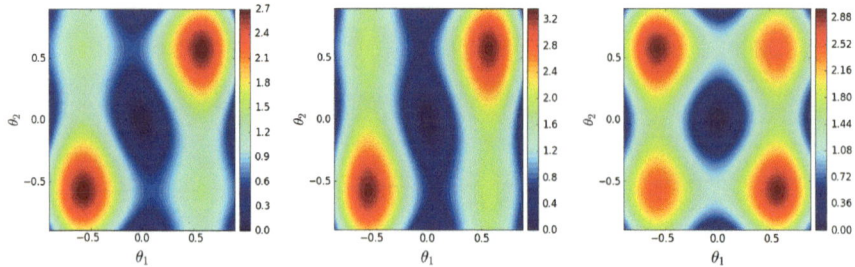

Fig. A.7 Contour plot of the mini-batch gradient descend function for mini-batches of size 2 (*left*), 10 (*middle*) and 40 (*right*)

rate is adjusted carefully, this causes the algorithm to jump over local minimums several times and it may take a longer time to the algorithm to converge. Adjusting learning rate in stochastic gradient descend is not trivial and for this reason stochastic gradient descend is not used in training neural networks. On the other hand, minimizing the vanilla gradient descend is not also tractable.

The trade-off between vanilla gradient descend and stochastic gradient descend is (A.7) that is called *mini-batch gradient descend*. In this method, the objective function is computed over a small batch of samples. The size of batch is much smaller than the size of samples in the training set. For example, k in this equation can be set to 64 showing a batch including 64 samples. We computed the error surface for mini-batches of size 2, 10, and 40 in our example. Figure A.7 shows the results.

We observe that a small mini-batch is not able to adequately represent the error surface. However, the error surface represented by larger mini-batches are more accurate. For this reason, we expect that the trajectory of mini-batch gradient descend becomes smoother as the size of mini-batch increases. Figure A.8 shows the trajectory of mini-batch gradients descend method for different batch sizes.

Depending of the error surface, accuracy of error surface is not improved significantly after a certain mini-batch size. In other words, using a mini-batch of size 50 may produce the same result as the mini-batch of size 200. However, the former size

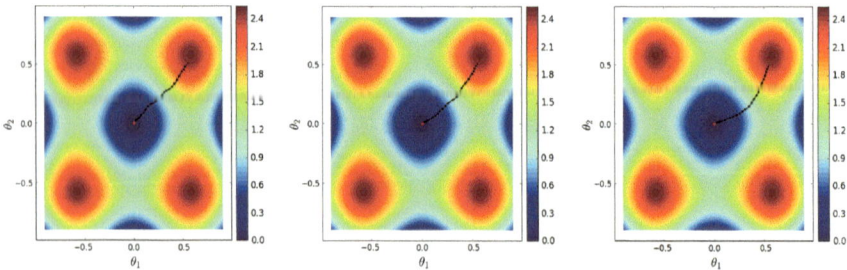

Fig. A.8 Trajectory of the mini-batch gradient descend function for mini-batches of size 2 (*left*), 10 (*middle*), and 40 (*right*)

is preferable since it converges faster. Currently, complex models such as neural networks are trained using mini-batch gradient descend. From statistical point of view, variance of gradient vector in mini-batch gradient descend is lower than stochastic gradient descend but it might be higher than batch gradient descend algorithm. The following Python script shows how to implement the mini-batch gradient descend algorithm in our example.

```
def J(x, w):                                                                                  1
e = (np.dot(x, w.transpose()) − np.dot(x, w.transpose() ** 3)) ** 2                            2
return np.mean(e, axis=0)                                                                      3
                                                                                              4
def dJ(x, w):                                                                                 5
    return np.mean((x−3*x*w.transpose()**2)*((x* w.transpose()) − (x* w.transpose             6
        () ** 3)), axis=0)                                                                    7
                                                                                              8
x1, x2 = np.meshgrid(np.linspace(−5,5,100),np.linspace(−5,5,100), indexing='ij')              9
x1x2 = np.stack((x1.flatten(), x2.flatten()), axis=1)                                         10
                                                                                              11
w1,w2 = np.meshgrid(np.linspace(−0.9,0.9,50),np.linspace(−0.9,0.9,50), indexing='ij')         12
w = np.stack((w1.flatten(), w2.flatten()), axis=1)                                            13
                                                                                              14
seed(1234)                                                                                    15
ind = range(x1x2.shape[0])                                                                    16
shuffle(ind)                                                                                  17
                                                                                              18
w_sol = np.asarray([0.55, 0.50])                                                              19
                                                                                              20
alpha = 0.02                                                                                  21
batch_size = 40                                                                               22
                                                                                              23
start_ind = 0                                                                                 24
                                                                                              25
for _ in xrange(50):                                                                          26
    end_ind = min(x1x2.shape[0], start_ind+batch_size)                                        27
    x = x1x2[ind[start_ind:end_ind], :]                                                       28
                                                                                              29
    if end_ind >= x1x2.shape[0]:                                                              30
        start_ind = 0                                                                         31
    else:                                                                                     32
        start_ind += batch_size                                                               33
                                                                                              34
    de = dJ(x, w_sol)                                                                         35
    w_sol = w_sol − alpha * de                                                                36
```

A.1 Momentum Gradient Descend

There are some variants of gradient descend algorithm to improve its convergence speed. Among them, *momentum gradient descend* is commonly used for training convolutional neural networks. The example that we have used in this chapter so far has a nice property. All elements of input x have the same scale. However, in practice, we usually deal with high-dimensional input vectors where elements of these vectors may not have the same scale. In this case, the error surface is a ravine where it is steeper in one direction than others. Figure A.9 shows a ravine surface and trajectory of mini-batch gradient descend on this surface.

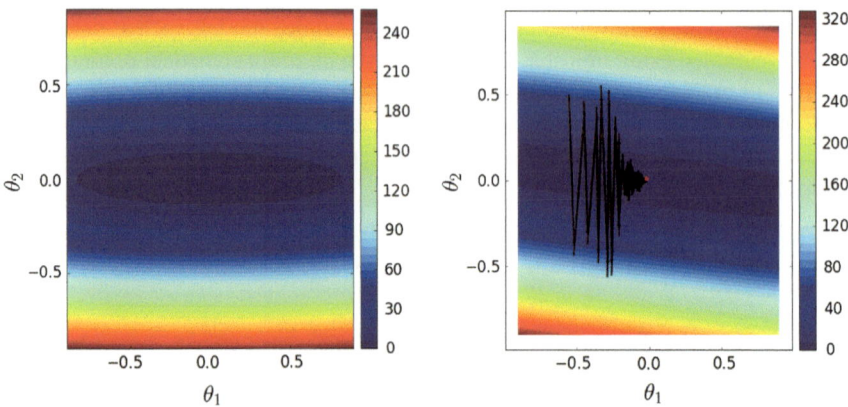

Fig. A.9 A ravine error surface and trajectory of mini-batch gradient descend on this surface

The algorithm oscillates many times until it converges to the local minimum. The reason is that because learning rate is high, the solution jumps over the local minimum after an update where the gradient varies significantly. In order to reduce the oscillation, the learning rate can be reduced. However, it is not easy to decide when to reduce the learning rate. If we set the learning rate to a very small value from beginning the algorithm may not converge in a an acceptable time period. If we set it to a high value it may oscillate a lot on the error surface.

Momentum gradient descend is a method to partially address this problem. It keeps history of gradient vector from previous steps and update the parameter vector θ based on the gradient of J with respect to the current mini-batch and its history on previous mini-batches. Formally,

$$v^t = \gamma v^{t-1} - \alpha \nabla J(\theta^t)$$
$$\theta^{t+1} = \theta^t + v^t. \tag{A.8}$$

Obviously, the vector v has the same dimension as $\alpha \nabla J(\theta^t)$. It is always initialized with zero. The hyperparameter $\gamma \in [0, 1)$ is a value between 0 and 1 (not included 1). It has to be smaller than one in order to make it possible that the algorithm forgets the gradient eventually. Sometimes the subtraction and addition operators are switched in these two equations. But switching the operators does not have any effect on the output. Figure A.10 shows the trajectory of the mini-batch gradient descend with $\gamma = 0.5$.

We see that the trajectory oscillates much less using the momentum. The momentum parameter γ is commonly set to 0.9 but smaller values can be also assigned to this hyperparameter. The following Python script shows how to create the ravine surface and implement momentum gradient descend. In the following script, the size of mini-batch is set to 2 but you can try with larges mini-batches as well.

Fig. A.10 Trajectory of
momentum gradient descend
on a ravine surface

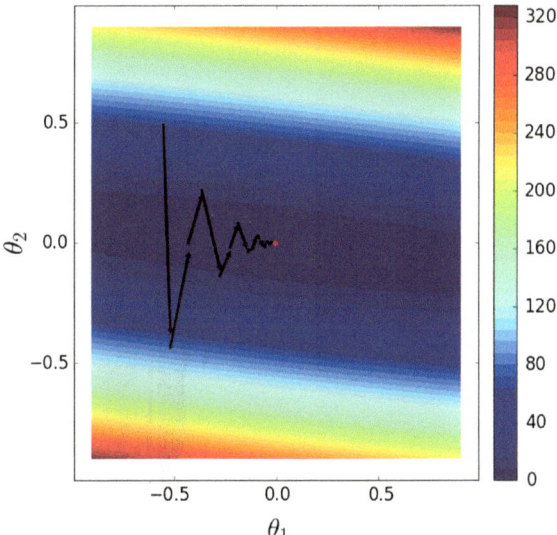

```
def J(x, w):                                                              1
    e = (np.dot(x, w.transpose())) ** 2                                  2
    return np.mean(e, axis = 0)                                          3
                                                                          4
def dJ(x, w):                                                            5
    return np.mean(x*(x* w.transpose()), axis = 0)                       6
                                                                          7
x1, x2 = np.meshgrid(np.linspace(-5, 5, 100),np.linspace(-20, -15, 100), indexing = 'ij')   8
x1x2 = np.stack((x1.flatten(), x2.flatten()), axis = 1)                  9
                                                                         10
w1,w2 = np.meshgrid(np.linspace(-0.9,0.9,50),np.linspace(-0.9, 0.9, 50), indexing = 'ij')   11
w = np.stack((w1.flatten(), w2.flatten()), axis = 1)                     12
                                                                         13
seed(1234)                                                               14
ind = range(x1x2.shape[0])                                               15
shuffle(ind)                                                             16
                                                                         17
w_sol = np.asarray([-0.55, 0.50])                                        18
                                                                         19
alpha = 0.0064                                                           20
batch_size = 2                                                           21
                                                                         22
start_ind = 0                                                            23
                                                                         24
momentum = 0.5                                                           25
momentum_vec = 0                                                         26
for _ in xrange(50):                                                     27
    end_ind = min(x1x2.shape[0], start_ind+batch_size)                  28
    x = x1x2[ind[start_ind:end_ind], :]                                  29
                                                                         30
    if end_ind >= x1x2.shape[0]:                                         31
        start_ind = 0                                                    32
    else:                                                                33
        start_ind += batch_size                                         34
                                                                         35
    de = dJ(x, w_sol)                                                    36
    momentum_vec = momentum_vec*momentum + alpha*de                      37
    w_sol = w_sol - momentum_vec                                         38
```

Fig. A.11 Problem of
momentum gradient descend

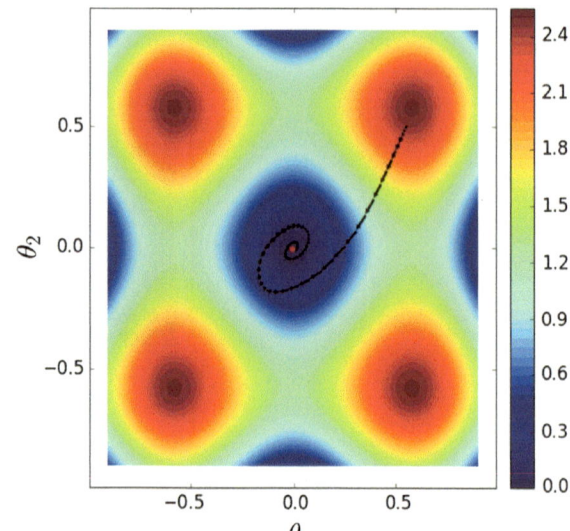

A.2 Nesterov Accelerated Gradients

One issue with momentum gradient descend is that when the algorithm is in the
path of steepest descend, the gradients are accumulated and momentum vector may
become bigger and bigger. It is like rolling a snow ball in a hill where it becomes
bigger and bigger. When the algorithm gets closer to the local minimum, it will
jump over the local minimum since the momentum has become very large. This is
where the algorithm takes a longer trajectory to reach to local minimum. This issue
is illustrated in Fig. A.11.

The above problem happened because the momentum gradient descend accumulates gradients blindly. It does not take into account what may happen in the next
steps. It realizes its mistake exactly in the next step and it tries to correct it after making a mistake. *Nesterov accelerated gradient* alleviates this problem by computing
the gradient of J with respect to the parameters in the next step. To be more specific,
$\theta^t + \gamma v^{t-1}$ approximately tells us where the next step is going to be. Based on this
idea, Nesterov accelerated gradient update rule is defined as

$$v^t = \gamma v^{t-1} - \alpha \nabla J(\theta^t + \gamma v^{t-1})$$
$$\theta^{t+1} = \theta^t + v^t.$$
(A.9)

By changing the update rule of vanilla momentum gradient descend to Nesterov
accelerated gradient, the algorithm has an idea about the next step and it corrects its
mistakes before happening. Figure A.12 shows the trajectory of the algorithm using
this method.

Fig. A.12 Nesterov
accelerated gradient tries to
correct the mistake by
looking at the gradient in the
next step

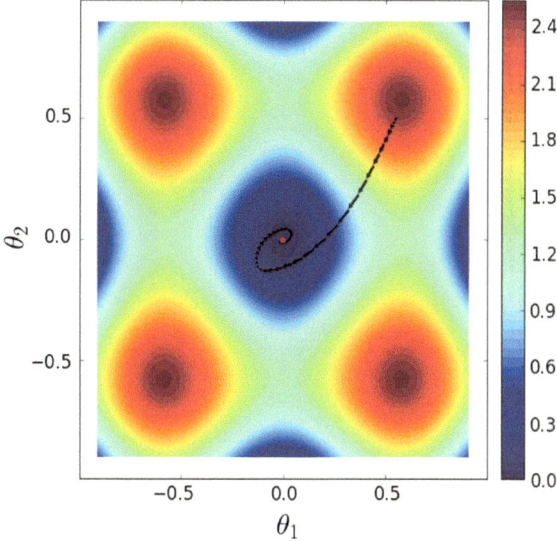

We see that the trajectory of Nesterov gradient descend is shorter than momentum gradient descend but it still has the same problem. Implementing the Nesterov accelerated gradient is simple. We only need to replace Lines 56–58 in the previous code with the following statements:

```
de_nes = dJ(x, w_sol−momentum_vec*momentum)                            1
momentum_vec = momentum_vec * momentum + alpha * de_nes                 2
w_sol = w_sol −momentum_vec                                             3
```

A.3 Adaptive Gradients (Adagrad)

The learning rate α is constant for all elements of ∇J. One of the problems in objective functions with ravine surfaces is that the learning rates of all elements are equal. However, elements analogous to steep dimensions have higher magnitudes and elements analogous to gentle dimensions have smaller magnitudes. When they are all updated with the same learning rate, the algorithm makes a larger step in direction of steep elements. For this reason, it oscillates on the error surface.

Adagrad is a method to adaptively assign a learning rate for each element in the gradient vector based on the gradient magnitude of each element in the past. Let ω_l denotes sum of square of gradients along the lth dimension in the gradient vector. Adagrad updates the parameter vector θ as

$$\theta_l^{t+1} = \theta_l^t - \frac{\alpha}{\sqrt{\omega_l} + \varepsilon} \frac{\delta J(\theta^t)}{\delta \theta_l^t}. \tag{A.10}$$

Fig. A.13 Trajectory of
Adagrad algorithm on a
ravine error surface

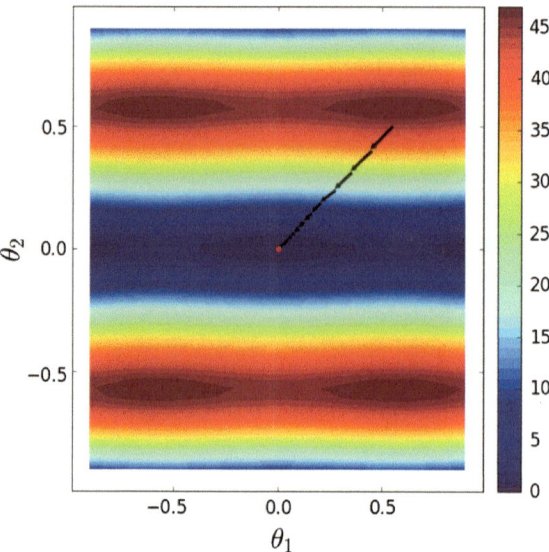

In this equation, θ_l shows the lth element in the parameter vector. We can replace Lines 56–58 in the previous script with the following statements:

```
de_nes = dJ(x, w_sol−momentum_vec∗momentum) momentum_vec =      1
momentum_vec ∗ momentum + alpha ∗ de_nes w_sol = w_sol −        2
momentum_vec                                                    3
```

The result of optimizing an objective function with a ravine surface is illustrated in Fig. A.13. In contrast to the other methods, Adagrad generates a short trajectory toward the local minimum.

The main restriction with the Adagrad algorithm is that the learning may rate rapidly drop after a few iterations. This makes it very difficult or even impossible for the algorithm to reach a local minimum in an acceptable time. This is due to the fact that the magnitudes of gradients are accumulated over time. Since the magnitude is obtained by computing the square of gradients, the value of ω_l will always increase at each iteration. As the result, the learning rate of each element will get smaller and smaller at each iteration since ω_l appears in the denominator. After certain iterations, the adaptive learning rate might be very small and for this reason the parameter updates will be negligible.

A.4   Root Mean Square Propagation (RMSProp)

Similar to Adagrad, *Root mean square propagation* which is commonly known as *RMSProp* is a method for adaptively changing the learning rate of each element in the gradient vector. However, in contrast to Adagrad where the magnitude of gradient is always accumulated over time, RMSProp has a forget rate in which the accumulated magnitudes of gradients are forgotten overtime. For this reason, the quantity ω_l is not always ascending but it may descend sometimes depending of the current gradients and forget rate. Formally, RMSProp algorithm update parameters as follows:

$$\omega_l^t = \gamma \omega_l^{t-1} + (1 - \gamma)\Big[\frac{J(\theta^t)}{\delta\theta_l^t}\Big]^2$$

$$\theta_l^{t+1} = \theta_l^t - \frac{\alpha}{\sqrt{\omega_l} + \varepsilon}\frac{\delta J(\theta^t)}{\delta\theta_l^t}.$$

(A.11)

In this equation, $\gamma \in [0, 1)$ is the forget rate and it is usually set to 0.9. This can be simply implemented by replacing Lines 56–58 in the above script with the following statements:

```
de_rmsprop = dJ(x, w_sol)                                                      1
rmsprop_vec = rmsprop_vec*rmsprop_gamma+(1-rmsprop_gamma)*de_rmsprop**2        2
w_sol = w_sol -(alpha/(np.sqrt(rmsprop_vec)))*de_rmsprop                       3
```

Figure A.14 shows the trajectory of RMSProp algorithm on a ravine error surface as well as nonlinear error surface. We see that the algorithm makes baby steps but it has a short trajectory toward the local minimums.

In practice, most of convolutional neural networks are trained using momentum batch gradient descend algorithm. But other algorithms that we mentioned in this section can be also used for training a neural network.

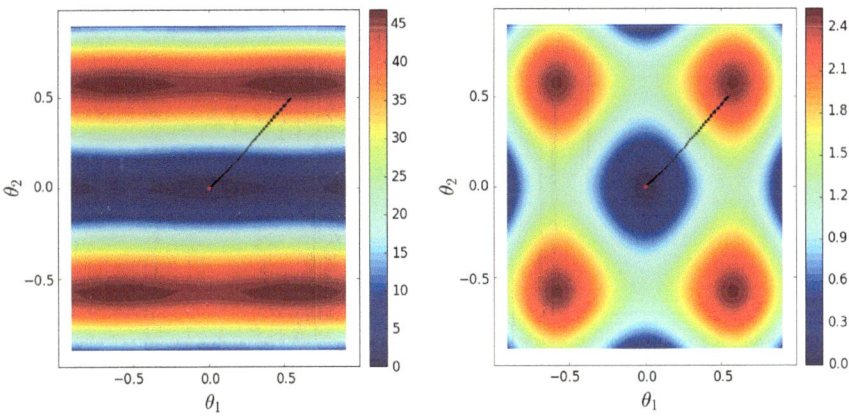

Fig. A.14 Trajectory of RMSProp on a ravine surface (*left*) and a nonlinear error surface (*right*) using mini-batch gradient descend

A.5 Shuffling

The gradient descend algorithm usually iterates over all samples several times before converging to a local minimum. One *epoch* refers to running the gradient descend algorithm on whole samples only one time. When mentioned that the error surface is always approximated using one sample (stochastic gradient descend) or a few samples (mini-batch gradient descend), assume the ith and $i + 1^{th}$ mini-batch. Samples in these two mini-batches have not changed compared to the previous epoch.

As the result, the error surface approximated by the ith in previous epoch is identical to the current epoch. Samples in one mini-batch might not be distributed in the input space properly and they may approximate the error surface poorly. Hence, the gradient descend algorithm may take a longer time to converge or it may not even converge in a tractable time.

Shuffling is a technique that shuffles all training samples at the end of one epoch. This way, the error surface approximated by the ith mini-batch will be different in two consecutive epochs. This in most cases improves the result of gradient descend algorithm. As it is suggested in Bengio (2012), shuffling may increase the convergence speed.

Glossary

Activation function   An artificial neuron applies a linear transformation on the input. In order to make the transformation nonlinear, a nonlinear function is applied on the output of the neuron. This nonlinear function is called activation function.

Adagrad   Adagrad is a method that is used in gradient descend algorithm to adaptively assign a distinct learning rate for each element in the feature vector. This is different from original gradient descend where all elements have the same learning rate. Adagrad computes a learning rate for each element by dividing a base learning rate by sum of square of gradients for each element.

Backpropagation   Computing gradient of complex functions such as neural networks is not tractable using multivariate chain rule. Backpropagation is an algorithm to efficiently computing gradient of a function with respect to its parameters using only one backward pass from the last node in computational graph to first node in the graph.

Batch gradient descend   Vanilla gradient descend which is also called batch gradient descend is a gradient-based method that computes the gradient of loss function using whole training samples. A main disadvantage of this method is that it is not computationally efficient on training sets with many samples.

Caffe   Caffe is a deep learning library written in C++ which is mainly developed for training convolutional neural network. It supports computations on CPU as well GPU. It also provides interfaces for Python and Matlab programming languages.

Classification score   A value computed by $wx + b$ in the classification layer. This score is related to the distance of the sample from decision boundary.

Decision boundary   In a binary classification model, decision boundary is a hypothetical boundary represented by the classification model in which points on one side of the boundary are classified as 1 and points on the other side of the boundary are classified as 0. This can be easily generalized to multiclass problems where the feature space is divided into several regions using decision boundaries.

Depth of network   Depth of deepest node in the corresponding computational graph of the network. Note that depth of a network is not always equal to the

© Springer International Publishing AG 2017
H. Habibi Aghdam and E. Jahani Heravi, *Guide to Convolutional Neural Networks*, DOI 10.1007/978-3-319-57550-6

number of layers (computational nodes). The reason is that in networks such as GoogleNet some of nodes have the same depth in computational graph.

Dropout Dropout is a simple but effective technique for regularizing a neural network. It works by randomly dropping a neuron from the network in each iteration of the training algorithm. This means that output and gradients of selected neurons are set to zero so they do not have any impact on forward and backward passes.

Early stopping It is a technique based on training and validation performance to detect overfitting during and stop the training algorithm. For this purpose, performance of the model is computed on a validation set as well as the training set. When their difference exceeds a certain threshold, the training algorithm stops. However, in some cases, even when the neural network starts to overfit on the training data, the validation accuracy might be still ascending. In that case, we may not stop the training algorithm.

Feature space A neural network can be thought as a composite feature transformation function. Given the input $x \in \mathbb{R}^p$, it transforms the input vector to a q-dimensional space. The q-dimensional space is called the feature space.

Generalization Ability of a model to accurately classify unseen samples is called generalization. A model is acceptable and reliable if it generalizes well on unseen samples.

Gradient check It is a numerical technique that is used during implementing the backpropagation algorithm. This technique ensures that gradient computation is done correctly by the implemented backpropagation algorithm.

Loss function Training a classification model is not possible unless there is an objective function that tells how good is the model on classifying training samples. This objective function is called loss function.

Mini-batch gradient descend Mini-batch gradient descend is an optimization technique which tries to solve the high variance issue of stochastic gradient descend and high computation of batch gradient descend. Instead of using only one sample (stochastic gradient descend) or whole samples of the dataset (batch gradient descend) it computes the gradient over a few samples (60 samples for instance) from training set.

Momentum gradient descend Momentum gradient descend is a variant of gradient descend where gradients are accumulated at each iteration. Parameters are updated based the accumulated gradients rather than the gradient in current iteration. Theoretically, it increases the convergence speed on ravine surfaces.

Nesterov accelerated gradient Main issue with momentum gradient descend is that it accumulates gradients blindly and it corrects its course after making a mistake. Nesterov gradient descend partially addresses this problem by computing the gradient on the next step rather than current step. In other words, it tries to correct its course before making a mistake.

Neuron activation The output computed by applying an activation function such as ReLU on the output of neuron.

Object detection The goal of object detection is to locate instances of a particular object such as traffic sign on an image.

Object classification   Object classification is usually the next step after object detection. Its aim is to categorize the image into one of object classes. For example, after detecting the location of traffic signs in an image, the traffic sign classifiers try to find the exact category of each sign.

Object recognition   It usually refers to detection and classification of objects in an image.

Overfitting   High nonlinear models such as neural network are able to model small deviations in feature space. In many cases, this causes that the model does not generalize well on unseen samples. This problem could be more sever if the number of training data is not high.

Receptive field   Each neuron in a convolutional neural network has a receptive field on input image. Receptive field of neuron z_i is analogous to the region on the input image in which changing the value of a pixel in that region will change the output of z_i. Denoting the input image using x, receptive field of z_i is the region on the image where $\frac{\delta z_i}{\delta x}$ is not zero. In general, a neuron with higher depth has usually a larger receptive on image.

Regularization   Highly nonlinear models are prone to overfit on data and they may not generalize on unseen samples especially when the number of training samples is not high. As the magnitude of weights of the model increases it become more and more nonlinear. Regularization is a technique to restrict magnitude of weights and keep them less than a specific value. Two commonly used regularization techniques are penalizing the loss function using L_1 or L_2 norms. Sometimes, combinations of these two norms are also used for penalizing a loss function.

RMSProp   The main problem with Adagrad method is that learning rates may drop in a few iterations and after that the parameters update might become very small or even negligible. RMSProp is a technique that alleviates the problem of Adagrad. It has a mechanism to forget the sum of square of gradient over time.

Stochastic gradient descend   Opposite of batch gradient descend is stochastic gradient descend. In this method, gradient of loss function is computed using only one sample from training set. The main disadvantage of this method is that the variance of gradients could be very high causing a jittery trajectory of parameter updates.

Time to completion   The total time that a model takes for a model to compute the output.

Vanishing gradients   This phenomena usually happens in deep networks with squashing activation functions such as hyperbolic tangent or sigmoid. Because gradient of squashing function become approximately zero as magnitude of x increases, the gradient will become smaller and smaller as the error is backpropagated to first layers. In most cases, gradient becomes very close to zero (vanishes) in which case the network does not learn anymore.

Width of network   Width of network is equal to number of feature maps produced in the same depth. Calculating width of network in architectures such as AlexNet is simple. But computing width of network in architectures such as GoogleNet is slightly harder since there are several layers in the same depth in its corresponding computational graph.

Reference

Bengio Y (2012) Practical recommendations for gradient-based training of deep architectures. Lecture notes in computer science, pp 437–478. doi:10.1007/978-3-642-35289-8-26, arXiv:1206.5533

Index

A
Accuracy, 124
Activation function, 62, 71, 142
AdaDelta, 154
Adagrad, 154, 271
Adam, 154
Advanced Driver Assistant System (ADAS),
        1, 167
AlexNet, 100
Artificial neura network, 61
Artificial neuron, 62
Augmenting dataset, 177
Average pooling, 97, 127, 144
Axon, 62
Axon terminals, 62

B
Backpropagation, 65, 68, 92
Backward pass, 68
Bagging, 201
Batch gradient descend, 263
Batch normalization, 127
Bernoulli distribution, 35
Bias, 20, 62
Binary classification, 16, 20, 41
Boosting, 201
Boutons, 62

C
Caffe, 105, 131
Classification, 16
Classification accuracy, 54, 106, 108, 159
Classification metric function, 106
Classification score, 22, 44
Class overlap, 16
Cluster based sampling, 175

Computational graph, 50, 59, 65
Confusion matrix, 108
Contrast normalization, 185
Contrast stretching, 185
Convolutional neural networks, 9, 63, 85
Convolution layer, 89
Covariance matrix, 8
Cross entropy, 65
Cross-entropy loss, 36, 147
Cross-validation, 176
    hold-out, 176
    K-fold, 177
CuDNN, 104, 131
Curse of dimensionality, 20

D
Dead neuron, 74
Decision boundary, 18, 21, 22, 30, 31, 116
Decision stump, 201
Deep neural network, 73
Dendrites, 62
Development set, 105
Directed acyclic graph, 102
Discriminant function, 20
Dot product, 21
Downsampling, 95
Dropout, 119, 146
Dropout ratio, 120
DUPLEX sampling, 175

E
Early stopping, 189
Eigenvalue, 8
Eigenvector, 8
Elastic net, 9, 119
Ensemble learning, 200

© Springer International Publishing AG 2017
H. Habibi Aghdam and E. Jahani Heravi, *Guide to Convolutional
Neural Networks*, DOI 10.1007/978-3-319-57550-6

Epoch, 139
Error function, 22
Euclidean distance, 5, 8, 175
Exploding gradient problem, 119
Exponential annealing, 123, 153
Exponential Linear Unit (ELU), 76, 142

F
F1-score, 111, 124, 161
False-negative, 108
False-positive, 2, 108
Feature extraction, 5, 57
Feature learning, 7
Feature maps, 90
Feature vector, 5, 57
Feedforward neural network, 63, 85
Fully connected, 81, 146

G
Gabor filter, 90
Gaussian smoothing, 178
Generalization, 116
Genetic algorithm, 203
Global pooling, 144
Gradient check, 131
Gradient descend, 24
Gradient vanishing problem, 119
GTSRB, 173

H
Hand-crafted feature, 58
Hand-crafted methods, 6
Hand-engineered feature, 58
Hidden layer, 63
Hierarchical clustering, 175
Hinge loss, 31, 32, 38, 39, 147
Histogram equalization, 185
Histogram of Oriented Gradients (HOG), 6,
        7, 57, 58, 80, 247
Hold-out cross-validation, 176
HSI color space, 6
HSV color space, 180
Hyperbolic tangent, 72, 190
Hyperparameter, 7, 64

I
ImageNet, 100
Imbalanced dataset, 45, 185
    downsampling, 186
    hybrid sampling, 186
    synthesizing data, 187
    upsampling, 186
    weighted loss function, 186
Imbalanced set, 107
Indegree, 68
Initialization
    MRSA, 142
    Xavier, 142
Intercept, 20, 62
Inverse annealing, 123, 153
Isomaps, 198

K
Keras, 104
K-fold cross-validation, 177
K-means, 175
K nearest neighbor, 17

L
L1 regularization, 118
L2 regularization, 118
Lasagne, 104
Leaky Rectified Linear Unit (Leaky ReLU),
        75, 142
Learning rate, 121
LeNet-5, 98, 150
Likelihood, 50
Linear model, 17
Linear separability, 16
Lipschitz constant, 227
Local binary pattern, 247
Local linear embedding, 198
Local response normalization, 101, 126
Log-likelihood, 50
Log-linear model, 48
Logistic loss, 38, 59
Logistic regression, 34, 52, 63
Loss function, 22
    0/1 loss, 23, 46
    cross-entropy loss, 36
    hinge loss, 31
    squared loss, 24

M
Majority voting, 42, 200
Margin, 30
Matching pursuit, 9
Max-norm regularization, 119, 154
Max-pooling, 95, 100, 127, 144
Mean square error, 25
Mean-variance normalization, 112, 114
Median filter, 179

Mini-batch gradient descend, 266
Mirroing, 182
Mixed pooling, 127
Model averaging, 200
Model bias, 117, 125
Model variance, 117, 125, 175
Modified Huber loss, 34
Momentum gradient descend, 267, 268
Motion blur, 178
MRSA initialization, 142
Multiclass classification, 41
Multiclass hinge, 47
Multiclass logistic loss, 190

N

Nesterov, 154
Nesterov accelerated gradients, 270
Neuron, 61
Nonparametric models, 17
Nucleus, 62
Numpy, 158

O

Objective function, 259
One versus one, 41
One versus rest, 44
Otsu thresholding, 6
Outdegree, 68
Outliers, 37
Output layer, 63
Overfit, 39, 116

P

Parameterized Rectified Linear Unit
        (PReLU), 142, 163
Parametric models, 17
Parametrized Rectified Linear Unit
        (PReLU), 75
Pooling, 90, 95
    average pooling, 144
    global pooling, 144
    max pooling, 144
    stochastic pooling, 144
Portable pixel map, 173
Posterior probability, 49
Precision, 110
Principal component analysis, 8
Protocol Buffers, 133

R

Random cropping, 180

Random forest, 52
Randomized Rectified Linear Unit
        (RReLU), 76
Random sampling, 175
Recall, 110
Receptive field, 88
Reconstruction error, 8
Rectified Linear Unit (ReLU), 74, 100, 117,
        142
Recurrent neural networks, 63
Reenforcement learning, 15
Regression, 16
Regularization, 117, 153
Reinforcement learning, 15
RMSProp, 154, 272

S

Sampling
    cluster based sampling, 175
    DUPLEX sampling, 175
    random sampling, 175
Saturated gradient, 29
Self organizing maps, 198
Shallow neural network, 73
Sharpening, 179
Shifted ReLU, 165
Sigmoid, 63, 71, 99, 117, 142
Sliding window, 238
Softmax, 49
Softplus, 60, 77, 162
Softsign, 73
Soma, 62
Sparse, 9
Sparse coding, 9
Spatial pyramid pooling, 127
Squared hinge loss, 33
Squared loss, 38
Squared loss function, 24
Step annealing, 123
Step-based annealing, 153
Stochastic gradient descend, 264
Stochastic pooling, 97, 144
Stride, 94
Supervised learning, 15

T

TanH, 142
T-distributed stochastic neighbor embed-
        ding, 198, 249
Template matching, 5
    cross correlation, 5

normalized cross correlation, 5
normalized sum of square differences, 5
sum of squared differences, 5
Tensor, 139
TensorFlow, 104
Test set, 105, 158, 174
Theano, 103
Time-to-completion, 207
Torch, 104
Training data, 8
Training set, 105, 158, 174
True-negative, 108
True-positive, 108
True-positive rate, 3

U
Universal approximator, 64
Unsupervised, 8
Unsupervised learning, 15

V
Validation set, 101, 106, 149, 158, 174
Vanishing gradient problem, 74
Vanishing gradients, 71, 119
Vienna convention on road traffic signs, 3
Visualization, 248

W
Weighted voting, 200
Weight sharing, 88

X
Xavier initialization, 113, 142

Z
Zero padding, 140, 144
Zero-centered, 8
Zero-one loss, 38